Kroc Foundation Series
Volume 13

ERYTHROCYTE MECHANICS AND BLOOD FLOW

Editors

Giles R. Cokelet, ScD
Departments of Radiation Biology and
Biophysics, and Chemical Engineering
University of Rochester
Rochester, New York

Herbert J. Meiselman, ScD
Department of Physiology and Biophysics
School of Medicine
University of Southern California
Los Angeles, California

Donald E. Brooks, PhD
Department of Pathology
University of British Columbia
Vancouver, British Columbia

ALAN R. LISS. INC. • NEW YORK

Library of Congress Cataloging in Publication Data

Main entry under title:
Erythrocyte mechanics and blood flow.
    (Kroc Foundation series; v. 13)
    Bibliography: p.
    Includes index.
    1. Blood flow — Congresses. 2. Erythrocytes — Congresses. 3. Rheology (Biology) —
    3. Rheology (Biology) — Congresses.
I. Cokelet, Giles R. II. Meiselman, Herbert J. III. Brooks, Donald E.
IV. Series: Kroc Foundation. Kroc Foundation series; v. 13.
QP105.E77   612'.111   79-5473
ISBN 0-8451-0303-2

This monograph includes the proceedings of a conference
sponsored by The Kroc Foundation, held in July 1978,
at its headquarters in the Santa Ynez Valley, California.

1st Row L to R: Peter Amacher and Robert L. Kroc; 2nd Row: Geoffrey V.F. Seaman, Herbert J. Meiselman Donald E. Brooks, Giles Cokelet, and Harry L. Goldsmith; 3rd Row: Lionel E. Sewchand, Paul L. LaCelle, Stephen Shohet, Robert M. Hochmuth, Denis Haydon, Richard Skalak, Evan A. Evans, Brian LaLone, Alan C. Groom, and Donald E. McMillan; 4th Row: Benjamin W. Zweifach, Jiri Palek, Walter Garey, Michael P. Sheetz, and Alfred C. Greenquist.

Contents

Participants and Contributors

Peter Amacher, Kroc Foundation, PO Box 547, Santa Ynez, California 93460

Donald E. Brooks, University of British Columbia, Department of Pathology, 2075 Westbrook Place, Vancouver, British Columbia, Canada V6T1W5   [XI, 119]

Giles Cokelet, University of Rochester Medical Center, Department of Radiation Biology and Biophysics, Rochester, New York   [XI, 141]

Evan A. Evans, Department of Biomedical Engineering, Duke University. Durham, North Carolina [31]

Harry L. Goldsmith, University Medical Clinic, Montreal General Hospital, 1650 Cedar Avenue, Montreal, Quebec, Canada [165]

Alfred C. Greenquist, Cancer Research Institute, UCSF Medical Center, San Francisco, California 94143

Walter Garey, Kroc Foundation, PO Box 547, Santa Ynez, California 93460

Russell G. Greig, Department of Pathology, University of British Columbia, Vancouver, British Columbia, Canada V6T 1W5 [119]

Alan C. Groom, Department of Biophysics, Health Science Center, University of Western Ontario, London, Ontario, Canada N6A 5C1 [229]

Denis Haydon, Physiological Laboratory, Downing Street, Cambridge, England CB2 3EG

Robert M. Hochmuth, Department of Chemical Engineering, Washington University, St. Louis, Missouri 63130 [57]

Johan Janzen, Department of Pathology, University of British Columbia, Vancouver, British Columbia, Canada V6T 1W5 [119]

Takeshi Karino, University Medical Clinic, Montreal General Hospital, Montreal, Quebec, Canada H3G 1A4   [165]

Paul L. La Celle, Department of Radiation Biology and Biophysics, University of Rochester School of Medicine and Dentistry, Rochester, New York 14620 [195]

Brian LaLone, Department of Physiology, University of Arizona College of Medicine, Tucson, Arizona 85724

S. C. Liu, Research Department, St. Elizabeth's Hospital, Boston, MA 02135 [15]

Donald E. McMillan, Sansum Medical Research Foundation, P.O. Drawer LL, Santa Barbara, California 93102 [211]

Herbert J. Meiselman, Department of Physiology, USC School of Medicine, 2025 Zonal Avenue, Los Angeles, California 90033   [XI, 75]

Jiri Palek, Hematology Research Laboratory, St. Elizabeth's Hospital, 736 Cambridge Street, Brighton, Massachusetts 02135 [15]

The number in bold type following a contributor's affiliation is the opening page number of that author's chapter.

Geoffrey V. F. Seaman, Division of Neurology, University of Oregon Medical School, Portland, Oregon 97201 [267]
Lionel E. Sewchand, Department of Physiology, USC School of Medicine, 2025 Zonal Avenue, Los Angeles, California 90033
Michael P. Sheetz, Department of Physiology, University of Connecticut School of Medicine, Farmington, Connecticut 06032 [1]
Stephen Shohet, Cancer Research Institute, UCSF-Moffitt Hospital, San Francisco, California 94143
Richard Skalak, Department of Civil Engineering and Engineering Mechanics, Columbia University, New York, New York 10027 [149]
Richard E. Waugh, Department of Radiation Biology and Biophysics, University of Rochester, Rochester, New York 14642 [31]
Benjamin W. Zweifach, Department of Bioengineering, University of California at San Diego, La Jolla, California 92037 [283]

Introduction

The ability of the erythrocyte to deform as it flows has long been recognized
as a sine qua non for effective blood circulation. As it passes through the capillaries
of the cardiovascular system the eight micron diameter red cell must negotiate three
to four micron diameter vessels and considerable deformation is involved. Beyond
this, erythrocyte deformability influences blood flow in all parts of the circulation,
since this cellular mechanical property influences whole blood viscosity.

In light of the importance of erythrocyte deformability, it is not surprising that
it has been the subject of much research. Considerable progress has been made in
several areas which touch on this property of the cell: 1) determination of the
composition, constituent conformation, and molecular structure of the red cell
membrane, 2) measurement of the mechanical properties of the erythrocyte membrane and development of constitutive equations for interrelating these properties,
3) interpretation of the macroscopic rheological properties of blood in terms of
erythrocyte deformation and aggregation, and 4) observation of the significance
of changes in erythrocyte deformability for in vitro and in vivo flow of red cells.
In spite of the obvious significance of work in each area to the understanding in
the other disciplines, information transfer and collaborative work between researchers in the different areas have been suboptimal; workers in one field are
often not cognizant of those problems in other areas which could be solved by
their contributions.

With these thoughts in mind, the Kroc Foundation sponsored a workshop en-
titled: "Red Blood Cell: Mechanics and Blood Flow" at its facilities in the Santa
Ynez Valley, California from July 17–21, 1978. Participants, drawn from the
research areas mentioned above, represented such varied research interests and
disciplines as biochemistry, colloid chemistry, engineering mechanics, fluid mechanics and rheology, physiology and hematology. The formal presentations at the
workshop fostered considerable informal discussion, much of it rather novel because it crossed disciplinary and research area boundaries.

An example of this interdisciplinary exchange centered around a current, first-
order model of the erythrocyte membrane, shown in Figure 1. The upper (outer)
sandwich-like structure represents the lipid-bilayer, with "membrane associated

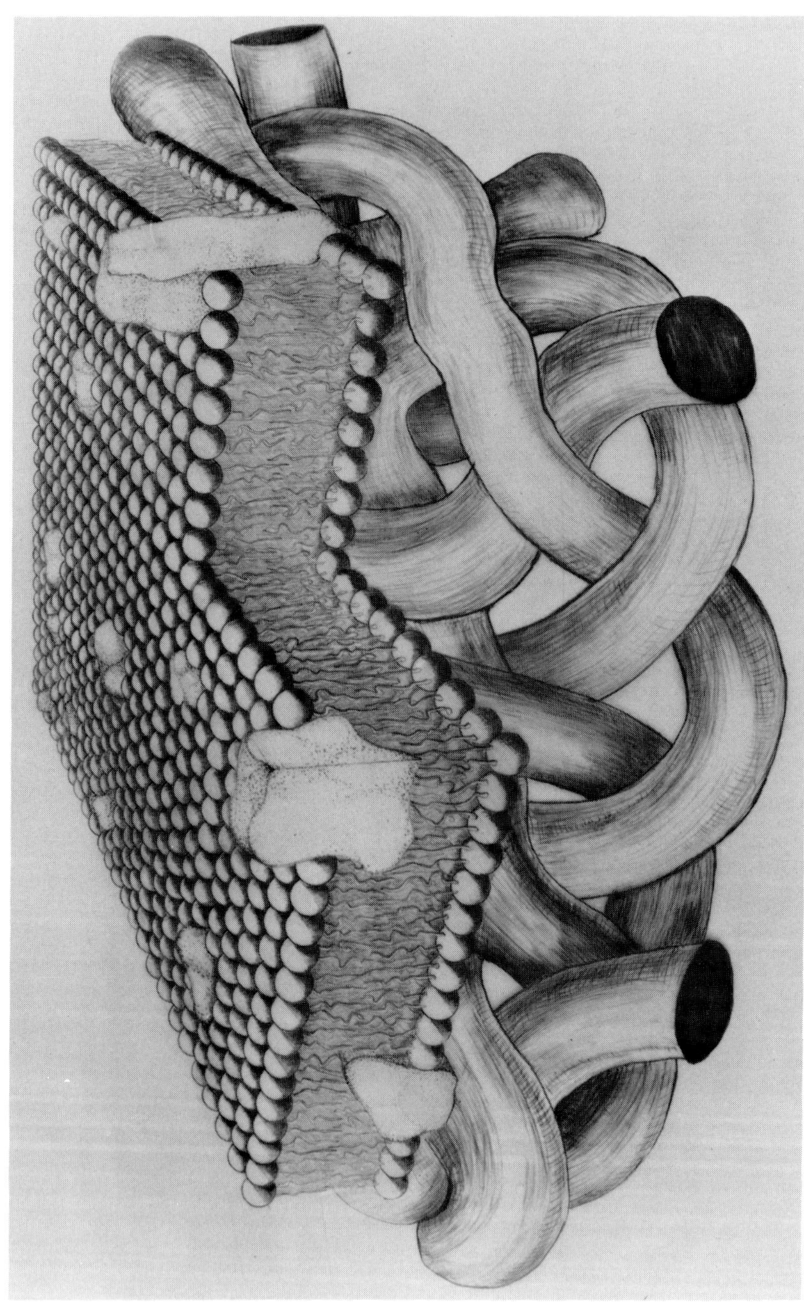

Fig 1. A schematic depiction of a composite model of the erythrocyte membrane, consisting of a lipid bilayer, integral proteins and structural network. [Reprinted with permission from E. Evans and R. Skalak, "Mechanics and Thermodynamics of Biomembranes," CRC Critical Reviews in Bioengineering (in press)].

proteins" (the globular elements) in this bilayer. The lower (inside) worm-like parts represent the spectrin cytoskeleton. This is, of course, just a crude model and is not intended to represent the details of the membrane. Nevertheless, this model was the framework for lively discussion, mainly between the membrane biochemists, those engaged in measuring the mechanical properties of this membrane, and those interested in relating the membrane properties and structure through the principles of engineering mechanics and fluid mechanics. It seemed to be generally agreed that the membrane shear elasticity (high compression modulus) arises from the lipid bilayer. The source of the membrane shear viscosity was judged to be unclear. The rheological role of the intramembraneous protein was also not known. There was general agreement that it could not be modeled until 1) its conformation was better understood (a "tree" structure embedded in the lipid could provide the observed slow diffusion in the bilayer); 2) the interactions between proteins were known; and 3) the bilayer viscosity was ascertained (perhaps from experiments on model phospholipid vesicles). Considerable discussion centered on questions about two-dimensional connections within the spectrin region and linkages between the spectrin and constituents in the lipid bilayer. These questions are important since they must be answered before a reasonable mathematical model can be constructed to relate membrane structure to mechanical properties (through the methods of engineering mechanics). Such models might be useful in investigating the possibility of fatigue being a failure mechanism determining the red cell life span, for instance. Similarly, informal discussions of these questions with respect to blood rheology and blood flow were extensive and very stimulating.

A general feeling, generated early in the workshop and present throughout the meeting, was that the interdisciplinary interactions developed in this diverse group of people were thought provoking and beneficial. With the hope that such interactions might be generated among other researchers as well, the presentations made at the meeting were collected together into this book. It is our hope that the reader will both benefit from reading these contributions and see possibilities for interdisciplinary efforts in this broad research area.

We wish to take this opportunity to thank all the participants for their contributions to the workshop and to this collection of viewpoints. Dr. Robert Kroc, President of the Kroc Foundation was both our gracious host and an active participant at the workshop. Dr. Peter Amacher, Director of Conference Programs for the Kroc Foundation, provided considerable experienced guidance to the cochairmen during all stages of the conference, skillfully leading us around most pitfalls and extracting us from those we persisted in falling into. To them and the Kroc Foundation, we extend our thanks for having made possible a very stimulating meeting.

Giles Cokelet, Herbert J. Meiselman, and Donald E. Brooks

1

The Bilayer and Shell Model of the Erythrocyte Membrane

Michael P. Sheetz

The shape and mechanical deformability of the human erythrocyte membrane are believed to be primary determinants of the rheological properties of normal erythrocyte suspensions. Inasmuch as the shape and deformability are properties of the structural organization of membrane proteins and lipids, the consideration of biological membrane structure is important in understanding these properties on a molecular level. I summarize the essence of our studies on erythrocyte membrane structure in the "bilayer and shell" model described herein. In this model the membrane bilayer is supported by a protein shell structurally distinct from the bilayer, with only protein links serving as structural contacts between the two layers.

Several recent reviews have described the organization of erythrocyte membrane components with regard to their asymmetry and their interactions with other components [1–3]. In these schemes, which are based upon the fluid mosaic model [4], the peripheral proteins are on the cytoplasmic membrane face and interact directly or indirectly with the integral glycoproteins that span the membrane. The exact nature of the linkages between peripheral and integral proteins is not understood and is the object of much current membrane research. Such varied processes as the capping and endocytosis of lectins bound to lymphocytes [5], histamine release from mast cells, [6], cell shape changes with oncogenic transformation, [7], and the ATP-dependent erythrocyte shape change [8] are all believed to be mediated by associations of peripheral proteins with the membrane. In the case of the ATP-dependent erythrocyte shape change, we have shown that the peripheral proteins spectrin and actin are involved, since antibody cross-linking of spectrin accelerates [9] and the addition of globular

Erythrocyte Mechanics and Blood Flow, pages 1–13
© 1980 Alan R. Liss, Inc., 150 Fifth Avenue, New York, NY 10011

actin inhibits the shape change [10]. The membrane lipid is probably involved in the ATP-dependent shape change because there is an asymmetric change in membrane surface area [11]. In order to define the molecular mechanism whereby spectrin and actin alter membrane surface area, the organization of the spectrin and actin complex and the interactions of that complex with the lipids need to be understood.

Spectrin and actin presumably form a network underlying the erythrocyte membrane. Such a network can be isolated from intact red cells or membranes by Triton extraction. We have designated the integrated structure left after Triton extraction the "erythrocyte cytoskeleton" [12]. Although the majority of the cellular spectrin and actin are contained in the cytoskeleton, many other components copurify with the complex [13]. We have shown that the cytoskeleton structure is destroyed when either spectrin [8] or actin [12] alone is solubilized from the complex, but other proteins may play essential roles in forming the cytoskeleton. Our studies have, therefore, centered on dissecting the cytoskeleton complex in order to define the component interactions under isotonic conditions. The majority of other studies have been performed on material prepared by hypotonic lysis of erythrocytes. Many of the protein interactions, however, are likely to be dependent upon ionic conditions. For example, small changes in divalent cation concentrations modulate microtubule polymerization [14], and small changes in pH can stimulate the polymerization of actin in the acrosomal process [15]. In the case of sperm, Triton can solubilize the membrane surrounding the acrosome without affecting the actin polymerization process or the factors controlling it. Triton was, therefore, used in these studies to remove the cytoplasm and the membrane permeability barrier under ionic conditions which approximate those in vivo. From the study of Triton cytoskeletons of intact erythrocytes we have derived the concept that a shell of peripheral cytoskeletal proteins forms an integrated structure independent of the lipid bilayer and integral membrane proteins.

MATERIALS AND METHODS

Protein was determined by the method of Lowry et al [16]. SDS polyacrylamide gel electrophoresis was performed according to the procedure of Fairbanks et al [17]. All chemicals were reagent grade or better.

Cytoskeleton Preparation

Fresh human blood was centrifuged (1,500g for 10 minutes), the supernatant and buffy coat were aspirated, and cells were suspended in an isotonic HEPES-Ringer solution. This procedure was repeated once, and then the suspended cells were incubated with 2 mM diisopropylfluorophosphate (DFP) in HEPES-Ringer for 30 minutes at 37°C. After DFP treatment, cells were pelleted and resuspended in Iso Tris (140 mM NaCl, 24 mM Tris pH 7.4) twice. A 40% suspension of red cells in Iso Tris was mixed at 4°C with an equal volume of a Triton X-100 suspension in the extraction buffer (140 mM KCl, 24 mM HEPES, 1.0 mM $MgCl_2$, 0.5 mM EGTA, 0.05 mM $CaCl_2$, and 2 mM reduced glutathione (pH 7.0). The mixture was applied to a linear sucrose gradient (10–60%) in the extraction buffer without Triton and centrifuged at 100,000g for 1 hour. The light-scattering band in the gradient was isolated, diluted in the extraction buffer, and centrifuged at 40,000g for 20 minutes. A portion of the diluted material was counted in a Petroff-Hausser bacteria counter after the addition of 0.001% uranyl acetate to enhance the contrast of the cytoskeletons. The pelleted cytoskeletons were resuspended in Iso Tris and the total protein content of the cytoskeletons was measured. The 0.6M KCl extraction of cytoskeletons was performed by adding 3M KCl to cytoskeletons in the extraction buffer and centrifuging at 40,000g for 20 minutes.

Light Microscopy

Cytoskeletons were fixed by mixing 1:1 with a 2% solution of glutaraldehyde in PBS (145 mM NaCl, 5 mM PO_4, pH 7.4). Cytoskeletons were viewed by dark-field illumination in an Olympus BH microscope.

RESULTS

When erythrocytes were treated with lytic concentrations of Triton X-100 under isotonic conditions, a complex of protein, lipid, and detergent could be isolated by sedimentation upon a linear sucrose density gradient. These complexes, termed "cytoskeletons," were prepared with different detergent concentrations to determine the effects of the detergent. Detergent concentrations were expressed in terms of amount of detergent per 10^{10} cells ($\sim$ 1 ml packed cells), since the properties of the complex correlated with changes in this parameter and not with changes in the percent of detergent in the extraction solution. The most notable change in the complex with increasing detergent concentration was the increase in buoyant density. Since an increase in density could result from a loss of detergent or lipid, both components were measured in cytoskeletons prepared with a low (30 mg Triton/10^{10} cells) or a high (300 mg Triton/10^{10} cells) concentration of detergent. The detergent represented less than 3% of the cytoskeleton mass and could not explain the density change (Table I). It is evident in Table I that the lipid content decreased dramatically with increasing detergent concentration, which could account for the increase in density.

When the higher concentration of detergent was used, the amount of protein present in the cytoskeleton band on the gradient was only about one-half of the protein present at the lower concentration. This could result from the specific extraction of particular lipid-associated proteins or the general loss of protein

TABLE I.

Triton concentration	Buoyant density (gm/cc)	Triton (mg/mg protein)[a]	Phospholipid[b] (mg/mg protein)	Cytoskeletal protein (10^{10} cells)
0	1.15^{c}	0	0.760^{c}	—
30 mg/10^{10} cells	1.17 ± 0.01	0.02 ± 0.005	0.50 ± 0.05	1.51 ± 0.2 mg
300 mg/10^{10} cells	1.27 ± 0.01	0.01 ± 0.003	0.03 ± 0.02	0.81 ± 0.2 mg

[a]Determined using H^3 Triton X-100.
[b]Computed from phosphate content of the shells [8].
[c]Taken from Steck [1].

from the cytoskeleton. When the proteins of the cytoskeleton were analyzed by SDS polyacrylamide gel electrophoresis, very little change was observed in the Coomassie blue staining pattern of proteins from the cytoskeletons prepared with 30 or 300 mg of Triton/10^{10} cells. Densitometer scans (Fig. 1) of the stained gels were integrated, and it was found that the percentage of stain in all the major protein bands except band 3 remained constant within the accuracy of the measurement. The amount of band 3 was decreased by 10–20% at the higher detergent concentration. When the gels were stained for carbohydrate with the periodic acid-Schiff reagent, no difference was noted between the high and low detergent samples. It is important to note that the SDS gels of whole erythrocyte membranes were significantly different from those of the cytoskeletons. Major portions of band 3, band 6, and band 8 were missing from the cytoskeletons. Since the extraction of band 3 alone cannot explain the loss of 50% of the cytoskeleton protein at the higher detergent concentration, it is apparent that at detergent concentrations greater than 30 mg Triton/10^{10} cells there is a rather nonspecific loss of protein components that accompanies lipid loss.

The loss of cytoskeleton material could possibly be visualized by dark field illumination as a decrease in the number or size of the cytoskeletons or as a decrease in the light scattered from them. At all detergent concentrations the number of cytoskeletons recovered from the gradient was 50–65% of the number of cells applied. Significant amounts of the cytoskeleton material adhered to the walls of the centrifuge tube and were lost when the gradients were fractionated.

Fig. 1. Densitometer scans at 600 mm of Coomassie blue stained SDS polyacrylamide gels (3.25% acrylamide) of a) whole erythrocyte membranes; b) cytoskeletons prepared with 30 mg Triton/10^{10} cells; and c) with 300 mg Triton/10^{10} cells.

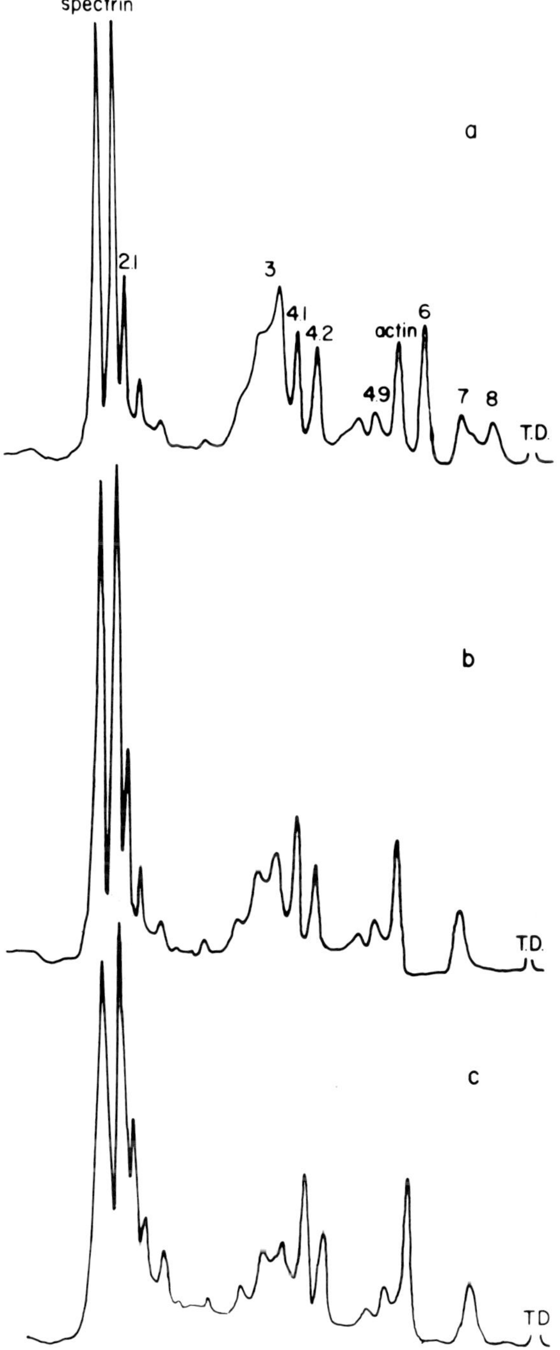
spectrin
a
2.1
3
4.1
4.2
actin
6
4.9
7
8
T.D.
b
T.D.
c
T.D.
1

As seen in Figure 2, the cytoskeletons prepared with 30 or 300 mg of Triton/10^{10} cells are 2.5–3.5 μ in diameter, but the amount of light scattered from the cytoskeletons is noticably decreased when the high Triton concentration is used. When the lower concentration of Triton is used, the surface of the cytoskeletons appears convoluted with thickenings or folds, which are not evident after treatment with the higher Triton concentration. Cytoskeleton size is independent of detergent concentration and correlates with the size of crenated ghosts. The loss of protein and lipid at higher detergent concentrations appears as a loss of foldings on the cytoskeleton surface.

Although the stoichiometry of proteins in the cytoskeleton is essentially constant with or without lipid present, many of the components may be bound to a

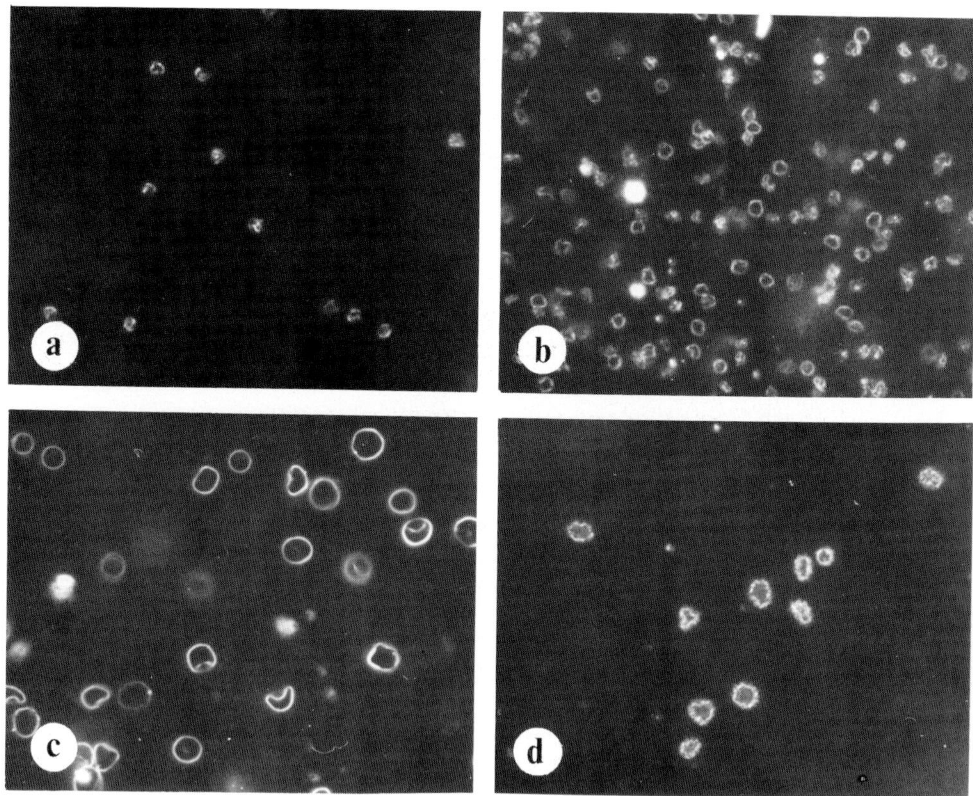

Fig. 2. Light micrographs taken under darkfield illumination of a) cytoskeletons prepared with 30 mg Triton/10^{10} cells; b) with 300 mg Triton/10^{10} cells; c) whole ghosts; and d) whole ghosts crenated by the addition of salt.

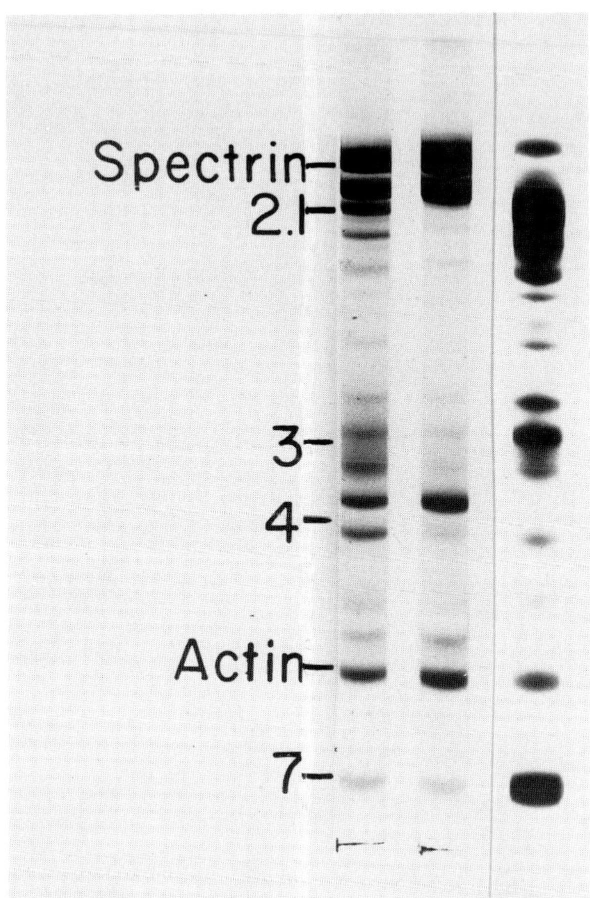

Fig. 3. Coomassie blue stained SDS polyacrylamide gels of cytoskeletons a) prepared with 300 mg Triton/10^{10} cells; b) after extraction with 0.6 M KCl; and c) the 0.6 M KCl extract.

structural protein core. Various ionic extractions were, therefore, performed to determine what components could be removed without destroying the integrity of the cytoskeleton. The most dramatic effect was obtained with 0.6 M KCl, which removed many of the components but left cytoskeletons that were visually unaltered. As seen in Figure 3, the 0.6 M KCl extracted cytoskeletons contained primarily spectrin, actin, and components 4.1 and 4.9. From densitometer scans of the gels it was determined that over 90% of the Coomassie blue stain was in those four bands. The 0.6 M KCl extract contained the components removed from the cytoskeletons, which indicates that proteolysis could not account for the changes.

DISCUSSION

Because we wished to study the organization of red cell membrane components in a native form, a procedure was devised for isolating at least the cytoskeletal components in an isotonic medium [8]. The procedure involves the extraction of intact erythrocytes with the nonionic detergent Triton X-100. Although recent reviews of detergent extraction of membranes classify Triton as a mild detergent that is not likely to denature proteins or disrupt protein-protein interactions [19, 20], the complex organization of components in the red cell membrane may be altered by Triton. Individual components such as spectrin or actin show no affinity for the detergent, but the removal of lipid from other proteins by detergent may alter their structure. Another more subtle effect of Triton on the organization of components is the effect Triton will have on cell shape just prior to lipid extraction. Our earlier studies of shape changes caused by amphophyllic compounds showed that morphological changes occurred at immeasurably short times after the addition of the compound to whole cells and ghosts [11, 21]. Therefore when Triton is added, cells may crenate dramatically before lipid is extracted. The small crenations could fall over, as in the spheroechinocyte, to form the apparent thickenings on the cytoskeletons prepared with the low detergent concentration. Since the removal of lipid would destabilize closely apposed regions of the cytoskeletons, higher detergent concentrations may cause budding off of folded regions of the cytoskeletons. In this way we could explain why the cytoskeletons have a size similar to highly crenated ghosts and lose protein nonselectively with removal of lipids. Although there are reservations about the effects of Triton and lipid removal itself, the procedure is unique in providing erythrocyte membrane proteins that have not been exposed to hypotonic lysis conditions or glycerol.

Our earlier studies of the organization of components in Triton cytoskeletons were hampered by proteases in the preparations. Specific proteases altered the polypeptide pattern of the SDS gels in that component 2.1 was absent and component 3 was increased in intensity in the gels [13]. DFP treatment of intact erythrocytes prior to extraction preserved component 2.1 and blocked time-dependent changes in the polypeptide pattern that had been observed [12]. It, therefore, appears that the cytoskeletons from DFP-treated cells contain proteins of the intact erythrocyte in an unproteolyzed form.

The Triton extraction procedure is quite selective with regard to the proteins that are removed. The major integral proteins, band 3 and glycophorin, are largely solubilized, as are band 6 and several other low molecular weight components. Otherwise, the SDS gels of the cytoskeletons and whole membranes appear remarkably similar. In the studies of Lutz et al [22] it was found that glycophorin and component 3 separate from the other membrane components in lipid-containing vesicles upon ATP depletion of erythrocytes. The budding off of vesicles from ATP-depleted cells and the solubilization of lipids with Triton are similar processes in that they separate lipids from the cell cytoskelton. Also, the same proteins coextract with the lipids in both cases. These studies would imply that the major portions of glycophorin and component 3 are not normally linked to the cytoskeleton. Indeed, measurements of band 3 mobility do not indicate that any direct interaction with spectrin exists [23]. The coaggregation of spectrin, band 3, and glycophorin observed by other investigators in erythrocyte ghosts can be explained by several more indirect mechanisms than an attachment of spectrin to the integral membrane proteins [24, 25]. Bourguignon and Singer [26] have suggested that a protein "X" links integral proteins with cytoplasmic actin and myosin in cultured fibroblasts, and a similar intermediary could enable spectrin to alter the lateral distribution of membrane glycoproteins. Thus, on a submicroscopic scale band 3 and glycophorin appear to be suspended in a lipid matrix, but constraints on their mobility may be imposed indirectly by spectrin

Extraction of lipid by Triton does not parallel the removal of band 3 and glycophorin. When only 40% of the phospholipid has been extracted, over 60% of band 3 and glycophorin has been removed. If the remaining 60% of the lipid is removed, less than 10% more of band 3 is lost. Either the detergent preferentially extracts the glycoproteins or the glycoproteins are sequestered in unstable lipid regions, which are preferentially released by detergent treatment. In ATP depletion, likewise, lipid is selectively removed along with just band 3 and glycophorin.

It is not clear how spectrin and the other peripheral proteins associate with the lipid bilayer. Direct lipid binding to spectrin has been observed in vitro [27] with hypotonically isolated spectrin, and from recent studies it has been suggested that spectrin controls the asymmetric distribution of red cell lipids in vivo [28]. The in vitro studies are of questionable significance since most proteins will associate with a hydrophobic-hydrophylic interface such as an air-water or lipid-water interface. Although the studies of lipid asymmetry changes with the modification of spectrin sulfhydryls imply that spectrin has a role in structuring lipids, spectrin's interaction with the lipids may be indirect through a binding protein [28]. In Triton extraction experiments sphingomyelin is selectively retained in the cytoskeletons [13]; however, phosphatidyl serine and ethanolamine are the lipids whose asymmetry in the red cell is presumably maintained by spectrin [28]. Although spectrin may be able to interact directly with lipid under certain conditions, studies with erythrocyte membrane vesicles show a much stronger interaction with protein components. Several laboratories have confirmed that the binding of spectrin to inside-out erythrocyte membrane vesicles is protein dependent and no binding occurs with right side-out vesicles [29–31].

Bennett has isolated a polypeptide of 72,000 daltons by proteolysis of inside-out vesicles [29]. This protein fragment binds to spectrin and prevents spectrin from binding to the inside-out vesicles. Antibodies raised against the fragment have recently been shown to bind to component 2.1 (Bennett, personal communication). These studies indicate that component 2.1 forms a link between spectrin and the membrane lipids. In the Triton extraction studies we find that when 2.1 is solubilized by 0.6 M KCl, components believed to be closely associated with lipids and integral membrane proteins also leave the cytoskeleton. For example, component 3, the putative anion channel, coextracts with 2.1. Component 2.1 itself has been labeled by impermeable reagents in intact cells and appears to extend through the bilayer [34]. The only proteins remaining in the cytoskeleton after 0.6 M KCl extraction are peripheral proteins which form stable aqueous solutions. In preliminary experiments we also find that 0.6 M KCl removes most of the lipids from cytoskeletons prepared with 300 mg Triton per 10^{10} cells. Thus it appears that when the putative spectrin-binding protein component 2.1 is removed, lipids and lipid-associated proteins are likewise extracted. This is consistent with the concept that peripheral proteins such as spectrin are linked to the lipids by proteins more integrally associated with the bilayer.

It is quite remarkable that neither the integrity nor the morphology of the cytoskeletons is altered by the extraction of the remaining lipids and lipid-associated proteins. On the other hand, the solubilization of either spectrin [8] or actin [12] alone will destroy the cytoskeleton structure. The lipids and lipid-associated proteins in the cytoskeleton are, therefore, not necessary structural components, and a peripheral protein complex or shell exists as an independent structure from the bilayer. Steck and Hainfeld [32] have also observed that a network of erythrocyte proteins remains intact after separation from the membrane bilayer. Thus, certain peripheral proteins in the erythrocyte may form a structure independent of the bilayer and equally as extensive as the bilayer.

In the current fluid mosaic model of membrane structure, the peripheral proteins are attached to the membrane lipids through integral proteins [4], but insufficient information has been available to formulate a detailed model of peripheral protein organization which would provide a molecular explanation of the physical properties of the cell. Because of its possible predictive value, we have formulated a detailed model of the erythrocyte membrane that is built upon three basic components: the membrane bilayer, the shell of peripheral proteins, and the links between the shell and the bilayer. In isolation the membrane bilayer is composed of a lipid bilayer and integral membrane proteins, which are freely mobile laterally. In the shell, on the other hand, the proteins are not able to reorganize rapidly, although the structure is deformable. When the shell is stressed for a prolonged period, it will reorganize to assume a new lower energy configuration. The linkages serve to join physically the two layers and to communicate certain signals between the layers. The cross-linking of membrane glycoproteins may affect the linkage proteins, which in turn could alter the ATPase in the shell. ATP hydrolysis and spectrin phosphorylation-dephosphory-lation in the shell could affect the linkage protein interactions with the bilayer surface and thereby alter membrane surface tension and membrane shape. We propose that the membrane bilayer is composed primarily of the membrane lipids, glycophorin, and band 3. We have defined the shell proteins earlier as spectrin, actin, and components 4.1 and 4.9. The major linkage protein is component 2.1, but other proteins such as band 4.1 or 4.2 may also physically bridge between the two layers.

We have designated this the "bilayer and shell" model in part because of the terms used in material science for similar two-layer structures composed of a shell and a covering or membrane [33]. A membrane is considered as a material that has very little rigidity or bending resistance but can resist lateral compression or expansion, as is the case for the membrane bilayer. On the other hand, a shell has rigidity and is needed to support the shape of the membrane. Although alternative models for the erythrocyte membrane structure are tenable, the "bilayer and shell" model provides a good fit of the existing biochemical and biophysical data. It also aids in the prediction of the effects of biochemical changes on the membrane physical properties.

ACKNOWLEDGMENTS

This project was supported by NIH grant HL-18317.

REFERENCES

1. Steck TL: The organization of proteins in the human red blood cell membrane. J Cell Biol 62:1–19, 1974.
2. Marchesi VT, Furthmayr H, Tomita M: The red cell membrane. Ann Rev Biochem 45:667–698, 1976.
3. Juliano RL: The proteins of the erythrocyte membrane. Biochim Biophys Acta 300: 341–378, 1973.
4. Singer SJ: Molecular biology of cellular membranes with applications to immunology. Adv Immunol 19:1–66, 1974.
5. Silverstein SC, Steinman RM, Cohn ZA: Endocytosis: Annu Rev Biochem 46:254–297, 1977.
6. Lawson D, Fewtull C, Lomperts B, Raff MC: Anti-immunoglobulin-induced histamine secretion by rat peritoneal mast cells studied by immunoferritin electron microscopy. J Exp Med 142:391–402, 1975.
7. Pastan I, Willingham M: Changes in cell shape upon transformation. Nature 274:550–555, 1978
8. Sheetz MP, Sawyer D, Jackowski S: The ATP-dependent red cell membrane shape change: A molecular explanation In Brewer G (ed): "The Red Cell." New York. Alan R. Liss, 1978, pp 431–450.
9. Sheetz MP, Singer SJ: On the mechanism of ATP-induced shape changes in human erythrocyte membranes. The role of the spectrin complex. J Cell Biol 73:638–646, 1977.

10. Birchmeier W, Singer SJ: Muscle G-actin is an inhibitor of ATP-induced erythrocyte ghost shape changes and endocytosis. Biochem Biophys Res Commun 77:1354–1361, 1977.

11. Sheetz MP, Singer SJ: Biological membranes as bilayer complex. A molecular mechanism of drug-erythrocyte interactions. Proc Natl Acad Sci USA 71:4457–4461, 1974.

12. Sheetz MP: DNAase I dependent dissociation of erythrocyte cytoskeletons. J Cell Biol 81:266–270, 1978.

13. Sheetz MP, Sawyer D: Triton shells of intact erythrocytes. J Supramol Struct 8:399–412, 1978.

14. Marcum JM, Dedman JR, Brinkley BR, Means AR: Control of microtubule assembly-disassembly by calcium-dependent regulator protein. Proc Natl Acad Sci USA 75:3771–3775, 1978.

15. Tilney LM: Stimulation of the acrosome actin polymerization by pH change. In Pepe FA (ed): "Cell Motility." New York: Academic Press (in press).

16. Lowry OH, Rosebrough NJ, Farr AF, Randall RT: Protein measurement with the Folin phenol reagent. J Biol Chem 193:265–275, 1975.

17. Fairbanks G, Steck TL, Wallach DFH: Electrophoretic analysis of the major polypeptides of the human erythrocyte membrane. Biochemistry 10:2606–2616, 1971.

18. Tanford C, Reynolds JA: Characterization of membrane proteins in detergent solutions. Biochim Biophys Acta 457:133–153, 1976.

19. Helenius A, Simons K: Solubilization of membranes by detergents. Biochim Biophys Acta 415:29–79, 1975.

20. Sheetz MP, Painter RG, Singer SJ: Biological membranes as bilayer couples. III. Compensatory shape changes induced in membranes. J Cell Biol 70:193–203, 1976.

21. Sheetz MP, Singer SJ: Equilibrium and kinetic effect of drugs on the shapes of human erythrocytes. J Cell Biol 70:247–251, 1976.

22. Lutz HU, Liu SC, Palek J: Release of spectrin-free vesicles from human erythrocytes during ATP depletion. I. Characterization of spectrin-free vesicles. J Cell Biol 73:548–560, 1977.

23. Cherry RJ, Burkli A, Busslinger I, Scheider G, Porish GR: Rotational diffusion of band 3 proteins in the human erythrocyte membrane. Nature 263:389–393, 1976.

24. Nicolson GL, Painter RG: Anionic sites of human erythrocyte membranes. J Cell Biol 59:395–406, 1973.

25. Shotton D, Thompson K, Wofsy L, Branton D: Appearance and distribution of surface proteins of the human erythrocyte membrane. An electron microscope and immuno-chemical labeling study. J Cell Biol 76:512–531, 1978.

26. Bourguignon LYW, Singer SJ: Transmembrane interactions and the mechanism of capping of surface receptors by their specific ligands. Proc Natl Acad Sci USA 74:5031–5035, 1977.

27. Mombers C, van Dijck PWM, van Deenan LLM, de Gier J, Verkley AJ: The interaction of spectrin-actin and synthetic phospholipids. Biochim Biophys Acta 470:152–160, 1977.

28. Haest CWM, Plasa G, Kamp D, Deuticke B: Spectrin as a stabilizer of the phospholipid asymmetry in the human erythrocyte membrane. Biochim Biophys Acta 509:21–32, 1978.

29. Bennett V: Purification of an active proteolytic fragment of the membrane attachment site for human erythrocyte spectrin. J Biol Chem 253:2292–2299, 1978.

30. Bennett V, Branton D: Selective association of spectrin with the cytoplasmic surface of human erythrocyte plasma membranes. J Biol Chem 252:2753–2763, 1977.

31. Litman D, Chen JH, Marchesi VT: Spectrin binds to the inner surface of the human red cell membrane via association with band 4.1–4.2 and 3. J Supramol Struct 8:209a, 1978.
32. Steck TL, Hainfeld JF: Protein ensembles in the human red-cell membrane. In Brinkley BR, Porter KR (eds): "International Cell Biology." New York: Rockefeller University Press, 1977, pp 6–14.
33. Naghdi IH: In Flugge (ed): "Encyclopedia of Physics." New York: Springer-Verlag, 1972, vol 6a part II, pp 425–440.
34. Staros JV, Richards FM: Photochemical labeling of the surface proteins of human erythrocytes. Biochemistry 13:2720–2776, 1974.

2
Protein-Protein Interactions in Red Cell Membranes

S. C. Liu and J. Palek

This article reviews our recent work in the following two areas. The first involves the ultrastructure of the spectrin-actin network in normal red cell membranes, combining the use of transmission electron microscopy and oxidative membrane protein cross-linkings. The second area is the effects of red cell energy metabolism, mainly red cell adenosine triphosphate (ATP) and reduced glutathione (GSH), on protein interactions in red cell membranes.

ULTRASTRUCTURE OF THE SPECTRIN-ACTIN NETWORK

Figure 1 depicts the methodology for studying the ultrastructure of the cytoskeletal network of red cell membranes. We prepared ghosts by osmotic hemolysis and immobilized them on carbon-coated grids by gently pressing the grid against the surface of a coverslip. These flattened ghosts were then subjected to Triton X-100 extraction to remove the integral membrane proteins and lipids [1], fixed with glutaraldehyde, stained with uranyl acetate, and subsequently examined by electron microscopy.

Figure 2 depicts ghosts after osmotic hemolysis and the cytoskeletal residues (Triton shells) after Triton X-100 extraction. The size of these shells was virtually identical with that of the original flattened ghosts. At a higher magnification, the network has a spongy structure and appears to consist of twisted fibers that are approximately 35 Å thick and resemble a string of beads in appearance. In addition, globular protrusions about 50 Å in diameter can be noted. Certain areas appear darker than the others, suggesting an overlap of the meshwork. Assuming that spectrin is a major constituent of this network and that each "fiber" represents double-standard spectrin (1+2) dimers (see cross linking data below), we calculate the length of the dimer fiber to be 750 Å, based on a molecular weight of 450,000, a protein density of 1.39/ml, and a fiber diameter of 35 Å.

Erythrocyte Mechanics and Blood Flow, pages 15–29
© 1980 Alan R. Liss, Inc., 150 Fifth Avenue, New York, NY 10011

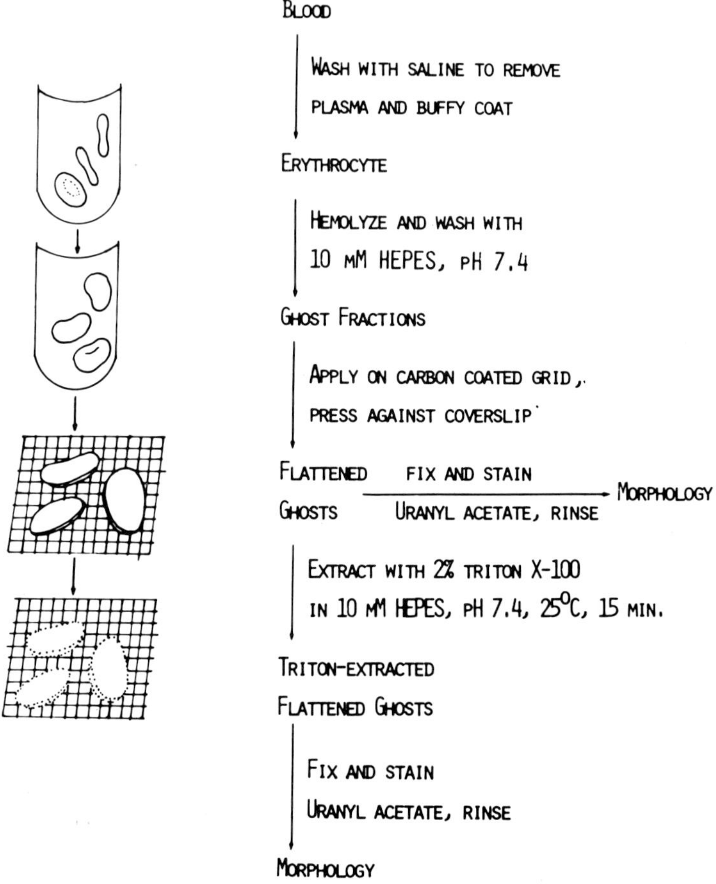

Fig. 1. Preparation of membrane cytoskeletons for transmission electron microscopy.

PROTEIN INTERACTIONS IN THE CYTOSKELETAL RESIDUES

In order to analyze the nearest protein neighbors in the cytoskeletal network, we cross linked the cytoskeletal proteins by intermolecular disulfide couplings with catalytic oxidation. Subsequently the residues were solubilized without a reducing agent and electrophoresed in polyacrylamide gels of a low concentration. This enabled us to separate the complexes of cross linked membrane proteins of molecular weight larger than those of spectrin. These were then subjected to electrophoresis in a second dimension in the presence of the reducing agent dithiothreitol (DTT). In the second dimension, the cross linked complexes

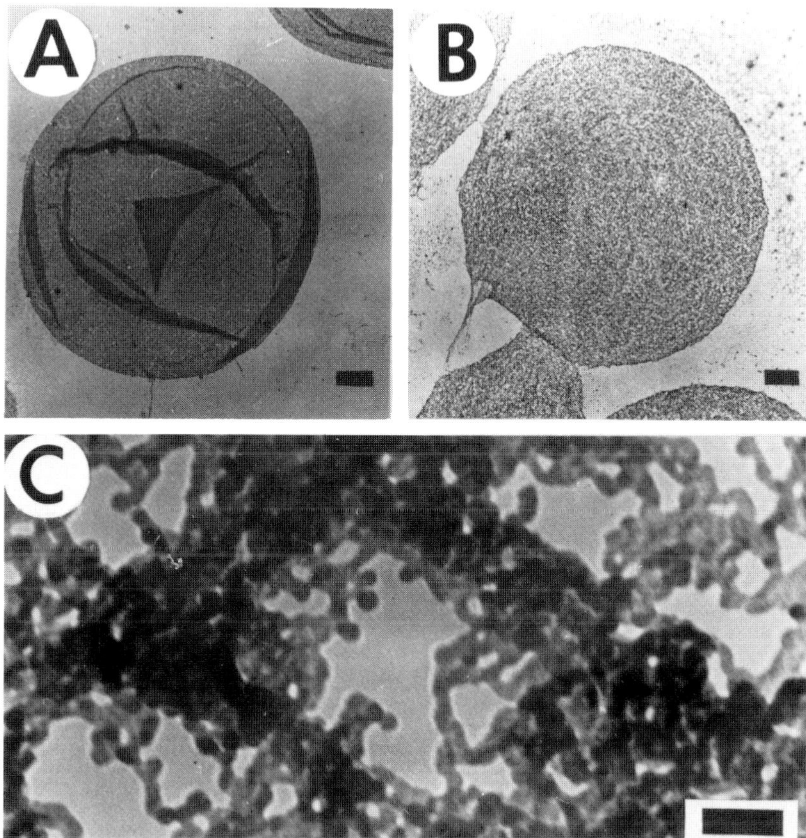

Fig. 2. Electron micrographs of flattened ghosts and Triton shells. (A) Flattened ghost. The diameter of the disc-shaped ghost is about 10 μm. (B) Triton residue of the flattened ghost. Samples were prepared as described in Figure 1. The Triton shell has essentially the same diameter as that in the flattened ghost. (C) Triton residue examined at high magnification. Calibration bars in (A) and (B) are 1 μ and in (C), 200 Å.

dissociated into the individual components, which were subsequently identified from their relative mobilities as described [2].

Figure 3 depicts the Coomassie blue stained proteins from untreated ghosts (first gel) and the "Triton shells" before and after cross-linking. As reported by others, the residues principally consisted of bands 1, 2, 4.1, 5, and residual amounts of band 3. After catalytic oxidation several new complexes were formed, namely, the 450K, 330K, 285K, 260K, and 175K dalton complexes.

The analysis of these complexes in the second dimension is shown in Figure 4. The uncross-linked proteins separated to form a skewed line. The off diagonal

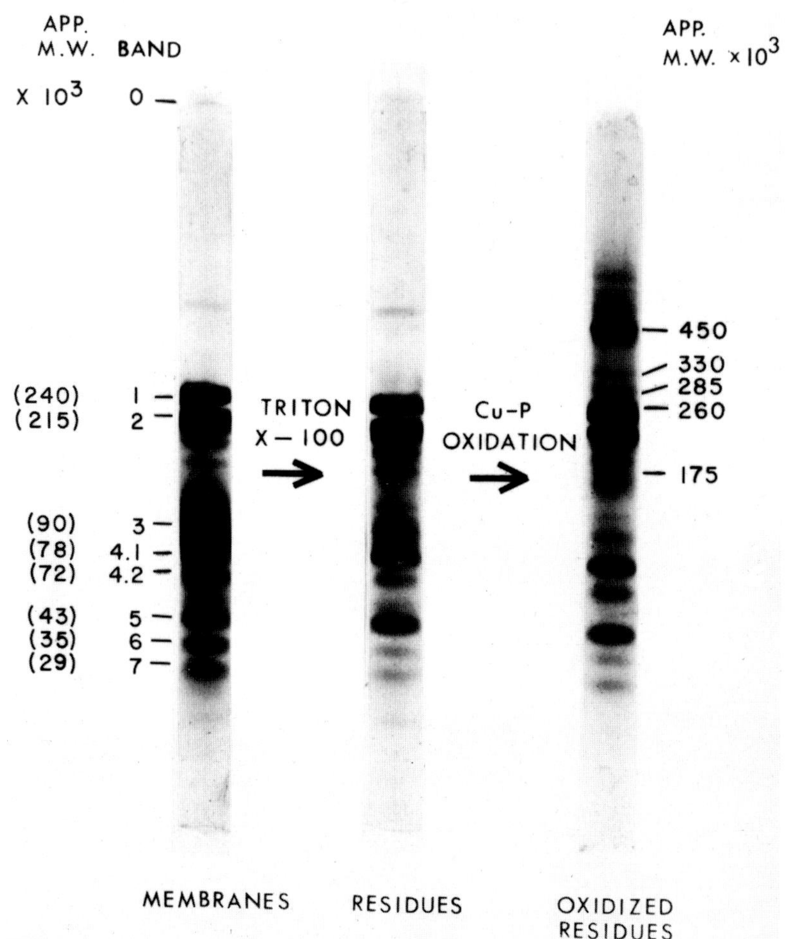

Fig. 3. Catalytic oxidation of Triton residues with $CuSO_4$/o-phenanthroline (CuP). Ghosts were treated with 2% Triton in the cold ($0°C$). $CuSO_4$ (10 μM) and o-phenanthroline (50 μM) were then added directly to the Triton shell suspension. The reaction was quenched by EDTA (2 mM, final concentration). Triton residues were collected by centrifugation (41,000 g for 30 minutes), dissolved in SDS concentrate, and electrophoresed in 2.5% acrylamide/0.3% agarose gels without reduction.

spots on the dotted line represent protein dimers such as 3_2 and 5_2. Additional complexes include 2 + 4.9, 1 + 5, and 1 + 2. The latter three complexes are better shown in the bottom half of Figure 4, in which the cross-linked sample was allowed to overrun in the first dimensional gel, so that it permitted a more detailed separation of the above complexes, which otherwise migrate closely with spectrin.

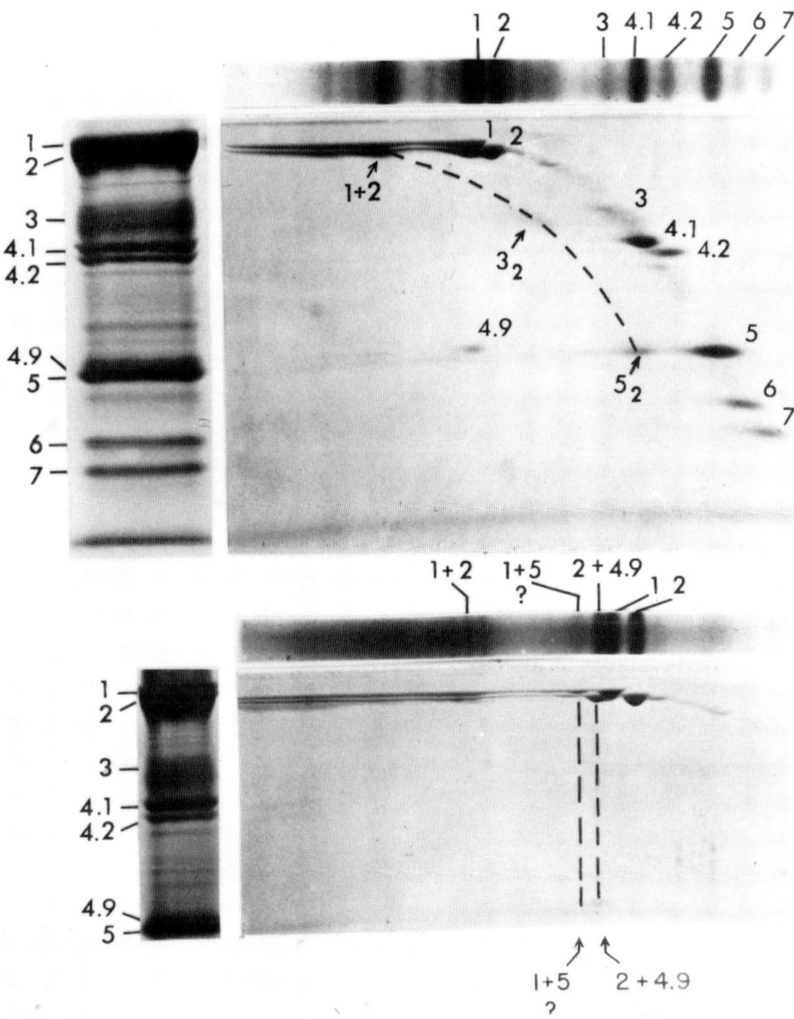

Fig. 4. Two dimensional analysis of protein cross-linkings in membrane cytoskeletons. Triton residues were prepared and catalytically oxidized as described in the legend of Figure 3. The samples were analyzed in the second dimension as described before [2]. The dotted curve in the upper gel represents the location of dimers of membrane proteins. The sample in the bottom gel was overrun in the first dimension, and only band 1 to 5 region was shown in the second dimension. Tentative identification of the released components from each complex was indicated.

The high yield of 1 + 2 complex in the Triton residues suggests a close association between these two spectrin chains. Multiple bands appeared in the spectrin dimer region (450K daltons), probably representing various Stoke's radii of the cross-linked dimers in SDS [2–5]. Presumably, this is due to multiple —SH groups which are involved in the intermolecular side-by-side contacts rather than single end-to-end contacts. The cross-linking between spectrin component 1 and band 5 (actin) (Fig. 4) was based primarily on the molecular weight estimations.

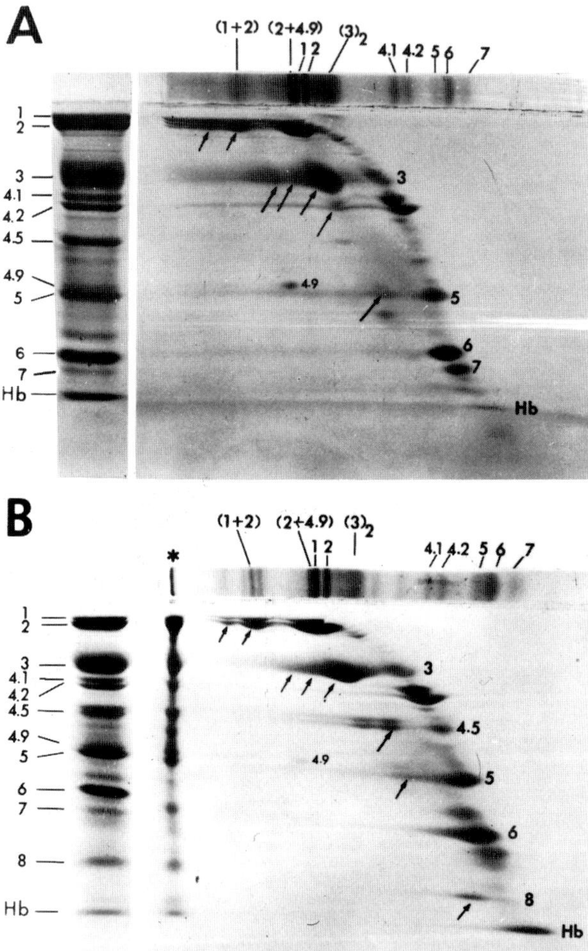

Fig. 5. Catalytic oxidation of membrane proteins in fresh and ATP-depleted erythrocytes. Ghosts from fresh (A) and anaerobic ATP-depleted (B) erythrocytes were catalytically oxidized with $CuSO_4$/o-phenanthroline at 25°C for ten minutes and electrophoresed as in Figure 4. Arrows indicate oligomers of spectrin, bands 3, 4, 5, and 8. Asterisk at the origin of (B) represents the large aggregate (> 1,000K daltons) in ghosts from ATP depleted erythrocytes. (Reprinted by copyright permission from "Erythrocyte Membranes: Recent Clinical and Experimental Advances." New York: Alan R. Liss, Inc., 1978, pp 75–88.

The off-diagonal spot of band 5 derived from the 285K complex was not well defined because of endogenous smearing of band 5, as shown on the second dimension. The composition of 330K complex is not clear because of its low yield in the second dimension, but it seems to involve band 1. (See bottom gel, Fig. 4.) Whether or not this complex represents 1 + 3, as recently observed by us in the isolated membranes [2, 6], remains to be shown.

Combining the ultrastructural information with the cross-linking data, we conclude that the cytoskeletal meshwork in red cell membranes is represented by rod-like spectrin dimers. Furthermore, the cross-linking data suggest the presence of oligomers of actin, complexes of 2 + 4.9 and small amounts of contacts between spectrin component 1 and actin. The question of whether the observation of actin dimers indicates some degree of polymerization of actin in the membrane is not known.

DEPENDENCE OF SPECTRIN ORGANIZATION ON CELL METABOLISM

In order to examine the changes in organization of spectrin and other proteins in red cell membranes as a function of intracellular ATP and GSH concentration, we employed cross-linking of the nearest membrane protein neighbors by intermolecular disulfide couplings, which were produced either by catalytic oxidation of ghosts or spontaneous oxidation of intact cells during aerobic incubation

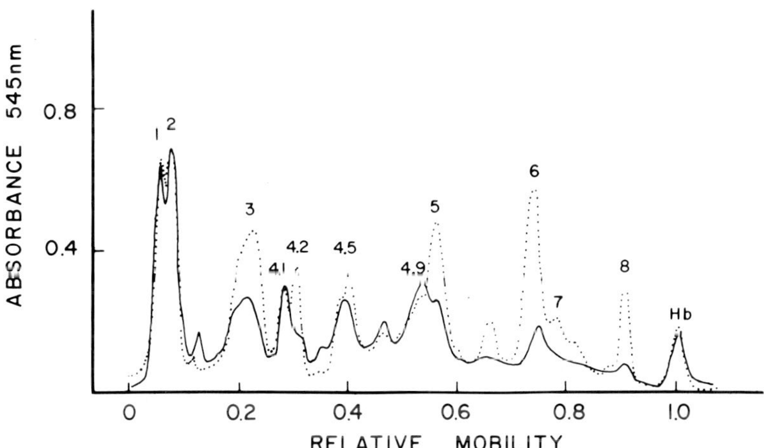

Fig. 6. Densitometric scans of Coomassie blue stained proteins released from the > 1,000K polymers after DTT reduction. This polymer is depicted in Figure 5B. Dotted line shows a reference membrane sample separated in an identical gel system. (Reprinted by copyright permission from "Erythrocyte Membranes: Recent Clinical and Experimental Advances." New York: Alan R. Liss Inc., 1978, pp 75–88.

[7, 8]. We analyzed the cross-linked complexes as described above for the Triton residues.

Figure 5 shows the catalytic oxidation of ghosts from fresh and ATP-depleted red cells. The ATP depletion was carried out anaerobically in a buffered solution without glucose for 22 hours at 37°C. Both fresh and ATP-depleted cells contained the complexes of 450K and 260K daltons, which resolved in the second dimension as 1 + 2 and 2 + 4.9, respectively. The major differences in the cross-linking pattern between ghosts derived from ATP-depleted cells in contrast to ghosts prepared from fresh red cells was a presence of a large complex at the start of the gel (mol wt > 1,000K). This complex is completely cleavable by DTT and can be separated into multiple components, which are further represented on a densitometric scan shown in Figure 6. If the scan is compared to

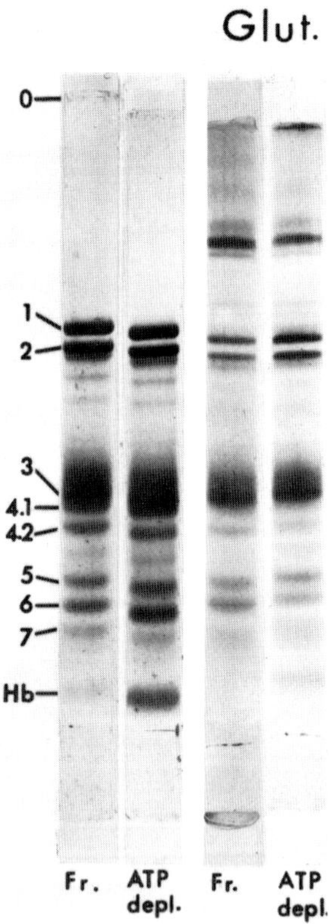

Fig. 7. Glutaraldehyde cross-linkings of isolated membranes from fresh and ATP-depleted erythrocytes. Ghosts were reacted with glutaraldehyde (2 mM) at 25°C for 30 minutes. Membrane proteins were solubilized in SDS and reduced by DTT (25 mM) before electrophoresis.

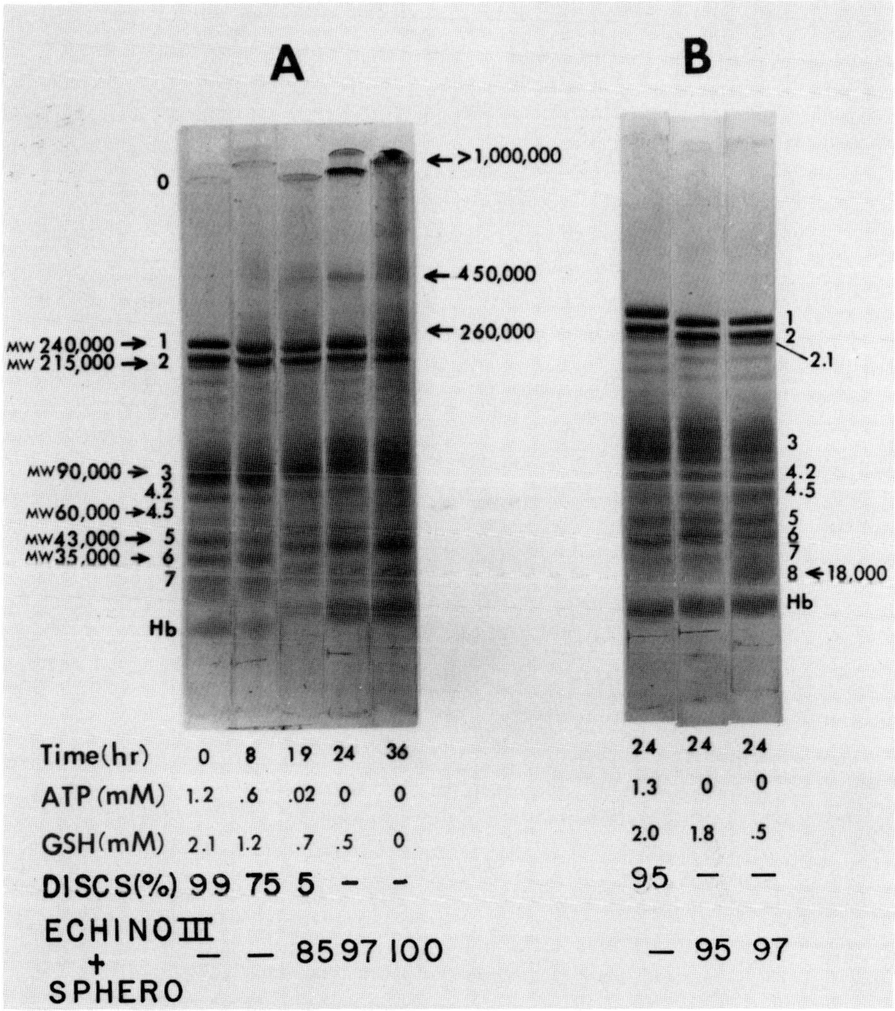

Fig. 8. Formation of large molecular weight complexes of membrane proteins during red cell ATP depletion in vitro. (A) Relationships between changes in red cell ATP and GSH levels, cell shapes, and membrane protein composition during aerobic incubation of red cells at 37°C without glucose. Membrane proteins were electrophoresed without DTT reduction. (B) Inhibition of the formation of large molecular weight complexes by maintenance of red cell ATP with adenine, inosine, and glucose (first gel), anaerobic incubation (second gel), or DTT reduction before electrophoresis (third gel). (Reprinted by copyright permission from Blood 51:387, 1978).

that of a control sample of normal membrane proteins separated in an identical gel system, it can be seen that this large mol wt complex is selectively enriched in spectrin; ie, the ratio of band 3 to spectrin or actin to spectrin is considerably lower than that in normal red cell membranes. This complex will be referred to as a spectrin-rich complex. Since intermolecular disulfide couplings require a close proximity (3–5 A°) of neighboring proteins, we concluded that in ATP-depleted red cells the proteins are rearranged into closer contacts, which results in an increase of their cross-linking.

A similar complex was formed spontaneously in the membrane without catalytic oxidation when red cells were incubated aerobically to deplete ATP [7]. These spectrin-rich polymers were apparently derived from the air oxidation of erythrocyte membrane proteins during incubation.

The next question was whether we could detect the difference of protein association between fresh and ATP-depleted red cells employing other cross-linking agents such as glutaraldehyde which cross-links the adjacent protein amino groups. Figure 7 illustrates that ghosts from ATP-depleted red cells had a large mol wt polymer (> 1,000K daltons) after glutaraldehyde cross-linking, whereas ghosts from fresh cells did not produce such a polymer. These findings support the hypothesis that the large mol wt spectrin-rich complex reflects a re-organization of spectrin into closer contacts within the membrane, thereby increasing the propensity of the individual spectrin subunits to be cross-linked to one another.

THE DEPENDENCE OF THE SPECTRIN-RICH POLYMER ON RED CELL ATP AND GSH CONTENT

We investigated the ATP, GSH dependence of the > 1,000K complex which took place spontaneously in intact red cells during their aerobic ATP depletion. Figure 8 depicts kinetics of changes in ATP, GSH content and membrane protein composition during aerobic incubation without glucose. Such cells exhibited a progressive increase of the > 1,000K complex at the start of the gel, as did the complexes of molecular weight of 450K and 260K. The formation of the > 1,000K complex coincided with shape transformation into type III echinocytes and spheroechinocytes classified according to Bessis [9]. The formation of the complexes further coincided with a progressive decrease in ATP and GSH levels. The occurrence of these protein complexes was completely prevented by maintenance red cell ATP by adenine and inosine, which sustained a normal GSH concentration as well. In cells previously aerobically ATP depleted, a subsequent restoration of ATP content and a recovery of biconcave shape resulted in a significant decrease of these complexes. However, under such conditions, significant fractions (75%) of GSH have been reduced as well. Thus the kinetic changes of the 1,000K complex could be related to both a restoration of ATP or GSH, or both. To differentiate the relative roles of GSH and ATP in the formation of the above protein complexes, we investigated glucose 6-phosphate dehydrogenase-deficient erythrocytes. These cells, when incubated at 37°C for three hours in the presence of an oxidant such as acetylphenylhydrazine, were able to maintain intracellular ATP concentrations while they exhibited a profound drop in GSH. The exposure

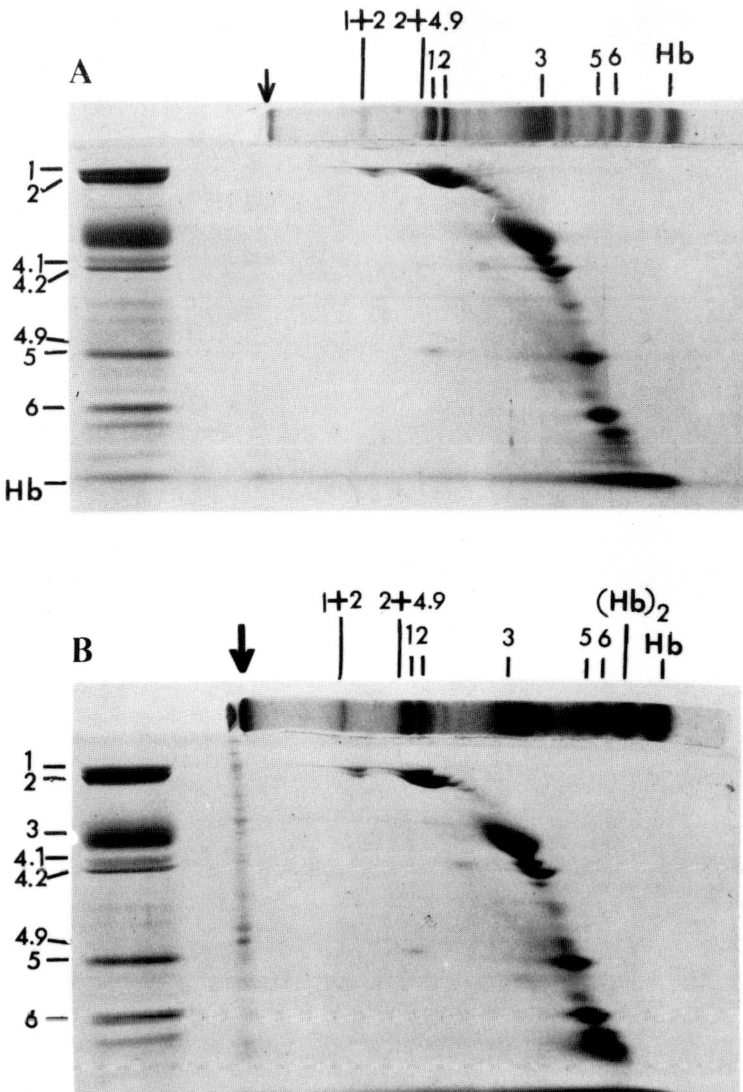

Fig. 9. Two-dimensional gel electrophorograms of membrane proteins of acetylphenylhydrazine (APH) treated normal and G6PD-deficient erythrocytes. Red cells were incubated under air with 4 mM APH for three hours. Arrows indicate the large aggregate at the origin of the first dimension. (A) Normal, (B) G6PD-deficient erythrocytes. (Reprinted by copyright permission from Blood 51:390, 1978.

of these cells to such oxidative stress produced a gel pattern, which is shown in Figure 9. It is apparent that the GSH-depleted cells with a normal ATP content did share with the aerobically ATP-depleted red cells in the presence of spectrin dimers and 2 + 4.9. In contrast to ATP-depleted red cells, the > 1,000K complex at the start of the gel of oxidized G6PD deficient erythrocytes consisted almost exclusively of globin subunits.

The above data suggest that the > 1,000K spectrin-rich polymer of ATP-depleted red cells is a consequence of two processes. First, spectrin, and to a lesser extent some other membrane proteins, were rearranged in closer contact with one another, presumably because of spectrin aggregation. Second, these aggregated proteins were subsequently coupled by disulfide cross-links which occur spontaneously when ATP depletion is performed under aerobic conditions or, in the case of anaerobically ATP-depleted red cells, after their exposure to catalytic oxidation. The second process can be reversed by reducing agents which dissociate

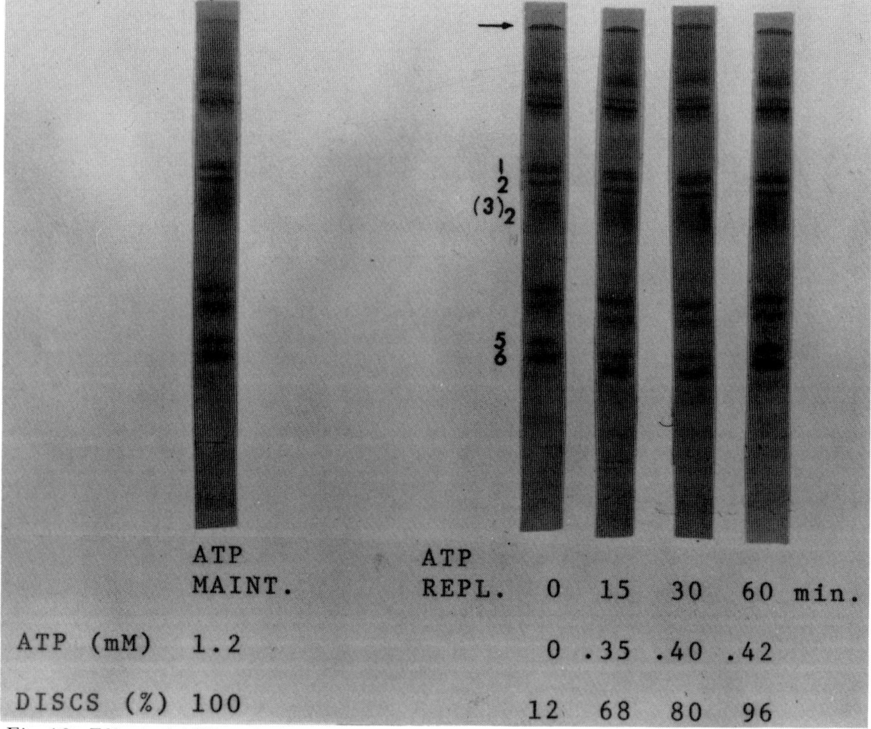

Fig. 10. Effect of ATP repletion on protein cross-linkings by $CuSO_4$/o-phenanthroline. Red cells were previously depleted in ATP by 20 hours, 37°C incubation under N_2 without glucose. ATP repletion of red cells was carried out at 37°C in buffer containing adenine (0.5 mM), inosine (12.7 mM), and glucose (2 gm/liter).

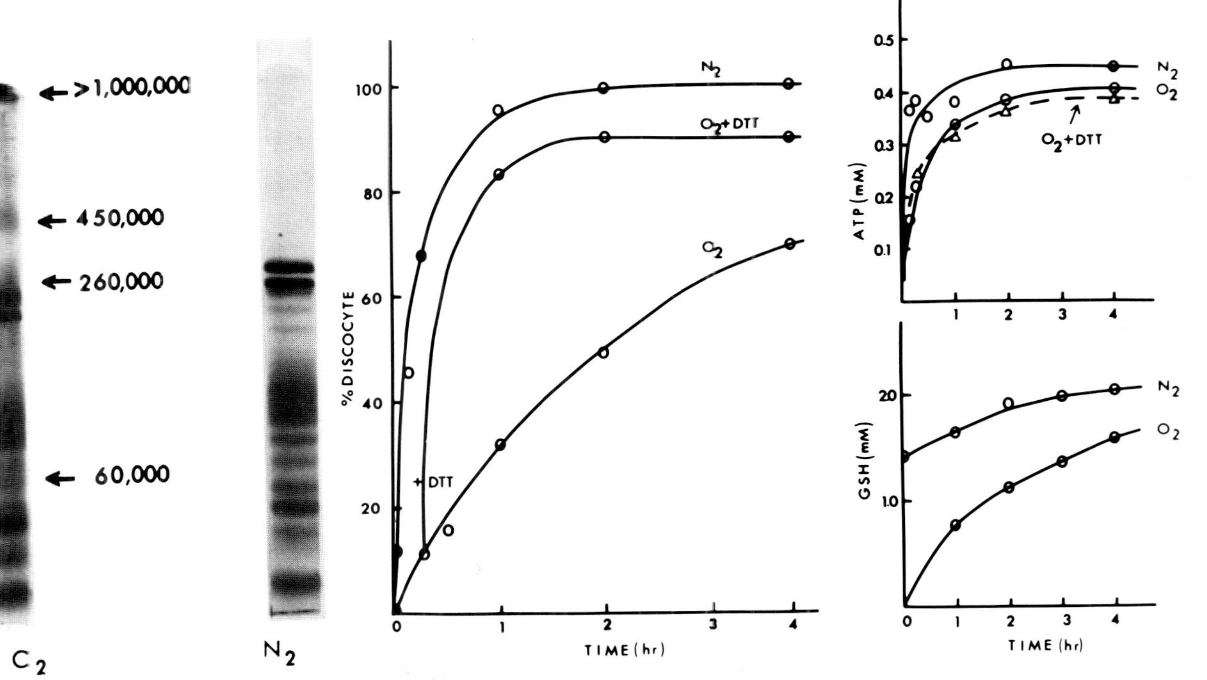

Fig. 11. Membrane protein composition, recovery of biconcave shape, ATP and GSH levels of ATP-depleted erythrocytes during a subsequent incubation with glucose, adenine, and inosine. Red cells were depleted by 19 hours of incubation (37°C) under N_2, O_2, or O_2 followed by an addition of 10 mM DTT at the onset of ATP rejuvenation. Percentage of discocytes, GSH, and ATP are plotted as a function of rejuvenation time. (Reprinted by copyright permission from Nature 274:505, 1978).

the disulfide cross-linkings. Furthermore, the first step, the rearrangement of spectrin, is probably an ATP-dependent process.

To investigate whether the second process is reversible with ATP repletion, we have depleted the red cell ATP under anaerobic condtions and subsequently restored ATP content with adenine and inosine. Ghosts from these cells were then subjected to catalytic oxidation. The results showed that the repleted cells were still able to produce this polymer, although they returned to their discoidal shape (Fig. 10). Thus, it appears that the rearrangment of spectrin during ATP depletion is an irreversible process; ie, once the aggregate is formed, it cannot be reversed by a subsequent ATP repletion. Also, it does not relate directly to shape change upon ATP repletion.

STABILIZATION OF CELL SHAPE

Cross-linking agents may be used as biological fixatives. Therefore, we inquired as to whether the cross-linking of reorganized spectrin through intermolecular disulfide bridges, which takes place spontaneously during ATP depletion under aerobic conditions, contributes to stabilization of the shape of these cells in their abnormal spheroechinocytic configuration [10].

Figure 11 indicates that aerobically ATP-depleted red cells which contain the spectrin-rich polymer, possess a markedly decreased propensity to restore their discoidal shape upon ATP repletion as compared to anaerobically prepared ATP-depleted cells which did not possess the spectrin-rich complex.

Additition of DTT, which completely dissociated the polymer, resulted in a marked improvement of restoration of the discoidal shape. In addition, there were no major differences in ATP content between the two cells types, suggesting that the delay in shape reversal was related to an oxidative protein cross-linking rather than to differences in kinetics of restoration of red cell ATP levels. This oxidative protein cross-linking of ATP-depleted red cells may act as an endogenous fixative responsible for stabilization of spheroechinocytic shape.

CONCLUSIONS

The ultrastructure of spectrin-actin meshwork was revealed in the membrane cytoskeleton, delineated by Triton X-100 extraction of erythrocyte ghosts. Based on the fibril diameter (35 Å), the calculated length of spectrin dimer is about 750 Å. The cross-linking data suggest multiple contacts between spectrin components 1 and 2 in the membrane cytoskeleton. Some structural associations between spectrin and other protein components such as 4.9 and actin are also indicated.

During aerobic depletion the erythrocytic membrane exhibits a major change characterized by a spontaneous formation of a large ($> 1{,}000$K) complex of proteins at the start of the gel. We have shown that this complex results from a rearrangement of spectrin to close contacts followed by intermolecular disulfide couplings. Although the disulfide couplings can be cleaved by restoration of the reducing power such as intercellular GSH, the rearrangement of spectrin in red cells cannot be reversed by a subsequent ATP repletion.

We have shown that the intermolecular disulfide couplings stabilize the shape of the cells. Furthermore, it is likely that these changes in spectrin organization in red cell membranes participate in regulation of cell shape, deformability, and stability, but the details of these interactions remain to be defined.

ACKNOWLEDGMENTS

This work was supported by NIH grant HL-15157 and by the Boston Sickle Cell Center. We appreciate the skilled technical help of Mr. Steven Sepe and Ms Maureen Monroe.

REFERENCES

1. Liu SC, Damico GA, Crusberg T, Palek J: Ultrastructure and protein organization of membrane skeleton in human erythrocytes. (Submitted for publication.)
2. Liu SC, Fairbanks G, Palek J: Spontaneous reversible crosslinking in the human erythrocyte membrane. Temperature and pH dependence. Biochemistry 16:4066, 1977.
3. Clarke M: Isolation and characterization of a water-soluble protein from bovine erythrocyte membranes. Biochem Biophys Res Commun 45:1063, 1971.
4. Steck TL: Crosslinking the major proteins of the isolated erythrocyte membrane. J Mol Biol 66:295, 1972.
5. Aizawa S, Kurimoto F, Yokono O: Crosslinking studies with different length dithiobisalkylimidates (1) solubilized erythrocyte spectrin. Biochem Biophys Res Commun 75:870, 1977.
6. Liu SC, Palek J: Crosslinking between spectrin and band 3 in human erythrocyte membranes (in preparation).
7. Palek J, Liu SC, Snyder LM: Metabolic dependence of protein arrangement in human erythrocyte membranes (1) Analysis of spectrin-rich complexes in ATP depleted red cells. Blood 51:385, 1978.
8. Palek J, Liu SC, Liu PA: Crosslinking of the nearest membrane protein neighbors in ATP depleted, calcium enriched and irreversible sickled red cells. In "Erythrocyte Membranes: Recent Clinical and Experimental Advances." New York: Alan R. Liss, Inc., 1978, p 75.
9. Bessis M: "An Illustrated Classification and Its Red Cell Shape." New York: Springer, 1973, p 1.
10. Palek J, Liu PA, Liu SC: Red cell membrane protein polymerization contributes to the irreversibility of spheroechinocytic shape. Nature 274:505, 1978.

3

Mechano-Chemical Study of Red Cell Membrane Structure In Situ

Evan A. Evans and Richard E. Waugh

Over the past decade, biochemists and ultrastructuralists have developed a view of the red cell membrane as a composite material [12–14]. The composite is made up of an amphiphilic bilayer "solvent" of lipids and cholesterol that contains integral proteins as "solutes." Associated with this two-dimensional fluid mixture are peripheral proteins either adsorbed or bound to the cytoplasmic face. From the correlation of red cell membrane material properties with the biochemical and ultrastructural results, the proposition has been made [15] that the membrane material can be represented as a solid-liquid composite. The solid-structural matrix is provided by the peripheral protein system, sometimes referred to as the "cytoskeleton" by biochemists and ultrastructuralists. This matrix or "scaffolding" gives the membrane its static resistance to extensional (shear) deformation at constant surface area. It is responsible for the elastic reversibility of red cell membrane extension. In addition, this component of the membrane is the primary determinant of the time-dependent extension recovery (viscoelasticity) and the fragmentation, or yield, behavior (viscoplasticity) of the red cell membrane [5, 15–17]. On the other hand, the liquid component (amphiphilic bilayer mixture of "solvent" and "solutes") provides the large resistance to increase or decrease in membrane area. This resistance to area dilation is four to five orders of magnitude greater than the resistance to extensional deformation at constant area [9, 10]; thus the red cell membrane can be considered as incompressible in two-dimensions for most cell deformations. When large transmembrane pressure differences are produced — eg, in osmotic swelling of red cell spheroids — then the resistance to area dilation is paramount and the shear rigidity can be neglected. The ratio of magnitudes — ie, $1:10^4$ or 10^5 — of the elastic modulus for extensional deformation at constant surface area to that for changes in surface area dilation or condensation is indicative of the vastly different chemical equilibrium in the solid matrix as compared to the liquid component.

Erythrocyte Mechanics and Blood Flow, pages 31–56

The discussion in this chapter focuses on the use of temperature-dependent mechanical experiments of red cell membrane elastic properties to study the thermodynamics of both membrane components in situ. The ultimate goals of this approach are to correlate the thermodynamic state of the membrane in situ with the studies performed by biochemists on molecular complexes extracted from the red cell membrane. Specifically, the origin of the weak extensional elastic behavior lies in the nature of conformational changes in the structural matrix balanced against hydrophobic and hydrophilic interactions (eg, hydrophobic effect, hydrogen bonding, electrostatic forces, etc). In addition, the basic topography of the structural matrix may depend on enzymatically mediated processes such as phosphorylation. By comparison, the strong cohesion of the amphiphilic bilayer mixture is primarily the result of hydrophobic effects; however the anisotropic solution chemistry of this mixture is not yet understood. The association between these two metaphorical components (solid matrix plus liquid mixture) is an important area for study by biochemists and biophysicists. Hopefully, the methods and preliminary results outlined here will provide insight into the physical chemistry of red cell membrane. Further details may be found in references [18–21].

MEMBRANE MECHANICS AND ELASTIC MODULI

The anisotropic or lamellar configuration of thin membrane structures like the red cell membrane is peculiar to the preferential assembly of amphiphilic molecules (eg, lipids, proteins, etc) into multicomponent mixtures. These mixtures can only be considered as continuous media in the sense of the two dimensions that describe the surface of the membrane. The very slow rate of exchange of membrane molecules between the encapsulating membrane surface and adjacent aqueous phases is evidence of the strongly anisotropic chemical behavior. Consequently, the encapsulating membrane of a biological cell or artificial lipid vesicle behaves as a closed system (ie, with fixed mass) for short periods of time (less than the order of hours). Mechanical experiments may be used to probe the intact structure of the membrane system by doing work on the structure for short periods. Mechanical properties of red cell membranes have been studied for many years; however, constitutive relations have only recently been developed that accurately represent the elastic behavior of the membrane, independent of the particular experiment [1, 2]. The surface elastic behavior of the membrane is characterized by two constants: The area compressibility modulus, K, which represents the elastic energy storage produced by area dilation or compression, and the shear modulus, μ, which represents the elastic energy storage produced by extension of the membrane in the surface plane without change in membrane area. These surface elastic moduli, with units of force per unit length, are the intrinsic determinants of membrane deformability; the additional effects of curvature elasticity or bending rigidity usually contribute negligibly to membrane resistance to deformation [3–5]. Direct measurements of the surface elastic moduli of red cell membrane have been reported [6–11] and are summarized in a recent review [5].

The intensive forces and moments supported by a thin membrane material are illustrated in Figure 1. These variables are the resultants of external forces that act on the membrane; they are intensively distributed per unit length along imaginary edges within the material surface [5, 19]. Specifically, the force per unit length along the edge is composed of two parts, one normal and one tangential

Fig. 1. Free body diagram for a thin multilayered membrane in which the adjacent layers are strongly associated (ie, coupled). Here, in addition to the shear resultant T_s and T_n shown in Figure 2, the membrane can have a moment resultant per unit length M (cgs units of dyne-centimeters per centimeter) and a transverse shear resultant Q_T (cgs units of dynes per centimeter) (taken from Evans and Hochmuth [5]).

to the edge. These are the tension, T_n, and shear resultant, T_s, respectively, with units of dynes/cm. Strongly associated layers of a composite membrane produce force resultant couples that give a moment resultant, M, which acts on the edge. In addition, a transverse shear, Q_T, is required to balance the surface gradient of the moment resultant. Because moment resultants are associated with force couples formed by very small interlayer distances (eg, on the order of molecular dimensions), the moment resultant and transverse shear contribute negligibly to the work of membrane deformation in micromechanical experiments. However, there are important situations in which moment resultants are essential for membrane stability.

Study of membrane material structure consists of an examination of the response of the membrane to the two force resultants, tension and shear. Since these force resultants act in the plane, it is appropriate to consider an element of the membrane surface that is chosen small enough to be essentially "flat" in comparison to the extrinsic curvature of the cell membrane capsule. In any deformation it is possible to select the orientation of this abstract element such that the element dimensions are simply extended or compressed; this orientation is called the "principal axes of deformation." The force resultants in the principal axes system are principal tensions, T_1 and T_2, which act along the edges of the material element without shear. Shear is present, however, but it acts maximally along lines at $\pm 45°$ to the principal axes in the material element. Consequently, the principal tensions are made up of two independent parts: 1) the mean or isotropic tension, $\bar{T}$; and 2) the deviatoric or maximum shear resultant T_s:

$$T_1 = \bar{T} + T_s$$

$$T_2 = \bar{T} - T_s$$

Figure 2 shows the principal axes deformation of a small membrane material element and the representation for the force resultant decomposition of the principal tensions into the mean or isotropic tension, plus the maximum shear resultant (with axes at $\pm 45°$ to the principal axes of extension). Implicit in Figure 2 is the association of isotropic tension with area change and shear resultant with extensional deformation at constant area. The intensive variables that describe the state of deformation of the membrane element are also shown in Figure 2: 1) α, the fractional change in element area; 2) $\tilde{\lambda}$, the element extension ratio at constant element area. The latter represents shear deformation, since the finite deformation shear strain is given by [19]

$$\epsilon_s = \frac{1}{4} \, |\tilde{\lambda}^2 - \tilde{\lambda}^{-2}|$$

for extension at constant element area. The equilibrium or static relationships between the intensive force resultants and intensive deformation variables are elastic constitutive equations — ie, membrane material equations of state at con-

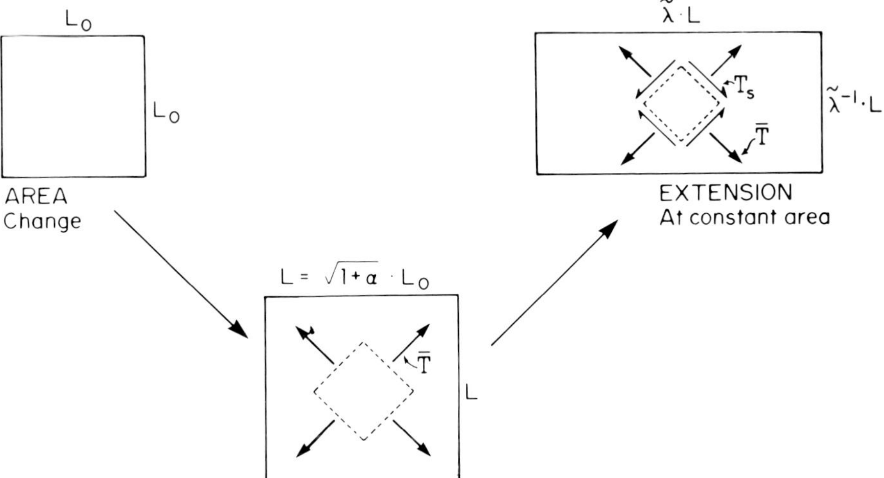

Fig. 2. Conceptual deformation of a square element of membrane surface in a reversible thermodynamic process. First, the area is increased; then, the element is extended at constant area. The intensive force resultants associated with material deformation are illustrated inside the element: $\bar{T}$ is the isotropic membrane tension; T_s is the deviatoric membrane tension – ie, the maximum membrane shear resultant.

stant temperature. The first order relations have been established and used to correlate micromechanical experiments on red cell membranes [see references 5 and 19 for extensive review] ; these relations are

$$\bar{T} = K \cdot \alpha \tag{1}$$

$$T_s = \frac{1}{2} \mu (\tilde{\lambda}^2 - \tilde{\lambda}^{-2}) = 2 \mu \epsilon_s \tag{2}$$

where K is the surface elastic modulus for area compressibility and μ is the surface elastic shear modulus. Both elastic moduli have units of dynes/cm.

The experimental methods and appropriate analyses will now be outlined for micromechanical determination of red cell membrane elastic moduli. Details of these approaches and cell preparation techniques are to be found in references [6, 8, 10, 11, 20] .

MECHANICAL DILATION OF RED CELL MEMBRANE AREA

The area compressibility modulus of red cell membrane is measured by aspiration of a pre-swollen cell into a micropipet as shown in Fig. 3. At a small initial pressure of 8.0–13.0 × 10³ dyne/cm² (6.0–10.0 mm Hg) the outer portion of the cell is spherical, which shows the dominance of isotropic tension. The pres-

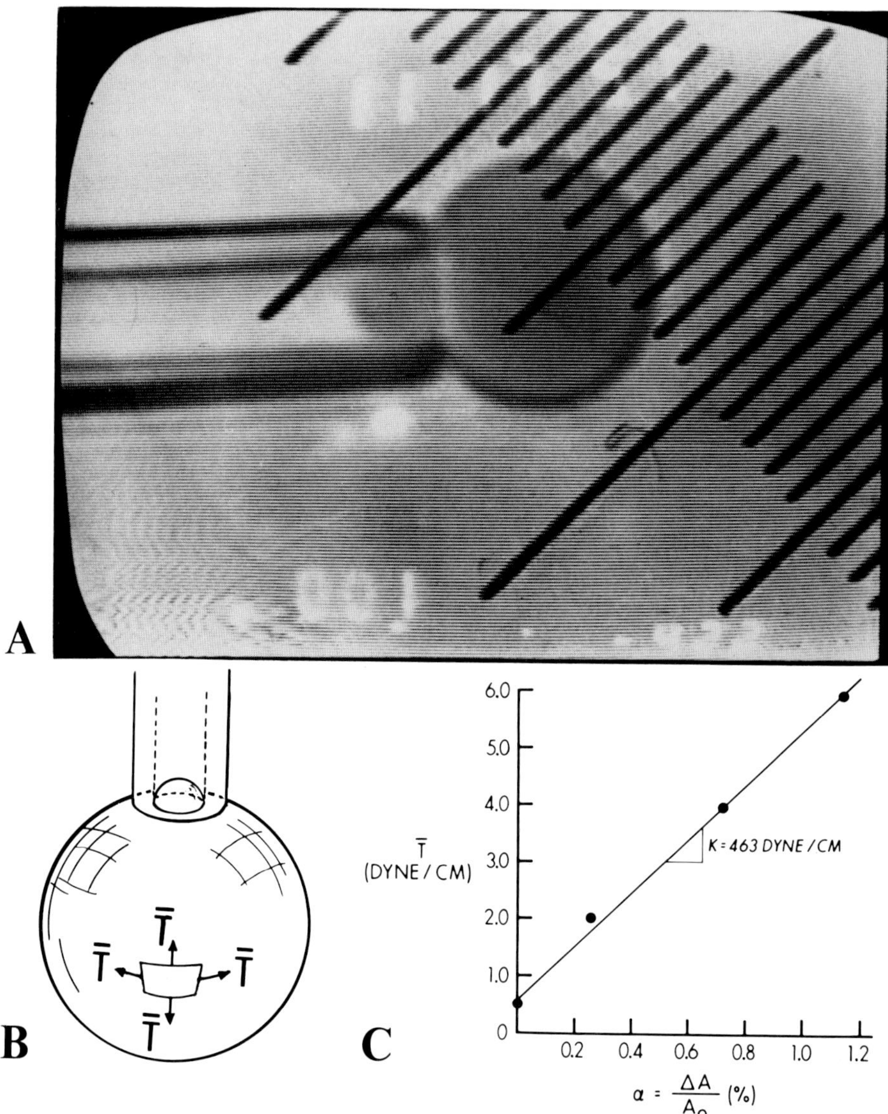

Fig. 3. Isotropic tension and area dilation produced by micropipet aspiration of osmotically swollen red cells. A videorecording of a single cell measurement (A) appears at the top the schematic of isotropic tension in the surface of an aspirated cell is at the (B) lower left; and the isotropic tension, $\bar{T}$, versus fractional change in membrane area, α is shown on the right (C) for a single cell experiment. The slope of the regression line is the isothermal elastic modulus for area compressibility, K.

sure is then increased by increments to a maximum value of $5.0–9.0 \times 10^4$ dyne/cm^2 (40.0–70.0 mm Hg). After the cell is released from the pipet, it is re-aspirated at the initial pressure to check for possible irrecoverable volume loss during the measurement. The experiment is recorded on videotape along with the time, temperature, and pressure for subsequent analysis. Hysteresis checks have been made, but none has been found within the limitations of experimental resolution. Each increment in applied pressure, ΔP, produces a change in the projection length, ΔL. The area compressibility modulus, K, is obtained from this data by calculating isotropic tension, $\overline{T}$, and fractional area change, α, for the data set (ΔP, ΔL).

The tension in the surface is obtained from the balance of forces on the surface. Provided that the frictional interaction between the cell and the pipet is negligible, the tension will be uniform throughout the surface. $\overline{T}$ can be calculated from the relationship

$$\overline{T} = \frac{\Delta P \cdot R_p}{2} \left(1 - \frac{R_p}{R}\right)^{-1} \tag{3}$$

where ΔP is the pressure difference between the pipet and the suspending medium, R_p is the radius of the pipet, and R is the radius of the outer spherical portion of the cell. The change in membrane area, ΔA, is related to the movement of the cell projection in the pipet, ΔL, and the change in cell volume, ΔV. Since the changes are small, the area change is given by the first order term in a Taylor series expansion

$$\Delta A \cong 2\pi \left(R_p \cdot \Delta L \cdot (1 - R_p/R) + \frac{\Delta V}{\pi R}\right) \tag{4}$$

All of the variables in Eq 4 can be measured directly in the experiment, except for ΔV. The change in volume must be determined by an independent relation because it is too small to be accurately measured.

It has been shown [11] that the reversible change in the cell volume is due to the pressure imposed on the cell by micropipet suction. Water is filtered out of the cell, but the number of ions in the cell remains essentially constant. The osmotic concentration within the cell increases to oppose water filtration. Steady state is reached when the *net* flow of water out of the cell is zero. The fractional change in volume due to this water loss is given by

$$\frac{\Delta V}{V_0} = \frac{-2R_w\overline{T}}{N_A kTc_i} \left(\frac{A_p/R_p + A_s/R}{A_p + A_s}\right) \tag{5}$$

where V_0 is the initial volume of the cell; R_w is the fraction of the cell volume that is osmotically active (~ 0.6); $\overline{T}$ is the isotropic tension in the membrane; $N_A \cdot k$ is the gas constant, T is absolute temperature; c_i is the osmotic concen-tration inside the cell; A_p is the area of the cell cap in the pipet; and A_s is the area of the cell in the suspending medium. The reversible volume change can be reduced in proportion to the increase in osmotic strength of the cell contents. In addition to reversible exchanges of water from the cell, sometimes an irrecoverable change in volume occurs during the measurement. These net changes are the result of small changes in the osmolarity of the solution near the pipet entrance. The irrecoverable changes can be essentially eliminated by closely matching the tonicity of the pipet and suspending solutions. This effect is also minimized at higher osmotic concentrations of the suspending media. The ionophore nystatin is used to load the red cells with high ionic strength potassium chloride (up to 300 millimolar). With Eq 5 for the volume change and Eq 4, the fractional change in area is calculated for the micropipet experiment. Since the slope of the isotropic tension versus fractional change in area is *independent* of the ionic strength, it is demonstrated that the effects of water volume exchange can be eliminated. Figure 3 shows a typical result for tension versus fractional change in area; the slope of the linear regression line is the surface elastic modulus of area compressibility, K, at a specific temperature. Figure 4 shows the experimental verification of the effects of water filtration from the cell and the volume correction. The surface elastic modulus of area compressibility ranges from 600 dyne/cm at 0°C to 300 dyne/cm at 50°C for red cell membrane [20].

MECHANICAL EXTENSION OF RED CELL MEMBRANE AT CONSTANT AREA

The elastic shear modulus of the membrane is measured by aspirating a flaccid, unswollen cell in the dimple region, as shown in Figure 5. From an initial pressure of 200–300 dyne/cm^2 ($\sim$ 2–3 mm H$_2$O), the pressure is increased in small increments. If the cell begins to fold or buckle, the measurement is terminated. Cells are then tested for hysteresis as the suction in the pipet is decreased in several increments to the initial pressure. The test is recorded on videotape with the temperature and pressures. Frictional effects from the pipet wall have been shown to be on the order of only 10% [20, 22], within the limits of experimental resolution, and they are minimized by averaging the loading and unloading phases of the aspiration experiment. The aspiration of the flaccid cell membrane requires suction pressures two orders of magnitude smaller than for the spheroidal cell experiments; consequently, the membrane area can be considered as constant — ie, an incompressible surface. This experiment produces membrane extension at constant area, as shown by the second step in Figure 2.

Membrane tensions produced by a small-caliber pipet are concentrated in, and near, the pipet mouth [6]. These tensions drop off inversely as the square of the distance from the pipet center, normalized by the pipet radius. Aspiration of the cell to such an extent that the membrane tensions at the cell rim or periphery are

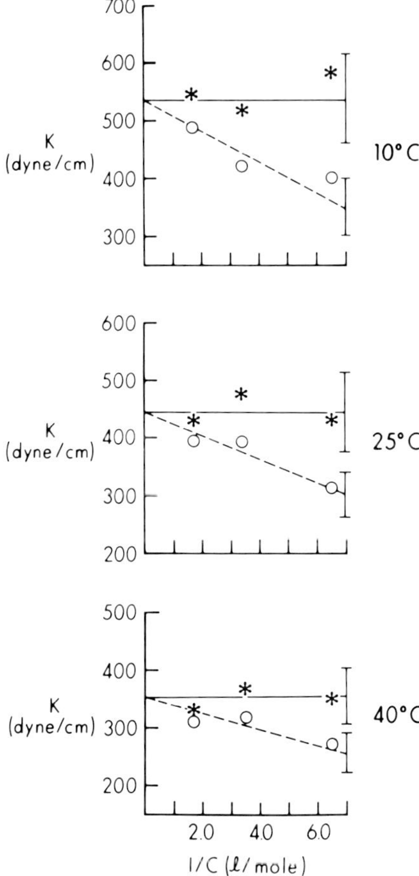

Fig. 4. The elastic area compressibility moduli of nystatin-treated cells are plotted versus one over the osmolarity in which the cells were swollen (in liters per mole); the data points are the average moduli for about 20 cells, calculated with and without the osmotic volume correction. Experiments are shown for three different temperatures. The solid line and dashed line in each plot represent the predicted behavior of the area compressibility modulus with and without the osmotic volume correction, respectively; these lines are calculated with the corrected and uncorrected average moduli obtained from measurements of several hundred untreated red cells (taken from Evans and Waugh [11]).

appreciable results in curvature changes at the cell rim, which are directly observable, thereby providing an experimental indicator for the limitation of the analysis. The analysis of the micropipet aspiration experiment involves two independent steps: First, the equations of equilibrium are satisfied for the forces that act on the membrane to give the intensive distribution of membrane force

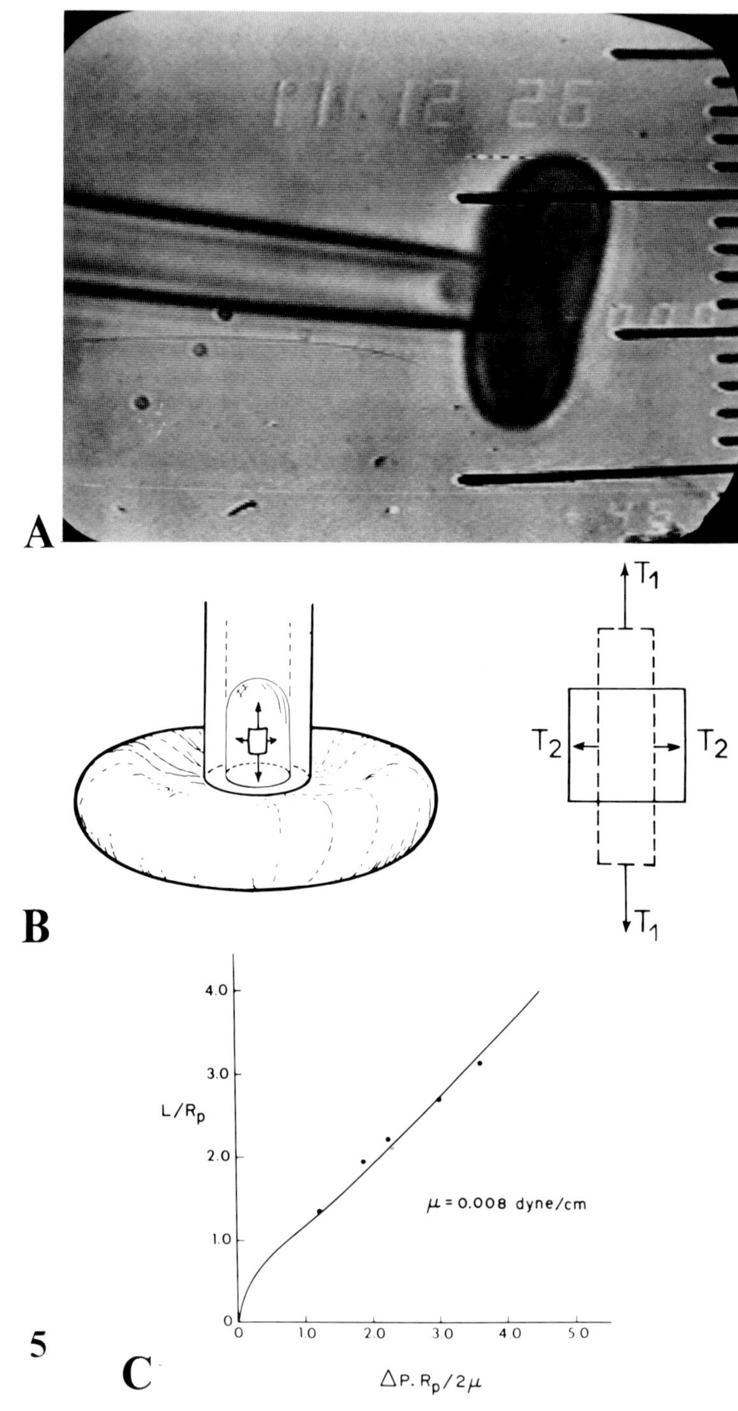

T₁
T₂
T₂
T₁
4.0
3.0
L/R_p
2.0
1.0
0
μ = 0.008 dyne/cm
0
1.0
2.0
3.0
4.0
5.0
ΔP·R_p/2μ
A
B
C
5

resultants; second, the deformation of the membrane surface is analyzed to determine the extension ratios of the material as a function of location. The result is the recipe for experimental determination of membrane shear force resultant in relation to membrane shear deformation. In turn, the experimental results are correlated with the elastic constitutive relation for membrane shear (Eq 2) to evaluate the membrane surface elastic shear modulus, μ.

The balance of forces and equations of equilibrium for the membrane require two critical assumptions: 1) the membrane is pushed tightly against the glass wall of the pipet, without wall friction; 2) only force resultants in the plane of the membrane are important in the equations of equilibrium; ie, contributions from bending moments are negligible. These assumptions are verified by experimental observations [19, 20]. The two assumptions permit the treatment of the principal force resultants (tensions) in the membrane as continuous from the aspirated region to the outer membrane surface. The exterior membrane surface is conceptualized as infinitely large in comparison to the pipet dimension; thus, the equations of equilibrium are integrated in the plane of the membrane from the pipet tip outward, to give

$$\Delta P = \frac{4}{R_p} \int_{R_p}^{\infty} \frac{T_s}{r} \, dr \tag{6}$$

The force resultants have been assumed to be negligible at distances far from the pipette (ie, at the periphery of the cell; this is consistent with the flaccid, essentially unchanged shape of the equatorial region of the red cell during the experiment). The intensive membrane shear force resultant is related to the externally applied pipet suction pressure by the integral equation (6).

The state of deformation in the membrane surface must be determined as a function of the length of the cell projection that is aspirated into the pipet. The material extension ratio is a function of radial position; it is used with the elastic constitutive relation (Eq 2) in Equation 6 to *predict* the suction pressure that is required to aspirate a membrane projection of length, L, given the pipet radius and material elastic shear modulus, μ. Correlation of the prediction and the observed relation between pressure and length provides the value for the elastic shear modulus. From the previous data for resistance to area dilation, it is

---

Fig. 5. Deviatoric tension (shear) and extensional deformation produced by micropipet aspiration of flaccid red cell disks. A videorecording of a single cell measurement (A) appears at the top; the schematic of principal tensions in the surface of an aspirated cell (B) is at the lower left; and the correlation of measured suction pressure, ΔP, and projection length, L, with the theoretical prediction (C) is shown on the right for a single cell experiment. The correlation is provided by the elastic shear modulus, μ, of the membrane.

recognized that the red cell membrane greatly resists changes in local surface area. Therefore, the surface can be assumed incompressible in the two dimensions that characterize the membrane surface. Only a single principal extension ratio, say along the radial direction, need be considered, since the other extension ratio is given by its reciprocal. Incompressibility or constant area implies that each element of area of the membrane surface remains constant in magnitude when deformed even though its shape has changed. Figure 6 illustrates the displacement and constant area deformation of an annular ring which results from the pipet aspiration. Two radial coordinates must be specified; r_0 is the initial radial position in the undeformed flat surface; r is the instantaneous radial position of the same point in the axisymmetric surface that comprises the projection and the surface exterior to the pipet entrance. The extension ratio at a point outside the pipet entrance is given by the ratio

$$\lambda_m = \frac{r_0}{r}$$

for the same material locations. The total areas from the origin ($r = r_0 = 0$) must be equal and are obtained by integration. This gives

$$r \geqslant R_p \qquad r_0^2 = \frac{A_{cap} + A_{cyl}}{\pi} + (r^2 - R_p^2)$$

where A_{cap} is the area of the spheroidal cap and A_{cyl} is the area of the cylindrical section. Consequently, the extension ratio is specified as

$$r \geqslant R_p \qquad \lambda_m^2 = \frac{A_{cap} + A_{cyl}}{\pi r^2} + \left(1 - \frac{R_p^2}{r^2}\right) \tag{7}$$

by division with the square of the radial coordinate, r. It is apparent that the square of the material extension ratio drops off as $1/r^2$ from the pipet entrance. The exact shape of the spheroidal cap can only be established by satisfying the equations of equilibrium for the membrane surface in this region subject to a hydrostatic pressure difference; this is not possible in closed form but can be done on a digitial computer. However, this region contributes minimally to the maximum degree of extension as the projection increases. Therefore, we simply use spherical surface segments to approximate the cap geometry as the projection length goes from $0 \leqslant L \leqslant R_p$ and consider as hemispherical for projection lengths greater than a pipet radius. With this approach, Eq 7 is given by

$$\lambda_m^2 = 1 + \left(\frac{R_p}{r}\right)^2 \left(\frac{2L}{R_p} - 1\right) \tag{7}$$

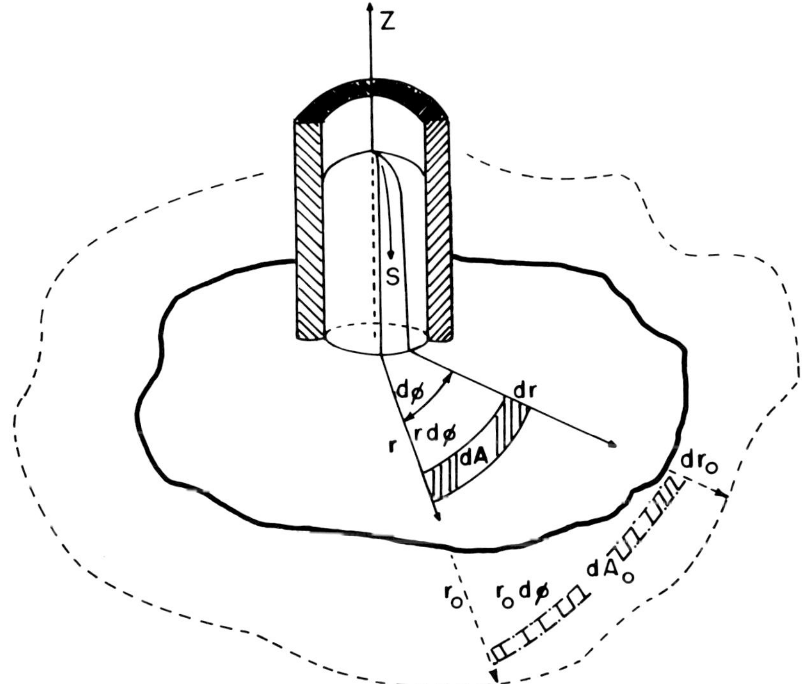

Fig. 6. Constant area deformation of a flat surface into a cylindrical pipet. The initial co-ordinate of a point in the surface is r_0, and its instantaneous location in space is given by r. The degree of extension along the radial direction is measured by the extension ratio, $\tilde{\lambda} = \dfrac{r_0}{r}$.

which is valid for projection lengths greater than one pipet radius. The maximum extension ratio in the membrane occurs at the pipet entrance and is equal to $\hat{\lambda} = \sqrt{2L/R_p}$.

With the first-order elastic constitutive relation for the shear resultant

$$T_s = \frac{\mu}{2}\left(\lambda_m^2 - \lambda_m^{-2}\right)$$

the integral relation (Eq 6) predicts the relation between the pipet suction pressure and the aspirated projection length [6]

$$\Delta P = \frac{2\mu}{R_p}\int_{R_p}^{\infty}(\lambda_m^2 - \lambda_m^{-2})\frac{dr}{r}$$

or

$$\Delta P = \left(\frac{2\mu}{R_p}\right)\left[\left(\frac{2L}{R_p} - 1\right) + \ln\left(\frac{2L}{R_p}\right)\right] \tag{8}$$

The relationship between ΔP and L is shown in Figure 5, along with data from a typical cell. The elastic shear modulus is determined by correlating the observed projection length versus aspiration pressure. The values for shear moduli of red cell membrane as a function of temperature are presented in Figure 7. The surface shear modulus ranges from 8×10^{-3} dyne/cm at 5°C to 5.5×10^{-3} dyne/cm at 45°C, with an apparent linear behavior. As previously stated, the elastic modulus for energy storage in extensional deformation of the red cell membrane is about five orders of magnitude smaller than the elastic modulus for energy storage in surface dilatation.

EXTRINSIC DEFORMABILITY OF RED CELLS

The major requirements for overall cell "deformability" in the circulation are obvious; these have been stated in many ways for over three decades since Eric Ponder's time: 1) surface area to volume ratio must exceed that for a sphere (a spherical membrane shell would act as a rigid particle); 2) membrane material must have a low resistance to extensional deformation (low shear modulus) but be strong enough not to fragment or yield in shear [5, 8, 17]. Deviations of the

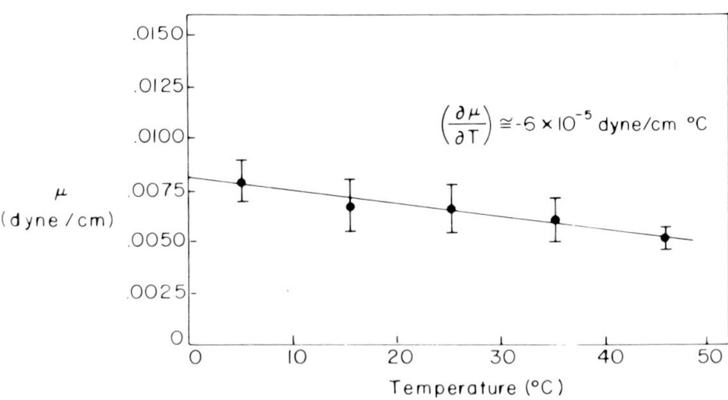

Fig. 7. The elastic shear modulus as a function of temperature. Points are the average of about 25 cell moduli at each of the five experimental temperatures; brackets are plus or minus one standard deviation. The solid line is the linear regression to the data (taken from Waugh and Evans [20]).

extrinsic form of the red cell from a discocyte (eg, to an echinocytic or stomatocytic form) can act to "stiffen" the cell. Such alterations can be produced by small changes in the chemical environment [3, 5]. Also, time-dependent or dissipative mechanisms [16] are important in the dynamics of cell deformation. The groundwork has been laid for further study in these areas; however, from the perspective of membrane science, an important question should now be addressed: Can we use measured material properties to provide information about the intact molecular structure of the red cell membrane? The answer is yes, but it requires an involvement with thermodynamics: The common "meeting place" for physics, chemistry, and life sciences.

MEMBRANE THERMOELASTICITY

The elastic properties of a closed membrane system are associated with reversible thermodynamic changes in the membrane; these changes are produced by deformation of the membrane structure. The elastic coefficients are derivatives of the free energy density at constant temperature (work per unit area of the membrane), taken with respect to intensive deformation. The mechanical work done on a material at constant temperature is the displacement of forces supported by the material. Mechanical work produces a coherent or deterministic change in the internal energy and heat content of the material, which is reversible for an elastic substance. Furthermore, the forces can be related to the material deformation and specified in both direction and magnitude. Temperature change alters the material internal energy and heat content in an incoherent manner without displacements of deterministic forces; it does so by random exchanges of momentum between the material surface and the adjacent environment and by radiation absorption and molecular excitation. For reversible processes, the two processes for changing internal energy and heat content are related by the combined first and second laws of thermodynamics with the material equations of state, or elastic constitutive relations. Experimentally, the interchange is evidenced, for example, in the reversible heat exchange to the environment measured by a calorimeter when work is done on the material at constant temperature *or* by the thermoelastic forces that are produced in a constrained material when its temperature is changed. Since it is impossible to construct a calorimeter to measure the reversible heat exchange that occurs during micromechanical experiments on cell membranes, thermoelastic behavior can be used to determine the relative changes in material internal energy and heat content that are produced by deformation. For example, it can be established whether the configurational state of molecular complexes is more *or* less ordered by the deformation and whether the energy of these complexes is appreciably changed by the deformation. For a biological membrane structure composed of amphiphilic components, thermoelasticity can provide a direct assessment of thermal repulsive forces in the membrane. In equilibrium, without externally applied forces, thermal repulsive forces in the membrane are balanced by the hydrophobic interaction between the adjacent aqueous phases in the membrane surfaces [18].

Thermoelastic behavior is predicted by the interrelation of the equations of state for the material with the combined first and second laws of thermodynamics for the closed material system [18]. The first equation of state is isotropic in the surface — ie, independent of surface coordinates; this equation relates the mean or isotropic tension (dyne/cm) to changes in temperature and fractional change in area. Differentially, this is written as

$$d\overline{T} \;=\; K d\alpha + \left(\frac{\delta\overline{T}}{\delta T}\right)_{\alpha} dT \tag{9}$$

where the membrane isotropic tension is $\overline{T}$; the fractional change in local area $(\Delta A/A_0)$ is α; and T is temperature ($^\circ$C). The coefficients in this differential form of the equation of state are the isothermal area compressibility modulus, K(dyne/cm), and the thermoelastic effect $(\delta\overline{T}/\delta T)_{\alpha}$. The first term in the differential form of the isotropic equation of state is the elastic constitutive behavior at constant temperature. The thermoelastic effect is the isotropic tension change that is produced by a change in temperature at constant surface area. This effect is given by the equation

$$\left(\frac{\delta\overline{T}}{\delta T}\right)_{\alpha} \;=\; -K \left(\frac{\delta\alpha}{\delta T}\right)_{\overline{T}} \tag{10}$$

in terms of the thermal area expansivity, $(\delta\alpha/\delta T)_{\overline{T}}$ ($^\circ$C^{-1}) [18]. The thermal area expansivity, like the area compressibility modulus, can be measured directly. It is the fractional change in area that is produced by a change in temperature at constant membrane tension.

For a solid membrane material (like red cell membrane), there exists an additional surface equation of state that relates the shear force resultant in the membrane, T_s (dyne/cm), to membrane extensional deformation at constant surface area and to temperature [18, 19]. Differentially, this equation of state (termed deviatoric because the maximum shear force resultant is given by the deviation between principal membrane tensions) is expressed as

$$dT_s \;=\; \left(\frac{\delta T_s}{\delta\widetilde{\lambda}}\right)_{T,\,\alpha} d\widetilde{\lambda} + \left(\frac{\delta T_s}{\delta T}\right)_{\widetilde{\lambda},\,\alpha} dT \tag{11}$$

where $\widetilde{\lambda}$ is the in-plane extension ratio of an element of membrane material at constant surface area (Fig. 2). The coefficients represent the elastic resistance to extension at constant temperature and surface area plus the thermoelastic shear resultant produced by temperature change at constant element extension, and surface area. Since the red cell membrane exhibits essentially first-order hyperelastic response to extension at constant surface area, the partial derivatives are given by

$$\left(\frac{\delta T_s}{\delta \widetilde{\lambda}}\right)_{T,\alpha} = (\widetilde{\lambda} + \widetilde{\lambda}^{-3})$$

and,

$$\left(\frac{\delta T_s}{\delta T}\right)_{\widetilde{\lambda},\alpha} = \left(\frac{d\mu}{dT}\right) \frac{(\widetilde{\lambda}^2 - \widetilde{\lambda}^{-2})}{2} \tag{12}$$

where the surface elastic shear modulus, μ (dyne/cm), is assumed to be independent of the membrane extension.

The coefficients in the isotropic and deviatoric equations of state are the two surface elastic moduli of the membrane, the thermal area expansivity, and the temperature derivative of the elastic shear modulus. These properties can be determined by direct mechanical experiment on vesicular membrane systems that encapsulate liquid interiors such as the red cell membrane. The temperature dependence of the elastic moduli of the red cell membrane have already been described.

TEMPERATURE DILATION OF RED CELL MEMBRANE AREA

The fractional change in red cell membrane area with temperature is determined by a procedure similar to that employed in the area dilation experiment. With micropipet aspiration of an osmotically swollen, nearly spherical red cell, the isotropic tension is easily controlled by the suction pressure. The cell is held at a constant suction pressure; the temperature change produces a reversible change in the length of the aspirated cell projection [20]. The length increase is proportional to the increase in area of the cell membrane. There is also a contribution to the movement of the cell projection due to reversible water exchange across the cell membrane. As discussed previously, the water exchange is inversely proportional to the osmotic strength of the red cell contents and, therefore, can be minimized by increasing the intracellular osmotic strength. This is accomplished by treating the cells with nystatin, loading them with potassium chloride, and washing away the nystatin, leaving red cells with high internal osmotic strength. If the change in volume due to water movement is eliminated, the

change in area of the red cell is simply a linear function of the projection length
for small area changes

$$\Delta A \cong 2\pi R_p \cdot \Delta L \cdot (1 - R_p/R)$$

From this relation, the change in membrane area is obtained as a function of
temperature at constant membrane tension. Figure 8 is an example of the area
change obtained for single cell experiments at different cellular osmotic strengths.
The fractional change in area is obtained by division of each experimental result
by the initial cell area at 25°C. The results for specific osmotic strengths are the
apparent fractional change in membrane area with temperature. The apparent
fractional change in membrane area is measured at different internal osmotic
strengths; thus, it is possible to obtain the actual fractional change in membrane
area with respect to temperature from the extrapolation to infinite internal
osmotic strength (zero water movement). The line in Figure 8 is the behavior for
cell area change with temperature determined by this procedure. The thermal
area expansivity of red cell membrane is found to be 1.2×10^{-3} $°C^{-1}$ with
a standard deviation of 0.2×10^{-3} $°C^{-1}$; ie,

$$\left(\frac{\delta\alpha}{\delta T}\right)_{\overline{T}=0} = 1.2 \times 10^{-3} \ °C^{-1}$$

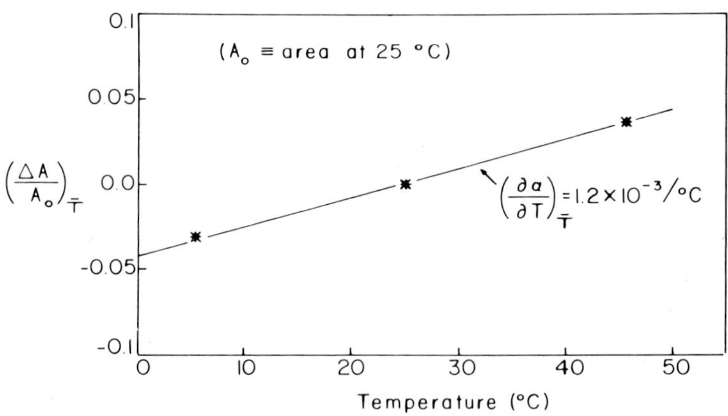

Fig. 8. Apparent area change of a single red cell relative to its area at 25°C versus temperature.
The apparent area change is calculated with the assumption that the cell volume is constant.
Data given here are for a cell swollen in 575 mOsm salt solution; thus, the effect of cell
volume change is less than 10% of the apparent change in area.

MEMBRANE THERMODYNAMICS AND MECHANICAL WORK

Reversible heat exchange and internal energy change are mechanochemical effects produced by membrane deformation. Prescriptions for measuring these effects are obtained from thermoelastic and thermodynamic relations [18]. For a reversible (elastic) process, the differential work is related to the combined first and second laws of thermodynamics,

$$d\widetilde{W} = d\widetilde{E} - Td\widetilde{S}$$

where $\widetilde{E}$ is the internal energy density (ergs/cm^2) of the membrane and $\widetilde{S}$ is the entropy density (ergs/cm^2/$^{\circ}$C) of the membrane material. The differential work done on the membrane per unit area by a mechanical experiment is expressed by the following equation [18]:

$$d\widetilde{W} = \bar{T} \cdot d\alpha + 2\,(1 + \alpha) \cdot T_s \cdot \frac{d\widetilde{\lambda}}{\widetilde{\lambda}}$$

where the first term on the right-hand side is the work associated with uniform dilation, or condensation, of the membrane; and the second term is the work of membrane extension at constant surface area. With cross-derivatives plus differential expansions of the thermodynamic state functions, $\widetilde{E}$ and $\widetilde{S}$, partial derivatives of internal energy density and entropy density can be related to the coefficients of the membrane equations of state [18].

The fractional change in internal energy density with expansion at constant temperature is given by

$$\left(\frac{\delta\widetilde{E}}{\delta\alpha}\right)_T = \bar{T} + T \cdot K \cdot \left(\frac{\delta\alpha}{\delta T}\right)_{\bar{T}} \tag{13}$$

and the internal energy density is

$$\left(\frac{\delta E}{\delta\widetilde{\lambda}}\right)_{T,\,\alpha} = \left(\mu - T\frac{d\mu}{dT}\right)\,(\widetilde{\lambda} - \widetilde{\lambda}^{-3}) \tag{14}$$

for extensional deformations at constant surface area. Likewise, the heats of expansion and extension at constant temperature are given by the fractional changes in entropy density produced by deformation; the heat of expansion is

$$T\left(\frac{\delta\widetilde{S}}{\delta\alpha}\right)_T - T \cdot K \cdot \left(\frac{\delta\alpha}{\delta T}\right)_{\bar{T}} \tag{15}$$

and the heat of extension is

$$T \left(\frac{\delta \widetilde{S}}{\delta \widetilde{\lambda}}\right)_{T,\alpha} = -T \cdot \frac{d\mu}{dT} \cdot (\widetilde{\lambda} - \widetilde{\lambda}^{-3}) \tag{16}$$

for constant surface area.

In the "natural" state, free of applied forces, the membrane isotropic tension is zero and the extension ratio is identically one. Therefore, the free energy density at constant temperature is a minimum, and

$$\left(\frac{\delta \widetilde{E}}{\delta \alpha}\right)_T - T\left(\frac{\delta \widetilde{S}}{\delta \alpha}\right)_T = 0$$

The heat of expansion at zero tension, $T \cdot K \cdot (\delta\alpha/\delta T)_{\overline{T} = 0}$, is opposed by the potential increase in internal energy density with expansion. The heat of expansion is a direct measure of the surface cohesion that resists the thermal expansion. Such a quantity is useful for determining the relative intrinsic resistance to lysis of a cell membrane, which often results in disease or with exposure to degrading chemical agents.

MECHANOCHEMISTRY OF RED CELL MEMBRANE

Equations 13–16 characterize the thermodynamic response of the membrane to deformation at constant temperature; the elastic free energy changes produced by area change or membrane extension are decomposed into internal energy and configurational entropy alterations which take the form of reversible heat exchange to the environment. With the results obtained for red cell membrane and calculations for a phospholipid bilayer, it is possible to form a comparison of a bilayer membrane to the red cell membrane. Above the phase transition temperature for ordered acyl chains, a phospholipid bilayer is a two-dimensional liquid without surface shear rigidity. Therefore, the reversible heat of extension and internal energy change in extensional deformation are zero for the bilayer membrane. However, along with cholesterol, it forms a "solvent" phase for integral proteins and is an important aspect of membrane chemistry. The heat of expansion, thermal area expansivity, and elastic area compressibility modulus are isotropic surface properties of a closed membrane system. As such, these properties characterize the changes in state of the membrane "mixture" with changes in surface area. As yet, essentially no data of this type exist for closed phospholipid bilayer systems — eg, single-walled vesicles. Unpublished x-ray data by Drs. Costello and Gulik-Krzywicki show that lamellar lecithin phases have a negative fractional change in thickness of about $2–2.7 \times 10^{-3}$ per degree centigrade. Since the volumetric thermal expansivity is probably much smaller, the thermal area expansivity for a phospholipid bilayer is at least this large, but of opposite sign [18]. This value is twice the measured thermal area expansivity of red cell membrane. Unfortunately, the elastic area compressibility modulus has not been measured for lipid bilayer membranes, but it is possible to estimate the area compressibility by idealizing the bilayer as two lipid monolayers above the phase transition for ordered acyl chains. If data are chosen from measurements for lecithin monolayers [23] the elastic area compressibility for a lecithin bilayer is estimated to range from 220 dyne/cm at $0°C$ to 150 dyne/cm at $50°C$ [20]. These values are about half the size of the elastic area compressibility modulus that has been measured for red cell membrane. These results lead to the deduction that the red cell membrane behaves like a mixed phase of compressible lipid-like liquid and much less compressible constituents. If such a simple model is postulated, then the area compressibility modulus would be elevated by an amount proportional to the total area divided by the area occupied by the compressible lipid phase and, likewise, the thermal area expansivity is expected to be lower by the inverse of this quotient. The reduced compressibility of the mixture could be the result of cholesterol or integral protein components, or both. Such deductions are *speculative* when dealing with heterogeneous mixtures but provide a basis for comparison of the well-defined mixtures with cell membrane materials. For example, the heat of expansion for the red cell membrane is determined to range from 197 ergs/cm^2 at $0°C$ to 116 ergs/cm^2 at $50°C$. The heat of expansion for the lecithin bilayer is calculated to be 125 ergs/cm^2. Above $45°–50°C$, the red cell membrane loses its structural rigidity and can be considered as a two-dimensional liquid mixture. It appears to have a heat of expansion that is expected for a lipid bilayer. Clearly, it is necessary to obtain direct experimental evidence on the state of pure lipid vesicle systems and on the state of lipid mixtures with specific components that make up a biological membrane (eg, the red cell membrane). These studies can be used to investigate changes in membrane structure that are produced by the biological chemistry of viral insemination and immune reaction, for example.

Consider now the thermoelastic behavior that is peculiar to solid membrane materials: The temperature dependence of the membrane shear modulus and the reversible heat of extension at constant surface area. Unlike the phospholipid component of a biological membrane, the composite red cell membrane exhibits solid elastic properties – ie, the ability to support membrane shear force resultants in proportion to shear deformation or uniaxial extension in the membrane plane. Similarly, we are able to neglect the small amount of work and thermoelastic change that is associated with the small surface area change. Because of the hyperelastic behavior exhibited by the red cell membrane and the magnitude of its shear modulus, it was anticipated that the membrane shear elasticity was due to the negative configurational entropy changes when the material was deformed — ie, ordering of the structural component. However, recent experiments have shown that the hyperelastic red cell membrane is not this type of simple entropic elastomer. The membrane elastic shear modulus *decreases* with temperature. The membrane configurational entropy is lowest in the undeformed state and *increases* with extension.

Biochemical and ultrastructural evidence strongly support the thesis that the spectrin protein material (adsorbed on or bound to the cytoplasmic face of the red cell membrane) is responsible for the shear elastic solid behavior of the red cell membrane. As yet, there is little known about the direct relation of spectrin to integral membrane proteins and the amphiphilic lipid components themselves. It is clear that spectrin prefers the interfacial "phase" adsorbed on the membrane surface as opposed to solution in an aqueous phase with moderate ionic strength (> 0.01 M). The natural state of spectrin in a normal red cell membrane is disrupted by temperatures in the range of $45°{-}50°$C, low pH (≈ 5), and low ionic strength aqueous media (eg distilled water).* At high temperature or low pH, spectrin still resists solution in the cytoplasmic phase and self-aggregates; it forms patches or clumps on the membrane surface. At low ionic strengths, spectrin will go into solution but some self-aggregation persists. The implication is that the conformation of the spectrin molecule is very sensitive to the balance of hydrophobic and electrostatic interactions. Furthermore, it appears that for the normal environment present in the red cell cytoplasm (pH and ionic composition), spectrin favors an adsorbed interfacial phase or layer configuration. Such a configuration represents a moderately ordered state because of restriction to essentially two dimensions. The thermoelastic measurements for membrane shear deformation show that the configurational entropy is increased by extension relative to the natural state; this indicates that extension forces the molecular complexes to disorder or utilize more degrees of freedom. If the concept of a spectrin layer is valid, it is expected that portions of the molecule would have to rotate out of the plane (eg, by transgauche rotation) when extended *with* both the constraint that the surface density remain constant and the limitations of molecular flexibility. Another aspect of extension at constant surface area is that material is compressed in the direction normal to the axis of extension; the compression or condensation of dense surface material like spectrin (while being extended in the other direction) could easily disrupt the organization or order of the molecular arrangement. In any case, such actions essentially force spectrin to partially go into solution in the cytoplasmic interior.

Recent evidence [21] indicates that spectrin may be bound to an integral membrane protein or fragment of protein. In addition, spectrin exists in at least

---

*ATP depletion also appears to affect the spectrin configuration but the evidence is deduced from red cell crenation (echinocyte formation) or shape change.

dimeric form as well as monomeric units when eluted from the red cell membrane and may be associated with other membrane proteins [14]. Consequently, extensional deformation of the red cell membrane (at constant surface area) could produce reversible (since it is elastic) changes in the association of these various membrane proteins by disrupting the normally ordered subsurface structure again consistent with the observed entropy increase.

The likelihood of spectrin as the primary structural element for surface shear rigidity and the plausibility of the previous arguments indicate that the heat of extension and internal energy density change should be considered on a per mole basis of spectrin material. The surface density of spectrin for the red cell membrane is on the order of 10^{-7} gm/cm^2, and its molecular weight is on the order of 10^5 daltons; therefore, the molar density is calculated to be 10^{-12} moles/cm^2. With this value and the measurements for the heat of extension of the red cell membrane and internal energy density change produced by extension, the values distributed per mole of spectrin are obtained. First, the reversible heat of extension is calculated from Eq 16 from the temperature gradient of the membrane surface elastic shear modulus, μ. The temperature gradient of the elastic shear modulus of the red cell membrane surface is -6×10^{-5} dyne/cm $-$ $^\circ$C [20]. Consequently,

$$T \left(\frac{\delta \bar{S}}{\delta \tilde{\lambda}} \right)_{T, \alpha} = 2 \times 10^{-2} \cdot (\tilde{\lambda} - \tilde{\lambda}^{-3}) \text{ ergs/cm}^2$$

Also, the reversible change in internal energy density with extension at constant area is given by Eq 14

$$\left(\frac{\delta \tilde{E}}{\delta \tilde{\lambda}} \right)_{T, \alpha} = 2.6 \times 10^{-2} \cdot (\tilde{\lambda} - \tilde{\lambda}^{-3}) \text{ ergs/cm}^2$$

The difference between these two relations is the free energy density change at constant temperature and density which is produced by extension

$$\left(\frac{\delta \tilde{F}}{\delta \tilde{\lambda}} \right)_{T,\alpha} = \left(\frac{\delta \tilde{E}}{\delta \tilde{\lambda}} \right)_{T,\alpha} - T \left(\frac{\delta \tilde{S}}{\delta \tilde{\lambda}} \right)_{T,\alpha} = \mu \left(\tilde{\lambda} - \tilde{\lambda}^{-3} \right)$$

where the coefficient is the surface elastic shear modulus. It is apparent that the small shear modulus is the difference between two contributions of nearly equal magnitude; consequently, the resolution of the temperature dependence of the shear modulus is critical in the determination of the free energy decomposition.

Per mole of spectrin material, these values become

$$ T \left(\frac{\delta \widetilde{S}}{\delta \widetilde{\lambda}} \right)_T \sim 2 \times 10^{10} \cdot (\widetilde{\lambda} - \widetilde{\lambda}^{-3}) \; \text{ergs/mole} $$

$$ \left(\frac{\delta \widetilde{E}}{\delta \widetilde{\lambda}} \right)_T \sim 2.5 - 2.8 \times 10^{10} \cdot (\widetilde{\lambda} - \widetilde{\lambda}^{-3}) \; \text{ergs/mole} $$

or, in terms of calories per mole,

$$ T \left(\frac{\delta \widetilde{S}}{\delta \widetilde{\lambda}} \right)_T \sim 500 \cdot (\widetilde{\lambda} - \widetilde{\lambda}^{-3}) \; \text{cal/mole} $$

$$ \left(\frac{\delta \widetilde{E}}{\delta \widetilde{\lambda}} \right)_T \sim 600 - 700 \cdot (\widetilde{\lambda} - \widetilde{\lambda}^{-3}) \; \text{cal/mole} $$

The increase in heat content due to the utilization of more degrees of freedom by the large molecule, when the interfacial phase is mixed with the adjacent aqueous phase, would be on the order of the gas constant, 8.3×10^7 erg/$^\circ$K/mole, times the absolute temperature; ie,

$$ N_A \cdot k \cdot T \sim 2.5 \times 10^{10} \; \text{ergs/mole} $$

$$ = 600 \; \text{cal/mole} $$

This is the same order as the measured value for the heat of extension per mole of spectrin material at constant surface density. As indicated originally, a natural tendency to increase entropy is represented by a positive heat of extension. In equilibrium, this is opposed by the internal energy or enthalpy required to "mix" two components, such as spectrin with the aqueous phase. The energy changes are expected to be the result of exposure of naturally sheltered hydrophobic regions of spectrin to the aqueous phase. Consequently, the following questions are asked: How much of the spectrin molecule is hydrophobic? How much is the free energy for transfer of a hydrophobic group to an aqueous phase? Thus, what proportion of the hydrophobic region would necessarily be exposed to account for the measured internal energy change that is produced by membrane extension?

Estimates of the number of hydrophobic groups are available (eg, reference 12), but the free energy of transfer from a lipid layer interface to an aqueous phase is not known. The free energy of transfer has been measured [22] for various amino acid residues from organic solvents like ethanol and dioxane to water. These results, combined with the composition data, give an estimate [19] of the total energy potentially available in the transfer of spectrin residues to the aqueous phase. It is about 300 Kcal/mole! This is a tremendously large value compared to that measured for red cell extensional deformation, 600–700 cal/mole. Consequently, it appears that a one-half percent increase in exposure of the hydrophobic portion of spectrin could account for the energy stored in shear deformation. This subtlety demonstrates the valuable insight that can be derived from thermomechanical experiments in conjunction with biochemical analysis. However, it is emphasized that extensive data must be obtained for well-defined membrane compositions in order to make reliable deductions. The previous discussion on spectrin and the red cell is clearly speculative, but, it is hoped, will stimulate further studies.

ACKNOWLEDGMENTS

E. Evans is supported by U.S.P.H.S. National Institutes of Health Research Career Development Award HL00063. In addition, this work was supported in part by NIH grant HL16711 and American Heart Association grant 761043. The assistance of Karen Buxbaum, who performed some of the thermal expansivity experiments and analysis, is appreciated.

REFERENCES

1. Evans EA: A new material concept for the red cell membrane. Biophys J 13:926–940, 1973.
2. Skalak R, Tozeren A, Zarda RP, Chien S: Strain energy function of red blood cell membranes. Biophys J 13:245–264, 1973.
3. Evans EA: Bending resistance and chemically-induced moments in membrane bilayers. Biophys J 14:923–931, 1974.
4. Zarda PR, Chien S, Skalak R: Elastic deformations of red blood cells. J Biomech 10:211–221, 1977.
5. Evans EA, Hochmuth RM: Mechanochemical properties of membranes. In Kleinzeller A and Bronner F (eds): "Current Topics in Membranes and Transport," Vol X. Academic Press, 1978, pp 1–64
6. Evans EA: A new membrane concept applied to the analysis of fluid shear- and micro-pipette-deformed red blood cells. Biophys J 13:941–954, 1973.
7. Hochmuth RM, Mohandas N, Blackshear PL Jr: Measurement of the elastic modulus for red cell membrane using a fluid mechanical technique. Biophys J 13:747–762, 1973.
8. Evans EA, LaCelle PL: Instrinsic material properties of the erythrocyte membrane indicated by mechanical analysis of deformation. Blood 45:29–43, 1975
9. Waugh R, Evans EA: Viscoelastic properties of erythrocyte membranes of different vertebrate animals. Microvasc Res 12:291–304, 1976.
10. Evans EA, Waugh R, Melnik L: Elastic area compressibility modulus of red cell membrane. Biophys J 16:585–595, 1976.
11. Evans EA, Waugh R: Osmotic correction to elastic area compressibility measurements on red cell membrane. Biophys J 20:307–313, 1977.

12. Marchesi VT, Steers E, Tillack TW, Marchesi SL: Properties of spectrin: A fibrous protein isolated from red cell membranes. In Jamison and Greenwalt (eds): "Red Cell Membrane." Philadelphia: 1969 Lippincott, pp 117–130.
13. Singer SJ: The molecular organization of membranes. Annu Rev Biochem 43:805–833, 1974.
14. Steck TL: The organization of proteins in the human red cell membrane. J Cell Biol 62:1–19, 1974.
15. Evans EA, Hochmuth RM: A solid-liquid composite model of the red cell membrane. J Membrane Biol 30:351–362, 1977.
16. Evans EA, Hochmuth RM: Membrane viscoelasticity. Biophys J 16:1–11, 1976.
17. Evans EA, Hochmuth RM: Membrane viscoplastic flow. Biophys J 16:13–25, 1976.
18. Evans EA, Waugh R: Mechanochemistry of closed, vesicular membrane systems. J Colloid Interface Sci 60:286–298, 1977.
19. Evans EA, Skalak R: "Mechanics and Thermodynamics of Biomembranes." CRC Press (in press).
20. Waugh R, Evans EA: Thermoelasticity of red blood cell membranes. Biophys J 26:115–131, 1979.
21. Bennett V, Branton D: Selective association of spectrin with the cytoplasmic surface of human erythrocyte plasma membranes. Quantitative determination with purified (32P) spectrin. J Biol Chem 252:2753–2763, 1977.
22. Nozaki Y, Tanford C: The solubility of amino acids and two glycine peptides in aqueous, ethanol and dioxane solutions. J Biol Chem 246:2211–2217, 1971.
23. Yue BY, Jackson CM, Taylor JAG, Mingins J, Pethica BA: Phospholipid monolayers at non-polar oil/water interfaces. Part 1. J Chem Soc (Lond), Faraday Transactions 1, 72:2685–2693, 1975.

4

Viscoelastic Solid Behavior of Red Cell Membrane

Robert M. Hochmuth

Because it is only a few molecules thick, human red cell membrane (and biomembrane in general) forms a "continuum" only in the plane of the surface. Hence, the continuum material properties of membranes are classified as two-dimensional surface properties that characterize reversible (elastic) and irreversible (viscous) deformations which occur in the membrane plane. If an external force applied to the membrane does not have a large isotropic component, then the membrane deforms at constant surface area; that is, the membrane undergoes only shear deformation. For an equal and opposite (ie, uniaxial) force applied to a rectangular surface element, the lines of maximum shear are at $\pm\,45°$ to the direction of application of the force, and the membrane's resistance to shear deformation and rate of shear deformation is characterized by two coefficients, a "shear modulus of surface elasticity" and a "coefficient of surface viscosity."

For the last decade, much of our research has concentrated on specification of the response of biomembranes (especially red blood cell membranes) to shear forces. We have observed that the red cell membrane can behave in a variety of ways that depend on the duration and magnitude of the applied force. For small forces and short times, the membrane behaves as a viscoelastic solid, capable of large elastic (recoverable) extensions with internal viscous dissipation [1, 2]. For small forces and long times the membrane exhibits "semi-solid" or relaxation behavior, in which the membrane material "creeps" and does not recover its original shape once the external force is removed [3, 4]. For large external forces, the membrane material yields and flows rapidly, experiencing a large degree of permanent (ie, plastic) deformation [5–7]. The fluid behavior in each of these regimes (solid, semi-solid, and plastic) is characterized by a different value for the coefficient of surface viscosity. In general, our observations support the view that the membrane is a composite material that can exhibit both solid and liquid characteristics [8].

Erythrocyte Mechanics and Blood Flow, pages 57–73
© 1980 Alan R. Liss, Inc., 150 Fifth Avenue, New York, NY 10011

In this paper, we concentrate on the description of the viscoelastic behavior of membranes in the solid regime. To illustrate the viscoelastic solid behavior of membranes and give the reader insight into the kinds of analyses we have performed [1–11], we describe the deformation and rate of deformation behavior of an idealized rectangular membrane strip that is subjected to an equal and opposite uniaxial force. More general developments involving complex force and moment resultants for different types of material behavior are given elsewhere [4, 9].

MEMBRANE DEFORMATION

In contrast to the phrases "cell deformation" and "cell deformability," which are qualitative descriptions of observed behavior and, as such, represent both intrinsic properties (eg, membrane and cytoplasm material properties) and extrinsic factors (eg, surface-to-volume ratio), the phrase "membrane deformation" can and must be defined precisely [4, 9]. To this end, consider a line element of length ds_0 between two points, P–P', on the surface of a membrane. When the membrane is deformed by an external force, the line element is stretched into a new line element, ds. The measure of membrane deformation is defined simply as

$$ds^2 - ds_0^2$$

We relate this deformation to a square membrane element with dimension, a, that is aligned with the direction of extension (ie, the principal axis system where the element dimensions are simply extended and compressed along the axes). If we deform the square at constant area into a rectangle with length λa and width $\lambda^{-1} a$, then the membrane deformation (Fig. 1) is given by

$$ds^2 - ds_0^2 = a^2 (\lambda^2 + \lambda^{-2} - 2)$$

or, since $ds_0^2 = 2a^2$,

$$\left(\frac{ds}{ds_0} \right)^2 - 1 = \frac{1}{2} (\lambda^2 + \lambda^{-2}) - 1$$

Thus, deformation is measured by a quadratic function of the extension or stretch ratio, λ, for extension at constant area. In general, we define [9] the intensive deformation variable, β, for deformation at constant area as

$$\beta \equiv \frac{1}{2} (\lambda^2 + \lambda^{-2}) - 1 \tag{1}$$

MEMBRANE FORCE RESULTANTS

External forces which act on the surface of the membrane cause the membrane to deform. For the example discussed in the previous section, a uniaxial force, f, applied to the boundary of the element causes the square element to be deformed into a rectangular strip. The expression for the intensive force resultant, T_1,* is defined simply by making an imaginary "cut" in the rectangular element (shown in Fig. 1), and then by distributing the force along the cut such that the external force is balanced (Fig. 2):

$$T_1 \frac{a}{\lambda} = f \tag{2}$$

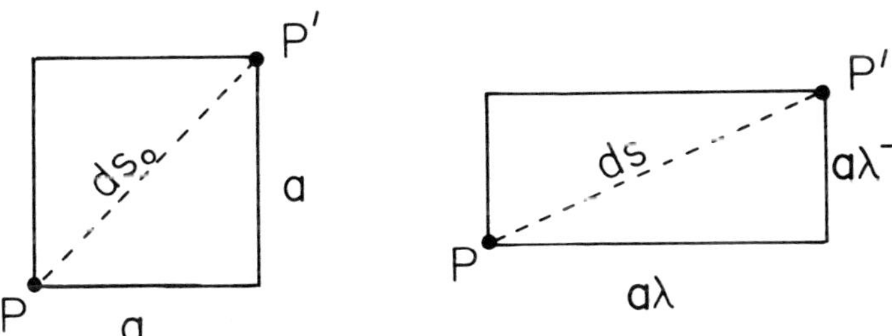

Fig. 1. Deformation at constant area of a membrane strip with unit length and unit width in the undeformed state. Note that a measure of the relative increase in distance between the two material points P–P' is given by $ds^2 - ds_0^2 = (\lambda^2 + \lambda^{-2} - 2)a^2$.

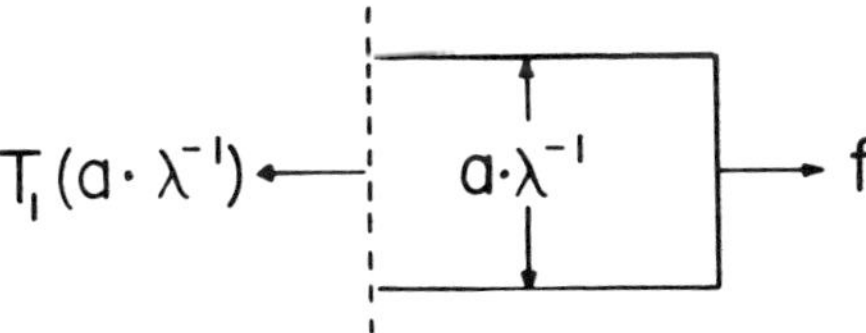

Fig. 2. Free-body diagram of the membrane strip shown in Figure 1. The extrinsic force is denoted by f and the intrinsic force resultant is denoted by T_1.

*Since the element is aligned with the principal axes, T_1 is a "principal membrane tension." The (tangential) shear resultant is zero on the faces of the element and is maximum along lines which are at $\pm 45°$ to the principal axes.

ELASTIC FREE ENERGY

Displacement of the force, f, through a small increment, $ad\lambda$, produces work on the membrane element shown in Figures 1 and 2. Thus,

$$dW = f(ad\lambda) = a^2 \ T_1\left(\frac{d\lambda}{\lambda}\right) \tag{3}$$

For a reversible (eg, quasiequilibrium) isothermal process, the work done on the membrane element is equal to the change in the Helmholtz free energy, F (the "total work function"), at constant temperature:

$$dW = (dF)_T$$

The isothermal process considered here takes place at constant area (surface density). Thus, we define an elastic potential energy density by division of the Helmholtz free energy by the fixed area, a^2

$$\widetilde{F} \equiv \frac{F}{a^2}$$

and we obtain the work per unit area

$$\frac{dW}{a^2} = \frac{(dF)_T}{a^2} = (d\widetilde{F})_T \tag{4}$$

At constant temperature and surface density, the free energy density will be a function only of the intensive deformation — ie, β (Eq 1). Therefore,

$$(d\widetilde{F})_T = \left(\frac{\delta\widetilde{F}}{\delta\beta}\right)_T d\beta = \left(\frac{\delta\widetilde{F}}{\delta\beta}\right) \cdot (\lambda^2 - \lambda^{-2}) \frac{d\lambda}{\lambda} \tag{5}$$

where β has been eliminated in favor of the extension ratio, λ, with the use of Eq 1. The direct comparison of Eqs 3 and 5 (via Eq 4) gives

$$T_1 = \left(\frac{\delta\widetilde{F}}{\delta\beta}\right)_T (\lambda^2 - \lambda^{-2}) \tag{6}$$

Thus, a thermodynamic equation of state (Eq 6) or elastic constitutive equation is specified at constant temperature for the force resultant, T_1, by

$$T_1 = \mu(\lambda^2 - \lambda^{-2}) \tag{7}$$

where a membrane elastic shear modulus is defined by

$$\mu \equiv \left(\frac{\delta \tilde{F}}{\delta \beta}\right)_{T}$$

If it is assumed that the free energy increases linearly with the deformation parameter, β, then μ is a constant of proportionality.

The maximum shear resultant or deviator, T_s, is the difference between principal tensions

$$T_s = \frac{T_1 - T_2}{2}$$

In the present example, $T_2 = 0$. Thus,

$$T_s = \frac{T_1}{2} = \frac{\mu}{2}\,(\lambda^2 - \lambda^{-2}) \tag{7a}$$

which is why we call μ the shear modulus of surface elasticity.

Equation 7 was first proposed by Evans in 1973 [10] to describe the hyperelastic behavior of two-dimensional membranes. (Note that in the original equation the value for μ was twice as large; ie, "$\mu_{original}$" = 2 "μ_{new}"). By substituting the expression for the external membrane force, f (Eq 2), into Eq 7, we demonstrate a fundamental feature that is observed in our experiments:

$$f = a\,\mu\,(\lambda - \lambda^{-3}) \tag{7b}$$

That is, the force-deformation behavior of the red cell membrane surface is essentially linear for large values of $\lambda(\lambda \approx 2)$, since the second term in Eq 7b can be neglected. We have observed this linear behavior repeatedly in our experiments to date, when the force is primarily uniaxial so that the isotropic force-resultant component can be neglected.

VISCOUS DISSIPATION

In the development of the previous section, the membrane was considered to behave as an elastic solid that exhibits reversible deformations. This consideration led to the development of a first-order constitutive equation at constant temperature (Eq 7). In contrast to a perfectly elastic material, the membrane could be represented as a two-dimensional viscous liquid such that all of the isothermal work performed on the membrane system is dissipated as heat to the environment at a lower temperature. This causes an increase in the entropy of the surroundings, with no change in the thermodynamic state of the membrane.

The rate of doing work on a membrane with uniaxial deformation is obtained simply by dividing the work increment, Eq 3, by a small time increment, dt:

$$\dot{W} = a^2 \left(T_1 \ \frac{d \ln \lambda}{dt} \right)$$

This is the mechanical power delivered to the membrane. For a liquid membrane deformed at constant temperature and surface density, the internal energy (and entropy, free energy, etc) remains constant; therefore, all of the work performed on the membrane is dissipated into heat, which is transferred to the surroundings. Thus, the rate of increase in the entropy of the surroundings, $(\dot{S})$, is given by

$$T_0 \ \dot{S} = a^2 \left(T_1 \ \frac{d \ln \lambda}{dt} \right)$$

where T_0 is the temperature of the surroundings. Since the entropy of the membrane system is constant, the entropy of the surroundings must increase according to Clausius' statement of the second law of thermodynamics. This is simply guaranteed by assuming that

$$T_1 \ \sim \frac{d \ln \lambda}{dt}$$

or

$$T_1 \ = \ 4\eta \ \frac{d \ln \lambda}{dt} \tag{8}$$

where η is called the coefficient of surface viscosity. (The factor of 4 in Eq 8 makes this expression consistent with "Newton's law of viscosity," which states that material force resultants are linearly related to a specific measure of the rate of deformation in the material. (For a general representation of Eq 8 developed from either a mechanics or an irreversible thermodynamics standpoint, see either reference 1 or reference 9.)

THE MEMBRANE AS A VISCOELASTIC SOLID

Equations 7 and 8 represent two ideal limits of material behavior. Equation 7 is an isothermal constitutive equation that represents the extension or shear deformation behavior of a two-dimensional hyperelastic solid membrane. Equation 8 is a first-order phenomenological relation which accounts for frictional forces (viscous dissipation) in a liquid membrane. We recognize that the actual red cell membrane is not a perfect solid or liquid; it exhibits manifold types of material behavior. In the domain where membrane deformations are recoverable, we idealize the material as a superposition of elastic and viscous components such that deformation in each component is the same. Hence, the total force resultant is simply the sum of the force resultants for each component, and Eqs 7 and 8 are summed to give

$$T_1 = \mu(\lambda^2 - \lambda^{-2}) + 4\eta \frac{d\,ln\,\lambda}{dt} \qquad (9)$$

The elastic term provides "memory," whereas the viscous term limits the rate of response of the membrane to an external force.

Equation 9 describes the membrane as a two-dimensional viscoelastic solid which is capable of large, recoverable deformations with internal viscous dissipation. For a very slow deformation process, the elastic term in Eq 9 will dominate (see Eq 7). On the other hand, for very rapid processes with large rates of deformation, the viscous term will dominate and the elastic term can be neglected (see Eq 8). In a particular type of experiment — that of membrane recovery after the external force is removed — the membrane is free of tension. Therefore, the elastic and viscous terms in Eq 9 are equal in magnitude and opposite in sign. To perform this experiment, the membrane is stretched slowly and then suddenly released; ie, T_1 is set equal to zero, and Eq 9 is integrated to give

$$\frac{\lambda^2 - 1}{\lambda^2 + 1} = \left(\frac{\lambda_m^2 - 1}{\lambda_m^2 + 1}\right) e^{-t/t_c} \qquad (10)$$

In Eq 10, λ_m is the maximum extension of a rectangular strip measured at the instant when T_1 is set equal to zero, and t_c is the time constant for recovery given by

$$t_c = \eta/\mu \qquad (11)$$

Thus, if we first measure a value for the elastic modulus, μ, in a "static" (quasiequilibrium) deformation experiment and then observe the recovery process (Eq 10) in order to obtain a value for t_c, we can determine a value for the membrane surface viscosity, η, from Eq 11.

EXPERIMENTS

For the past several years we have used three different types of static deformation experiments for the determination of a value for the shear modulus of surface elasticity, μ. They are 1) fluid shear deformation of red cells, 2) micropipet aspiration of red cell membrane, and 3) micropipet extension of whole cells.

Hochmuth and Mohandas [11] and Hochmuth et al [5] used fluid flow through a parallel plate flow channel to deform disc-shaped red cells that were attached at a single point to one of the surfaces of the channel. With a force-deformation relation similar to Eq 7b, analysis of the cellular deformation process in this experiment [5, 10–12] provided values for μ that are on the order of 10^{-2} dyne/cm.* The three-dimensional value of 10^4 dynes/cm^2 given by Hochmuth et al [5, 11] was obtained by dividing the surface value of 10^{-2} dyne/cm by an arbitrary membrane thickness of 10^{-6} cm.

Evans [10], Evans and LaCelle [3], and Waugh [13] have aspirated portions of red cell membranes, generally from the "dimple" region of flaccid cells, into micropipets with internal diameters of about 1 μm. Here the membrane is stretched along the direction of aspiration as it is drawn symmetrically into the mouth of the pipet by the suction force. The force is created by the negative pressure in the pipet relative to that in the reservoir. Again, the analysis of the force-deformation behavior of membrane (eg, Eq 7b) in these experiments [10] provided values for μ on the order of 10^{-2} dyne/cm. Recent accurate measurements by Waugh [13] indicate that $\mu \approx 0.006$ dyne/cm.

Although the micropipet technique demands a greater degree of skill in comparison to the flow channel technique, it allows the force on the membrane to be applied locally and, along with the deformation, to be determined accurately. In the flow channel technique, the fluid shear force is distributed over the surface of the membrane, and its value can only be estimated by multiplying the fluid shear stress at the surface of the channel by the projected area of the red cell as if the cell were a perfectly flat disc on the surface of the channel.

Recently we have developed the technique of "micropipet extension of whole cells" [14], which is easier to perform than micropipet aspiration and which provides a more direct measure of force than fluid shear deformation. In this technique, cell deformation is produced by aspirating a small portion of the rim of a red cell into the tip of a micropipet and then gently withdrawing the micropipet as illustrated in Figure 3. Upon release from the micropipet, the cell returns to its original biconcave shape. (Both deformation and recovery phases are recorded on videotape.) It should be noted that this particular experiment closely resembles the "idealized experiment" presented earlier, which involved the deformation (Eq 7) and recovery (Eq 10) of a rectangular membrane strip. In addition, the analytical model of this experiment is conceptually simple (compared to the micropipet aspiration and fluid shear deformation experiments) and, thus, is given here to illustrate the analytical process.

---

*Evans, in his analysis of the fluid shear deformation behavior of red cells [10], applied Eq 7b to a "membrane disc" such that the width, a, was a function of the axial distance measured in the direction of flow. Hochmuth et al [5, 11] used "Hooke's law" (the linearized form for Eq 7b) in their analysis, and Skalak et al [12] used a force-deformation relation, which was essentially cubic in the deformation parameter, λ.

For the "whole cell micropipet extension" experiment, the net force on the cell can be calculated by multiplying the pressure drop across the membrane tongue by the cross-sectional area of the pipet:*

$$f = (\Delta p)\,\frac{\pi}{4}\,D_p^2$$

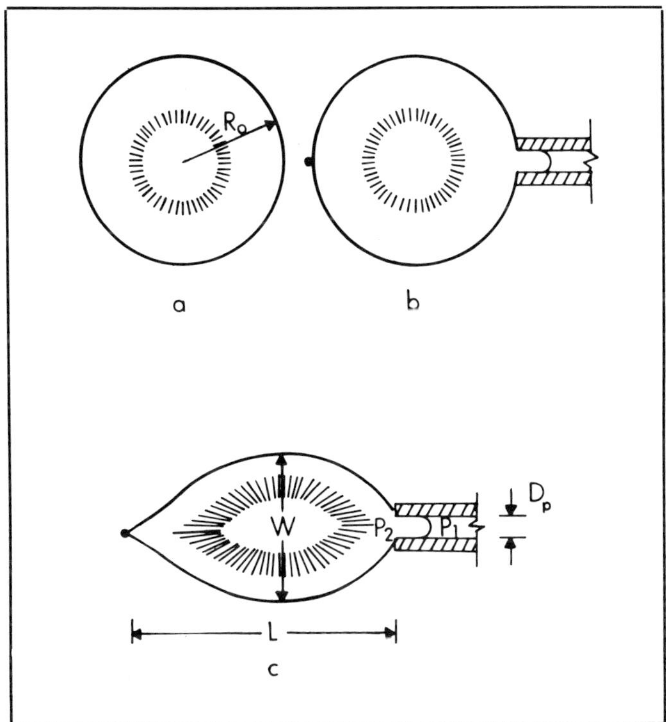

Fig. 3. Whole cell deformation with a micropipet where the cell remains attached to the surface at a single "point." The extension of the membrane into the micropipet (the "tongue" length) is held constant during the deformation process. Upon release from the micropipet the cell recovers its biconcave shape. During the recovery process the maximum length L and width W are functions of time.

*This assumes that the cell does not "round up" and, thus, the pressure inside the cell is the same as the atmospheric pressure in the reservoir. At very large extensions, the cell will begin to round-up because of an internal pressure, which is balanced by an isotropic tension in the membrane.

where f is the force acting on the cell, Δp is the pressure drop across the membrane tongue, and D_p is the diameter of the pipet. The equilibrium deformation for uniaxial extension of the two membrane surfaces (upper and lower) of the red cell is analyzed with the following set of equations:

$$\text{Axial Force Balance (2)} \qquad f = 2T_X\, w(x) \tag{12}$$

$$\text{Constitutive Equation (7)} \qquad T_X = \mu(\lambda^2 - \lambda^{-2}) \tag{13}$$

$$\text{Constant Area Deformation } \lambda \equiv \frac{dx}{dx_0} = \frac{w_0(x_0)}{w(x)} \tag{14}$$

The location of the "cut" for the axial force balance (Eq 12) is given by the position, x, which is the distance from the center of the cell as measured along the force axis. In Eqs 12–14, T_X is the axial "tension" in the membrane, μ is the shear modulus of surface elasticity, w(x) is the width in the deformed state, λ is the axial extension in the deformed state relative to the undeformed state, and w_0 is the width in the undeformed state; ie,

$$x_0^2 + \frac{w_0^2}{4} = R_0^2 \tag{15}$$

where R_0 is the radius of the (undeformed) membrane disc. We combine Eqs 12–15 to obtain

$$\lambda^4 = 1 + \frac{\bar{f}}{(1 - \bar{x}_0^2)^{1/2}}\, \lambda^3 \tag{16}$$

where $\bar{f} = f/4\mu R_0$ and $\bar{x}_0 = x_0/R_0$. Equation 16 gives the extension ratio, λ, for a given $\bar{f}$ and $\bar{x}_0$. The numerical integration of Eq 14 with the use of Eq 16 permits a calculation for the overall length, L, and maximum width, W:

$$\frac{L}{2} = \int_0^{\frac{L}{2}} dx = \int_0^{R_0} \lambda\, dx_0 \tag{17}$$

$$\frac{W}{2} = \frac{R_0}{\lambda_{x=0}}$$

For the range $1.25 \leqslant \sqrt{L/W} \leqslant 1.55$, the result for $\bar{f}$ vs $\sqrt{L/W}$ ($\sqrt{L/W}$ is a measure of λ for uniform extension of a rectangular strip) is essentially linear. In

Eq 17 the integration was terminated at the "point" of attachment (left-hand side in Fig. 3); the size of the point was assumed to be 0.1 μm relative to an 8 μm cell. The integration on the other side of the membrane was terminated at the location appropriate to a pipet-to-cell diameter of 0.10 (0.8 μm pipet and 8 μm cell). The slope of the $\bar{f}$ vs $\sqrt{L/W}$ line is not very sensitive to the selection of these cut-off points. In terms of the slope of the line, the force increment is approximately

$$\Delta \bar{f} = 1.6 \, \Delta (\sqrt{L/W}) \tag{18}$$

for the range of cell extensions given by $1.25 \leqslant \sqrt{L/W} \leqslant 1.55$.

A total of 47 deformation experiments has been performed [14] with the result that

$$\mu = 0.0045 \pm 0.0011 \text{ dyne/cm}$$

After the static deformation is observed (for cell extensions where $\sqrt{L/W} \leqslant 1.6$), the cell is released and allowed to recover its original biconcave shape. Even though Eq 10 applies only to uniform viscoelastic recovery of a rectangular strip, the time-dependent recovery process of the red cell is modeled closely by Eq 10 [2] as long as λ is defined as

$$\lambda = \sqrt{L/W} \, / \, \sqrt{L_0/W_0}$$

This is the same as saying that the strip is not a perfect square in the undeformed state, but a rectangle with length L_0 and width W_0. With λ defined in this way, Eq 10 becomes

$$\frac{\left(\frac{L}{W}\right) - \left(\frac{L}{W}\right)_0}{\left(\frac{L}{W}\right) + \left(\frac{L}{W}\right)_0} = \frac{\left(\frac{L}{W}\right)_m - \left(\frac{L}{W}\right)_0}{\left(\frac{L}{W}\right)_m + \left(\frac{L}{W}\right)_0} \, e^{-t/t_c} \tag{10a}$$

where $(L/W)_m$ is the maximum length-to-width ratio of the cell (at $t = 0$). Equation 10a is used to correlate the experimental results in terms of L/W vs t with $(L/W)_0$ and t_c chosen to give a best (least-squares) fit to the experimental data. Data from four separate recovery experiments and Eq 10a are shown in Figure 4.

Since $\eta = t_c \cdot \mu$, a measurement for t_c and μ for a single cell gives a value for the coefficient of surface viscosity, η, as measured by viscoelastic recovery for that cell. The results to date [14] are summarized in Table I.

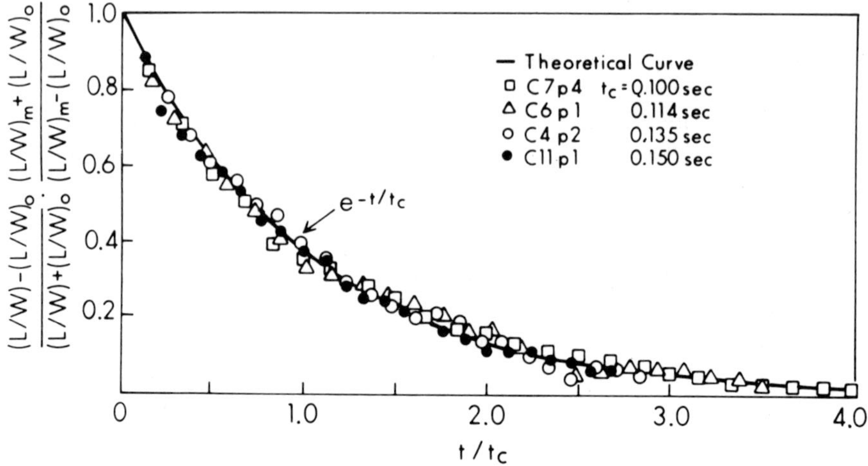

Fig. 4. Plot of cell recovery data according to Eq 10a for four different recovery experiments. Note that all of the data "collapse" onto a single curve.

TABLE I. Measurement of the Shear Modulus of Surface Elasticity, μ, and Recovery Time Constant, t_c, for Individual Cells*

$\mu \times 10^3$ (dyne/cm)	t_c(sec)	$\eta \times 10^4$ (dyne sec/cm)
4.93	0.111	5.47
5.81	0.135	7.84
3.15	0.140	4.41
4.60	0.114	5.24
3.17	0.090	2.85
5.72	0.105	6.01
4.59	0.100	4.59
5.58	0.121	6.75
3.94	0.142	5.59
4.46	0.152	6.78
3.24	0.145	4.70
3.28	0.125	4.20
2.96	0.150	4.44
4.34	0.113	4.90
4.44	0.104	3.52
4.28 ± 0.97	0.123 ± 0.020	5.15 ± 1.31

*For each cell, $\eta = t_c \cdot \mu$.

DISCUSSION OF RESULTS

It is significant that the three different experimental techniques discussed in the previous section give nearly the same value for the shear modulus of surface elasticity, μ. Although each experiment has a different distribution of membrane force resultants and deformations, the results are consistent because 1) membrane deformation as defined in Eq 1 can be quantified in each experiment; 2) the extensive membrane forces and intensive force resultants are determined (eg, Eq 2); and 3) the elastic constitutive equation (Eq 7) is an appropriate representation of the static force-deformation behavior of the membrane. Thus, we are confident that μ is a true intrinsic material property of the red cell membrane and not a subjective measure used to characterize red cell "deformability" in a particular experiment.

The mean value for μ obtained in the whole cell micropipet deformation experiments (0.0045 dyne/cm) is about 30% less than the mean value obtained by Waugh [13] with the micropipet aspiration technique analyzed by Evans [10]. These values are remarkably close if we consider the number of possible contributions to error in both experiments and analyses; eg, pipet wall friction [13], optical measurement errors, pressure uncertainties, membrane curvature elasticity (bending rigidity), buckling or membrane folding, curvature effects on deformation distribution, membrane yield at the attachment location, biaxial tension, etc. These errors are not large and are carefully considered in the experimental procedures. Many potential sources for error can be detected and avoided.

The values for the time constants shown in Table I agree (within 30%) with our previous values [2, 15] and those of Chien et al [16, 17]. In Chien's experiment a portion of the red cell membrane was aspirated into a pipet and then the time-dependent recovery within the pipet of the aspirated portion was recorded on videotape upon removal of the suction pressure. Chien's experiment gives a value for the recovery time constant of 0.13 ± 0.05 sec, which is in excellent agreement with the data presented in this study. Since the results obtained by the two different mechanical experiments when analyzed by the viscoelastic constitutive equation (Eq 9) give the same result, we conclude that the recovery time constant is an intrinsic property of the red cell membrane material, independent of the cell geometry. In addition, the experimental observation (Fig. 4) shows that the viscoelastic constitutive relation (Eq 9) adequately represents the recovery process for the large range of material extension involved during the recovery phase. Therefore, from the recovery time we can determine the coefficient for membrane surface viscosity in this solid material domain. For the 15 cells in Table I, this value is about 5×10^{-4} dyne sec/cm (poise $\cdot$ cm or surface poise) with a maximum range on a cell-by-cell basis of approximately $3-7 \times 10^{-4}$ dyne sec/cm.

COMMENTS ON SURFACE VISCOSITY AND PARTICLE DIFFUSIVITY

It is interesting to speculate on a possible correspondence of our surface viscosity measurement to the measurements made by other membrane scientists of surface diffusivity of various marker particles in biomembranes and phospholipid vesicles. We emphasize that viscosity is a continuum property and, thus, is a measure of the dissipative interactions among many molecules averaged over a "large" area. On the other hand lateral, or surface, diffusivity (diffusion of particular "marker particles" in the plane of the membrane) is related not only to the aggregate properties of the membrane (eg, surface viscosity) but also to the particular properties of the tagged molecule(s). Properties such as size, shape, location in the membrane, and degree of association with membrane proteins will affect the mobility of the marker particles. This is why published values for the "coefficient of surface diffusivity" vary from 10^{-7} cm^2/sec to the limits of resolution of the experiments (ca 10^{-12} cm^2/sec). However, surface diffusivity measurements do provide a means of "probing" the molecular structure of the membrane with particular marker particles. Also, we know that surface diffusivity is qualitatively related to membrane viscosity in that larger values for the membrane surface viscosity will produce smaller values for the coefficient of surface diffusivity, all other things remaining the same.

In general, diffusivity studies have been undertaken in both pure two-dimensional fluids (phospholipid bilayers and multilayers) and living membranes such as the erythrocyte membrane. We will refer to both as "biomembranes."

The original studies of Frye and Edidin [18] show that mouse and human cells can fuse, forming "heterokaryons," in the presence of Sendai virus. Rough estimates from their data give values for the diffusivity of 10^{-10} cm^2/sec. The elegant studies of Poo and Cone [19] and Liebman and Entine [20] on the self-diffusion of rhodopsin in disc membranes give values of about 0.5×10^{-8} cm^2/sec. Diffusivity in cultured rat myoblasts with two different tags have values of 2×10^{-10} cm^2/sec and 9×10^{-9} cm^2/sec [21]. Similar values were obtained in cultured mouse fibroblasts by Jacobson and co-workers [22]. The recent work of Fowler and Branton [23] indicates that surface diffusion can occur in human red cell membranes and that the values for diffusivity in red cells can be highly temperature sensitive (0.5×10^{-10} cm^2/sec at 37°C and 0.6×10^{-11} cm^2/sec at 23°C). In summary, it appears that a value for the surface diffusivity in disc membranes is about 10^{-9} cm^2/sec; in myoblasts, fibroblasts, and heterokaryons it is about 10^{-10} cm^2/sec; and in human red cell membrane it is about 10^{-11} cm^2/sec.

Values for diffusivity in relatively simple systems can also vary. Even for the same lipid multibilayer with two very similar fluorescent lipid analogs, Jacobson and coworkers [24] found that diffusivity values can differ by a factor of 4 (from 4×10^{-8} to 1.6×10^{-7} cm^2/sec at 25°C and 8×10^{-8} to 3×10^{-7} at 37°C). Also, these investigators [24] demonstrated that surface diffusivity is strongly dependent on temperature for lipids above the phase transition. (The order—

disorder transition for the hydrocarbon chains.) The strong temperature dependence suggests that the surface diffusivity should be strongly dependent on surface density (since the coefficient of volumetric thermal expansivity is small for lipids [25]).

From the previous discussion, it is apparent that structural material associated with the bilayer component of a cell membrane greatly influences the fluidity estimated by probe diffusivity. In such heterogeneous membrane materials, it is not possible to provide a simple kinetic relation between diffusivity, temperature, surface viscosity, and particle size, as in ordinary gases. However, diffusion is limited by dissipative interactions between the particle and the material environment, and surface viscosity is the measure of dissipation in the material produced by finite rates of deformation. Consequently, we anticipate that surface diffusivity and surface viscosity are correlated. For instance, we can estimate the surface diffusivity for a two-dimensional liquid abstraction of the red cell membrane — ie, a homogeneous liquid surface with a surface viscosity equal to the value measured from our red cell viscoelastic recovery experiments. In order to make this estimate, we use the surface mobility relation derived by Saffman [26] for drag on a particle restricted to motion in the surface plane of the liquid sheet:

$$b = \frac{1}{4\,\pi\eta}\left[\ln\left(\frac{\eta}{\tilde{\eta}\cdot a}\right) + 0.5772\right]$$

where b is the two-dimensional mobility, a is the radius of the diffusing particle, η is the surface viscosity of the anisotropic (two-dimensionally isotropic) liquid sheet, and $\tilde{\eta}$ is the viscosity of the adjacent aqueous media. Because of the logarithmic function, the surface mobility is extremely insensitive to particle radius and viscosity of the adjacent media. With the Einstein relation for Brownian motion,

$$D = k \cdot T \cdot b$$

the surface diffusivity is determined by the absolute temperature, T, Boltzmann's constant, k, and the mobility. With values of $T = 298°K$, $\tilde{\eta} = 10^{-2}$ poise for water, and $a = 2 \times 10^{-7}$ cm for the particle radius, we show a relation between D and η (Fig. 5.), and we calculate that the surface diffusivity would be on the order of 7×10^{-11} cm^2/sec for the particle diffusing in a liquid surface with a surface viscosity of 6×10^{-4} poise·cm. This value is in the same range as measurements of surface diffusivity of large protein markers in cell membrane surfaces [18, 21, 23]. This calculation is provided only to illustrate the dominant contribution that dissipation in structural material associated with the membrane lipid component can make to limiting the kinetic motion of large integral molecules as well as limiting the material response to rate of deformation on a macroscopic (continuum) scale.

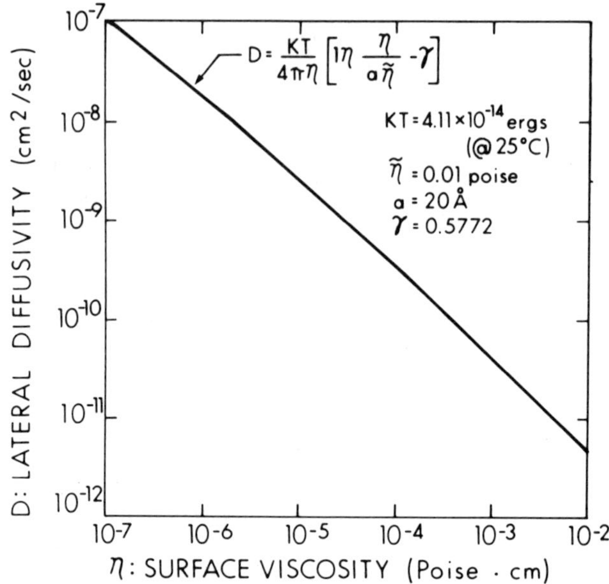

Fig. 5. The relation between the membrane surface viscosity, η, and the laterial diffusivity, D, according to Saffman's relation for particle mobility, b, and Einstein's relation for diffusivity, D = kTb. Here k = Boltzmann's constant, T = absolute temperature, a = particle radius, $\tilde{\eta}$ = viscosity of the adjacent aqueous media, and γ = Euler's constant.

ACKNOWLEDGMENTS

I thank my colleague at Duke University, Dr. Evan Evans, for suggesting the method of "micropipet extension of whole cells" for the measurement of membrane surface elasticity and for giving me many ideas and suggestions along the way. Also, Dr. Evans assisted greatly in the preparation of the last section of this paper.

In the course of this work I was supported by an NIH Research Career Development Award (HL70612). The various NIH grants which supported this research include HL21803, HL12839, HL2024, and HL16711. In addition I thank Karen Buxbaum, Lou Hampel, David Markle, Sid Smith, and Pat Worthy for assisting in various parts of this research.

REFERENCES

1. Evans EA, Hochmuth RM: Membrane viscoelasticity. Biophys J 16:1–11, 1976.
2. Hochmuth RM, Worthy PR, Evans EA: Red cell extensional recovery and the determination of membrane viscosity. Biophys J 26:101–114, 1979.

3. Evans EA, LaCelle PL: Intrinsic material properties of the erythrocyte membrane indicated by mechanical analysis of deformation. Blood 45:29–43, 1975.

4. Evans EA, Skalak R: "Mechanics and Thermodynamics of Biomembranes." CRC Press (in press).

5. Hochmuth RM, Mohandas N, Blackshear PL Jr: Measurement of the elastic modulus for red cell membrane using a fluid mechanical technique. Biophys J 13:747–762, 1973.

6. Evans EA, Hochmuth RM: Membrane viscoplastic flow. Biophys J 16:13–25, 1976.

7. Hochmuth RM, Evans EA, Colvard DF: Viscosity of human red cell membrane in plastic flow. Microvasc Res 11:155–159, 1976.

8. Evans EA, Hochmuth RM: A solid-liquid composite model of the red cell membrane. J Membr Biol 30:351–362, 1977.

9. Evans EA, Hochmuth RM: Mechanochemical properties of membranes. In Bronner F, Kleinzeller A (eds): "Current Topics in Membranes and Transport," Vol X. New York: Academic Press, 1978, pp 1–64.

10. Evans EA: New membrane concept applied to the analysis of fluid shear and micro-pipette-deformed red blood cells. Biophys J 13:941–954, 1973.

11. Hochmuth RM, Mohandas N: Uniaxial loading of the red cell membrane. J Biomech 5:501–509, 1972.

12. Skalak R, Tözeren A, Zarda RP, Chien S: Strain energy function of red blood cell membranes. Biophys J 13:245–264, 1973.

13. Waugh RE: "Temperature Dependence of the Elastic Properties of Red Blood Cell Membrane." PhD Thesis, Duke University, 1977.

14. Hochmuth RM, Hampel WL III: Surface elasticity and viscosity of red cell membrane. Paper presented at the 71st Am Inst Chem Engr Meeting Miami, November, 1978.

15. Hochmuth RM, Worthy PR, Evans EA: "Surface Viscosity of Red Cell Membrane." AIChE Symposium Series 74: No 182; 1–3, 1978.

16. Chien S, Sung K-LP, Skalak R, Tözeren A: Theoretical and experimental studies on viscoelastic properties of red cell membrane." Biophys J 24:463–487, 1978.

17. Sung PL, Chien S: "Viscous and Elastic Properties of Human Red Cell Membrane." AIChE Symposium Series 17: No 182;81–84, 1978

18. Frye LD, Edidin M: The rapid intermixing of cell surface antigens after formation of mouse-human heterokaryons. J Cell Sci 7:319–335, 1970.

19. Poo M, Cone RA: Lateral diffusion of rhodopsin in the photoreceptor membrane. Nature 247:438–441, 1974.

20. Liebman PA, Entine G: Lateral diffusion of visual pigment in photoreceptor disk membranes. Science 185:457–460, 1974.

21. Schlessinger J, Axelrod D, Koppel DE, Webb WW, Elson EL: Lateral transport of a lipid probe and labeled proteins on a cell membrane. Science 195:307–309, 1976.

22. Jacobson K, Wu E, Poste G: Measurement of the translational mobility of concanavalin A in glycerol-saline solutions and on the cell surface by fluorescence recovery after photobleaching. Biochem Biophys Acta 433:215–222, 1976.

23. Fowler V, Branton D: Lateral mobility of human erythrocyte integral membrane proteins. Nature 268:23–26, 1977.

24. Wu E-S, Jacobson K, Papahadjopoulos D: Lateral diffusion in phospholipid multilayers measured by fluorescence recovery after photobleaching. Biochemistry 16:3936–3940, 1977.

25. Liu N-I, Kay RL: Redetermination of the pressure dependence of the lipid bilayer phase transition. Biochemistry 16:3484–3486, 1977.

26. Saffman PG: Brownian motion in thin sheets of viscous fluid. J Fluid Mech 73:593–606, 1976.

5

Measures of Blood Rheology and Erythrocyte Mechanics

Herbert J. Meiselman

The study of hemorheology (ie, the flow and deformation of blood) is based on the desire to understand quantitatively both the macroscopic and the microscopic behavior of blood when it flows through vessels of the circulatory system. Until a few years ago, most efforts were directed toward determining the macroscopic rheological behavior of normal blood and plasma; these studies were prompted by the observed non-Newtonian (ie, apparent viscosity being a function of shear rate) flow properties of whole blood. More recent studies have attempted to relate these overall properties to the behavior of the individual constituents of blood. These later studies are thus microrheological in nature and involve examining such factors as red cell aggregation, plasma viscosity, red cell deformability, membrane mechanical properties, and the rheology of cellular contents.

This presentation is divided into three main areas: 1) the macroscopic flow behavior of blood and red blood cells; 2) the microrheological behavior of red blood cells; and 3) a consideration of the relations between the metabolic state and the mechanical properties of human erythrocytes. Inasmuch as several reviews already exist [eg, 1–11], the present discussion is intentionally limited in both breadth and depth; the author pleads nolo contendere to any charges of errors of omission!

MACROSCOPIC FLOW BEHAVIOR

Overview

Initial interest in blood rheology appears to have been prompted by its remarkable fluidity when compared to other model particle suspensions. Figure 1 compares the relative viscosity, η_r (viscosity of suspension/viscosity of suspending phase) of a number of model suspensions with that of human blood as a function of c, the particle volume concentration. It is obvious that at any given c, the relative viscosity is considerably lower in the case of blood. At a hematocrit (ie, volume frac-

Erythrocyte Mechanics and Blood Flow, pages 75–117
© 1980 Alan R. Liss, Inc., 150 Fifth Avenue, New York, NY 10011

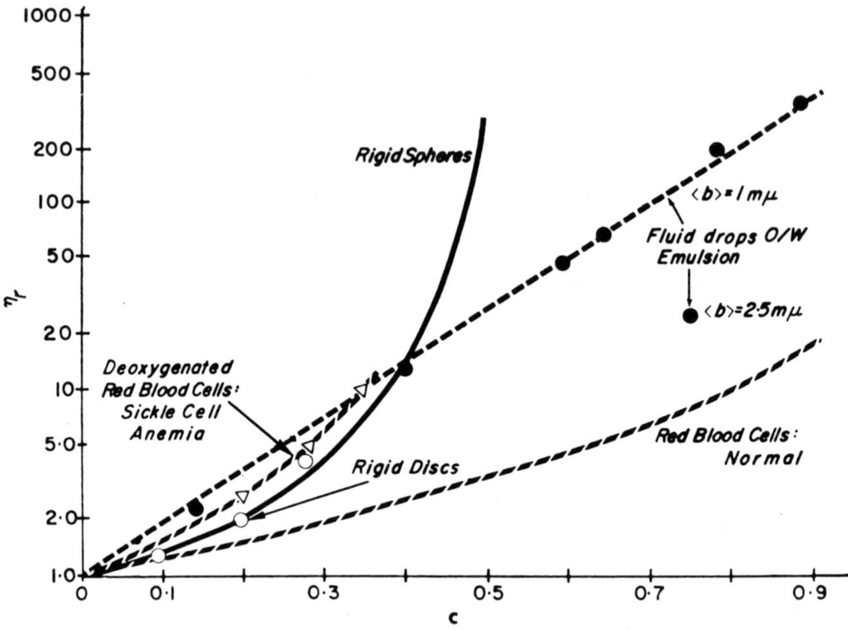

Fig. 1. Comparison of the relative viscosity-volume concentration behavior of normal human red cells in plasma with that of model particle suspensions. Note the lower relative viscosity, at all particle concentrations, shown by normal blood; the effect of cell shape and rigidity is demonstrated by the points indicated for sickle cell anemia. Normal human red cell–plasma data were obtained at shear rates sufficient to abolish cellular aggregation (from [42]).

tion of red cells) of 0.45, the relative viscosity of human blood is only about 3, whereas in suspensions of rigid spheres or discs at the same volume fraction, η_r exceeds 200. Although deformability of the red blood cell (RBC) is important for the relative ease of blood flow, its viscosity is not comparable to that of an emulsion of deformable droplets. As shown in Figure 1, η_r for an oil-in-water emulsion having a mean droplet radius of one micron is significantly greater than that of blood for all values of c; oil-in-water emulsions reach a quasi-solid state for c > 0.75. Altered shape and deformability (sickle cell data, Fig. 1) or altered deformability without shape changes produced by aldehyde treatment [1, 6, 12] do, however, markedly elevate the relative viscosity.

Instrumentation

The instrumentation used to measure the flow properties of RBC suspensions can be grouped into two general categories: 1) rotational viscometers, in which the suspension is sheared between a fixed and a moving surface, and 2) tube visco-

meters, in which the suspension is sheared by flow past the stationary inside wall
of the tube.

Couette viscometer. The instrument that subjects the sample under study most
closely to simple, steady, laminar shear is the rotating cylinder viscometer, usually
known as a Couette viscometer (Fig. 2). In these instruments, either the inner or
outer cylinder is rotated at a steady rate, and the shear stress is transmitted through
the layer of fluid between the cylinders; the resulting torque is usually measured
at the stationary cylinder. A significant number of blood rheology studies have
been carried out using the Couette viscometer, with many of them employing the
so-called GDM viscometer developed by P. J. Gilinson, C. R. Dauwalter, and E. W.
Merrill [13]. This GDM instrument uses a friction-free air bearing to support the
torque-measuring cylinder and allows measurement over a shear rate range from
about 0.1 to 100–300 inverse seconds.

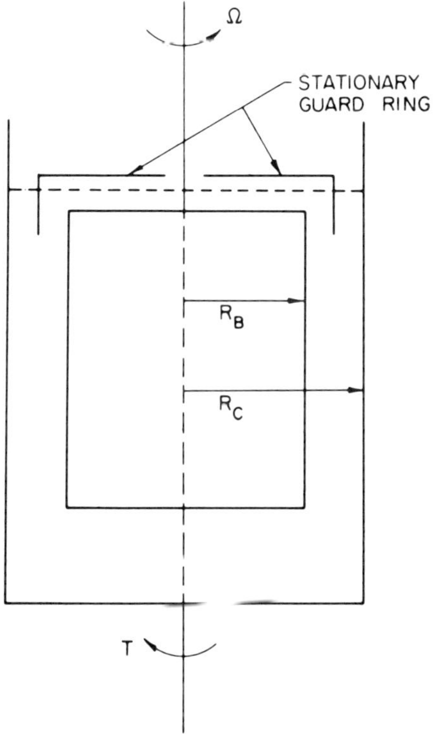

Fig. 2. Couette viscometer equipped with a stationary guard ring in order to eliminate surface
tension artifacts; this guard ring can be applied only to systems in which the surface used to
measure the torque (shown here as the outer cylinder) is held almost stationary. Horizontal
dashed line indicates nominal sample level (from [10]).

A feature of the true Couette viscometer that is useful for data reduction is the fact that the total shear torque (T) is constant across the gap, so that the shear stress, neglecting end effects, is inversely related to the square of the radius to the point of measurement:

$$\tau = \frac{T}{2\pi LR^2} \; ; L = \text{length of bob, R = cylinder radius}$$

The shear rate is not constant across the gap, but for a Newtonian fluid it also varies inversely with the square of the radius. For this specific Newtonian case, the shear rate is given by

$$\dot{\gamma} = \frac{-2\omega}{R^2} \left[\frac{R_c^2 R_B^2}{R_c^2 - R^2_B} \right]$$

where R_c and R_B are the radius of the cup and bob and ω is the angular speed of the rotating element [10]. In the case of non-Newtonian fluids, in which the viscosity depends on the rate of shear, the variation of shear rate across the viscometer gap will depend on the shear stress–shear rate relationship for the fluid being studied; a numerical solution has been obtained that allows calculation of shear rates for some non-Newtonian fluids [10, 14].

Because of the air–fluid interface at the upper end of a coaxial cylinder viscometer, extreme care must be taken to avoid surface tension artifacts. For the simple, Newtonian fluid, water, the high surface tension can introduce serious errors in the measurement of torques if there is a free meniscus in the gap between the two cylinders or between the bob support and the outer cylinder surface. Macromolecular solutions often show marked surface viscosity, so that at low shear rates a substantial fraction of the total torque may be transmitted through the surface layer.

This artifact has been the cause of serious problems in the field of blood rheology, especially when the low shear behavior of plasma has been studied. In an early article, Wells and Merrill [15] reported non-Newtonian behavior for plasma. Since that time, many other studies have reported similar results, using the earlier data [15] to substantiate their findings. The error in this early publication [15], due to serious surface tension artifacts caused by a stiff, wax-like surface film of lipids and lipoproteins, has been pointed out by Merrill [16].

These surface effects have been eliminated in Couette viscometers [1, 10, 16] by the use of a guard ring such as that shown in Figure 2. This stationary guard ring can only be applied to systems in which the surface used to measure the torque is held almost stationary. Therefore it is not applicable to viscometers in which the torque is measured on the rotating element or to instruments with soft tension strips or wires that permit large deflections.

Additional potential experimental problems associated with Couette viscometry include end corrections, sedimentation, wall exclusion effects, temperature effects, flow stability considerations, and torque–time relations at low, steady rates of shear. A detailed examination of these problems and some suggestions for resolving them has been presented elsewhere [10].

Cone-plate viscometer. This type of instrument consists of a flat plate and a cone with a very obtuse angle (Fig. 3). The apex of the cone just fails to touch the plate surface, and the fluid under study fills the narrow gap formed by the cone and the plate. The gap at the apex of the cone is usually very small, and the gap width increases linearly with radial distance, as does the linear velocity of the rotating element. Therefore, if the flow is strictly tangential, the shear rate in a Newtonian fluid is constant at all points in the gap. For the case of a rotating cone with a very small angle between the two elements (less than 4°), the shear stress and shear rate for a Newtonian fluid are given by:

$$\tau = \frac{3T}{2\pi R^3} \qquad \dot{\gamma} = \frac{\omega}{\tan \alpha}$$

where R = cone radius, T = measured torque on the cone, ω = cone rotational speed, α = angle between cone and plate.

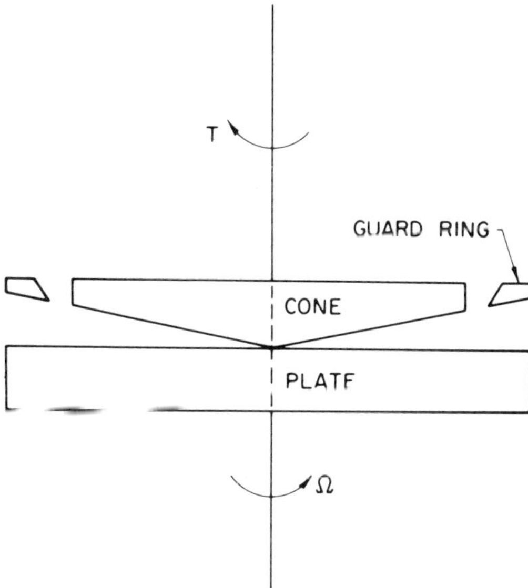

Fig. 3. Cross-sectional view of a cone and plate viscometer; the guard ring arrangement shown is not suitable if the cone is rotated and the resulting torque on it is measured. The material under study fills the gap between the cone and plate and the guard ring projects below the surface of the sample (from [10]).

Cone and plate viscometers, when constructed with a small angle between the two surfaces, require small samples and are thus frequently encountered in the study of blood rheology. The surface tension of the sample is used to keep the fluid in the gap. At very high rotational speeds, however, centrifugal forces will eventually override these surface tension forces and the fluid will be ejected from the gap; it is quite probable that secondary flows will occur before this happens.

Most of the sources of error discussed above for the Couette viscometer are applicable to the cone and plate instrument. Sedimentation of RBC away from the upper surface can be even more serious in this configuration, however, since the entire conical surface may be working against a cell-free, or at least a cell-poor, liquid sheath. Torque—time effects at low shear rates should also be important, but there appear to be no systematic studies of this problem. Surface tension artifacts at the meniscus can be extremely large in cone and plate instruments because the effect of this surface film is acting at the maximum radius, thus providing a large moment arm. The magnitude of this effect and the manner in which it may be eliminated have been demonstrated by Evans and co-workers [17]. In their system, the plate is rotated and the torque is measured on an essentially stationary cone. With this system, a simple guard ring as shown in Figure 3 has proved to be

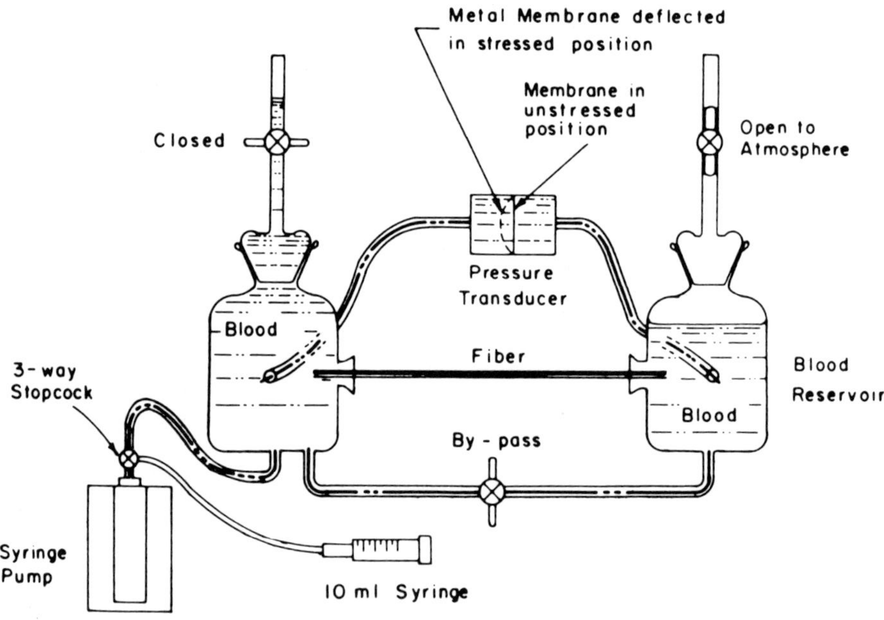

Fig. 4. Schematic drawing of a tube viscometer. Volumetric flow rate through the tube (fiber) is determined by the syringe pump and the resulting pressure drop is measured via the liquid-coupled pressure transducer (from [19]).

adequate to reduce surface film effects to an acceptable level. This guard ring arrangement would not be helpful if large deflections of the cone occurred or if the cone were rotated and the resulting torque on it was measured.

Tube viscometer. Tube or capillary viscometry is perhaps the most well-known method for studying the flow characteristics of a liquid or suspension. Figure 4 shows a typical arrangement for such a viscometer, in which fluid is pumped from one reservoir, through the tube, into the other reservoir. At each steady flow rate, the pressure drop across the tube is determined, usually via an electronic pressure transducer.

Although commonly used, the tube viscometer is one of the most difficult to interpret, especially for non-Newtonian materials. There are at least four major complications: 1) the shear rate varies from zero at the center of the tube to a maximum at the wall; 2) entrance conditions can affect the character of flow for some distance downstream of the inlet, and may influence the particle concentration distribution and actual concentration in the case of a suspension; 3) meniscus and surface tension artifacts may be extremely important unless the entire pressure-measuring portion of the system is liquid filled as shown in Figure 4; 4) at high flow rates, a portion of the driving force may remain in the fluid as kinetic energy, rather than being totally dissipated by viscous friction.

For the case of simple, laminar flow in a tube, the following equations are valid for the calculation of shear stress and shear rate (both at the wall) from experimental pressure drop-flow rate data [10].

$$\tau = \frac{(\Delta P)(D)}{(4)(L)}$$

$$\dot{\gamma} = \frac{3}{4}\left(\frac{8\overline{V}}{D}\right) + \frac{1}{4}\left(\frac{8\overline{V}}{D}\right)\frac{d\ln\left(\frac{8\overline{V}}{D}\right)}{d\ln\left(\frac{\Delta PD}{4L}\right)}$$

where D = tube diameter, ΔP = pressure drop along tube, L = tube length, and V = $4\dot{Q}/\pi D^2$ = bulk average velocity. Note that this shear rate equation reduces to the well-known expression below for the case of a Newtonian fluid, since the value of the derivative becomes unity.

$$\dot{\gamma}, \text{ wall, Newtonian} = \frac{(32)(\dot{Q})}{\pi D^3}$$

where $\dot{Q}$ = volumetric flow rate. Note also that as an index to the wall shear rate, many investigators use the reduced average velocity, $\overline{U}$, in tube diameters per second ($\overline{U} = \overline{V}/D = 4\dot{Q}/\pi D^3$). Thus, $\overline{U}$ is equal to one-eighth the Newtonian shear rate at the wall.

All of the above-mentioned potential artifacts associated with the Couette and cone and plate instruments are also possible with the tube viscometer. In particular, sedimentation effects can cause serious problems in horizontal instruments due to the possibility of a non-homogeneous red cell concentration across the tube diameter. This would be particularly serious at low flow rates in long tubes because of the extended residence time of the blood in the tube.

By microscopic observation with both visual and photographic techniques [18], blood has been shown to behave as a homogeneous suspension in horizontal tubes when the value of the reduced average velocity $\overline{U}$ was about $0.5-1.0$ sec^{-1} or greater. Below this value, sedimentation of red cells away from the top inside wall of the tube was observed; increased sedimentation was noted at progressively lower flow rates. Measurements were also made of the actual rate of sedimentation in horizontal tubes; in a 100 micron ID tube, using normal RBC in plasma at a volume fraction of 0.39, the red cell pack was observed to fall nearly one-half the tube diameter within three minutes of cessation of flow. Sedimentation may be dependent on tube length and is directly related to the degree of red cell aggregation, but a complete study of variables to predict the onset of phase separation has yet to be made. It does appear, however, that the operation of horizontal tube viscometers with blood below a $\overline{U}$ of about 1 sec^{-1} could lead to serious problems.

Salient Results

In this section, the macroscopic flow properties of blood and cell suspensions will be examined briefly. However, before beginning this analysis, it is important to note that, for a given blood sample, the results from these various types of instruments are in excellent agreement. This agreement holds true as long as the characteristic dimension of the instrument (eg, gap width in the Couette or cone and plate viscometer, tube diameter in the tube viscometer) is large compared to the eight micron diameter of the RBC. The ultimate selection of the appropriate instrument thus relates to such factors as ease of use, sample size, required shear rate range, and, of course, cost of the apparatus.

Typical rheological data for RBC in plasma, obtained in a tube viscometer, are shown in Figure 5. In this figure, the wall shear stress (τ_w) is plotted against $\overline{U}$, where $\overline{U}$ is defined as the average blood flow velocity divided by the tube diameter; for a given tube, wall shear stress is proportional to the pressure drop along the tube, whereas $\overline{U}$ is proportional to the volumetric flow rate. Note that two tube diameters were employed (523 and 850 microns ID) and that for each tube, RBC–plasma suspensions having RBC volume fractions of 0.201 and 0.393 were measured.

Examining first the data represented by the solid points (RBC volume fraction of 0.393), it is obvious that the curve representing these data points is not straight, especially at low values of $\bar{U}$. Further, even in those $\bar{U}$ regions where a straight line could be fitted to that portion of the data, the geometric slope is not unity; the slope does, however, approach 1.0 in the region of $\bar{U} = 500–1,000$ sec^{-1}. This nonlinear behavior means that the suspension behaves as a non-Newtonian fluid (ie, apparent viscosity decreasing with increasing shear rate), since on the double logarithmic coordinate system used in Figure 5, a Newtonian fluid would be represented by a straight line with a slope of unity [19].

The open symbols in Figure 5 were obtained for an RBC–plasma suspension with an RBC volume fraction of 0.201. Note the similar nonlinear behavior, with curvature at the lower values of $\bar{U}$; the degree of curvature and the departure from linear behavior are not, however, as marked as with the 0.393 volume fraction sample. In addition, note that for any given volumetric flow rate (as $\bar{U}$), the pressure drop (as τ_w) is greater for the suspension with the higher RBC volume fraction. Indeed, hematocrit (ie, RBC volume fraction) is probably the most significant determinant of the apparent viscosity of blood or RBC suspensions [1, 2, 4, 8]. As

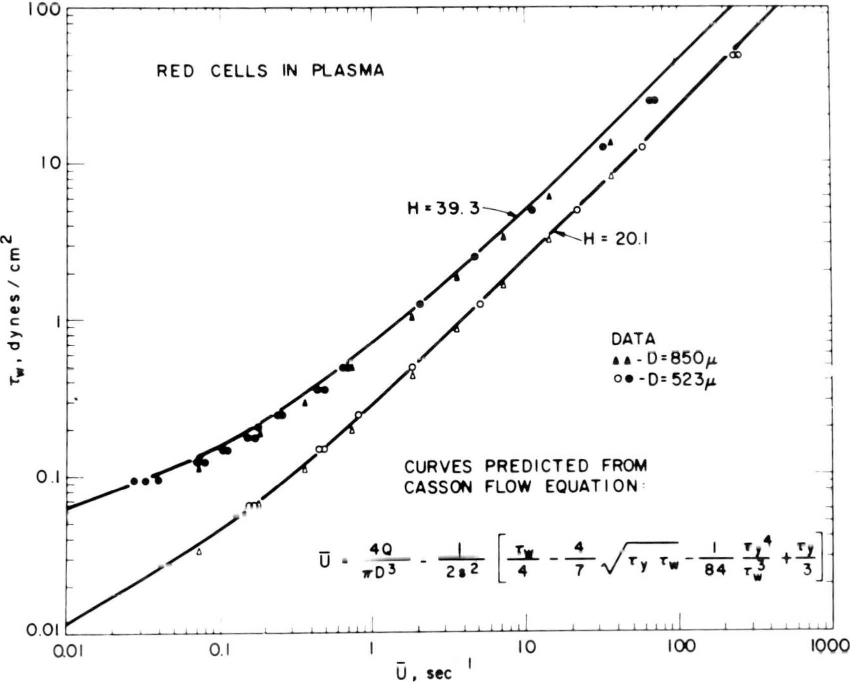

$$\bar{U} = \frac{4Q}{\pi D^3} - \frac{1}{2s^2}\left[\frac{\tau_w}{4} - \frac{4}{7}\sqrt{\tau_y\,\tau_w} - \frac{1}{84}\frac{\tau_y^4}{\tau_w^3} + \frac{\tau_y}{3}\right]$$

Fig. 5. Pressure–flow data (as wall shear stress–reduced average velocity $\bar{U}$) for normal human red cells in plasma. Data for two tube diameters (523 and 850 microns) and two hematocrits (0.201 and 0.393) are shown. The solid lines are the predicted results based on rheologic data obtained with the two samples in a co-axial viscometer (from [19]).

a rough approximation, there is a 4% increase in the apparent viscosity for each unit increase in hematocrit. That is, if the hematocrit of a suspension is increased from 0.45 to 0.46, the apparent viscosity will be elevated by 4%. This approximation is limited to medium to high shear rates (say above 10–50 sec^{-1}), since at lower shear rates the viscosity of RBC–plasma suspensions is even more sensitive to the volume fraction of RBC [1, 8].

Figure 6 shows an alternate method of plotting rheological data for RBC suspensions; the data in this figure were obtained using a Couette rotational viscometer [13]. The data are plotted on a square-root coordinate system, which is sometimes referred to as a Casson plot [1, 16]. This choice of coordinate system has at least two advantages: 1) the very low shear rate region (say 0–2 sec^{-1}) is expanded, allowing accurate data placement; 2) the data for blood or RBC–plasma suspensions appear linear or nearly linear in this low shear rate region, thus allowing extrapolation to the shear stress axis. Also shown in Figure 6 are dashed lines indicating loci of constant viscosity (eg, 5, 10, and 100 centipoise). These dashed lines allow visual appreciation of apparent viscosity changes with alteration of experimental variables such as shear rate, hematocrit, and suspending medium.

Data for two RBC suspensions, both at a hematocrit of 0.40, are shown in Figure 6. Dealing first with those for RBC suspended in isotonic saline (ie, 0.9% NaCl), note that the apparent viscosity is always less than 5 centipoise *and* that the curve extrapolates to the origin of the plot. Conversely, the data for RBC in plasma show increasing viscosity with decreasing shear rate (5 centipoise at 12 sec^{-1}, 10 centipoise at 2 sec^{-1}, and 100 centipoise at about 0.1 sec^{-1}) *and* a finite, nonzero intercept on the shear stress axis. This intercept is usually referred to as the yield shear stress of the suspension, and it represents the minimal fluid shear stress necessary to initiate flow. The existence and magnitude of this yield stress as well as the increased low shear rate apparent viscosity of the RBC–plasma suspension is due to the aggregation of RBC (ie, rouleaux formation) at low shear rates [1, 2, 5, 16, 19]. In plasma, this aggregation is induced by the protein fibrinogen, but in nonprotein suspending media it can be caused by the presence of macromolecules of appropriate molecular weight [2, 20]. Finally, for comparison purposes, the lines for plasma and saline alone are shown in Figure 6; these fluids are Newtonian in nature and thus plot as straight lines with a zero–zero intercept [16].

It is interesting to note that there has been considerable debate in the literature regarding the existence of a yield shear stress for normal blood or RBC–plasma suspensions; the debate seems to wax and wane with a periodicity of three to five years! Although many indirect methods for obtaining a measure of the yield stress have been reported, the most direct and convincing measurements have been carried out by Benis and coworkers [19]. With a tube viscometer similar to that shown in Figure 4, a motor-driven syringe was used to pump blood from a feed reservoir, through the tube, and into a receiving reservoir. After steady flow had been established and a steady pressure drop across the tube was recorded, the

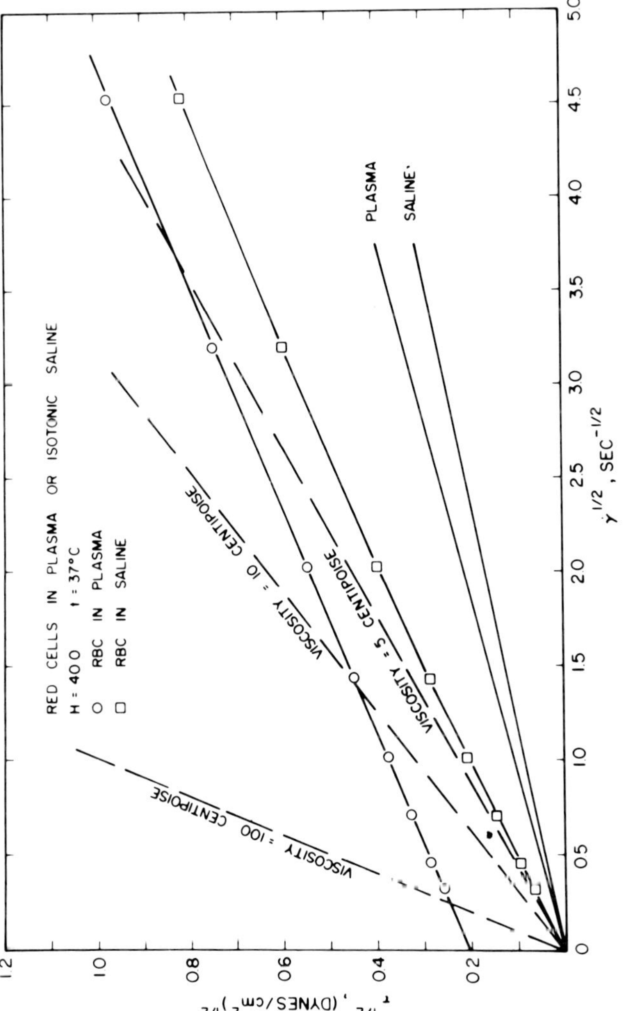

Fig. 6. Low shear rate behavior of human red cells in either plasma or isotonic saline at a constant hematocrit of 0.40. The data are plotted on a square-root coordinate system and, for reference, the dashed lines indicate the behavior for Newtonian fluids of several constant viscosities. Note that normal human plasma, like saline, is a Newtonian liquid (from [86]).

syringe pump was stopped and isolated from the system, thus allowing the flow to return toward zero. Pressure drop across the tube was then recorded as a function of time via the electronic pressure transducer. Volumetric flow through the tube was also recorded via the transducer by precalibration of the volume-pressure behavior (ie, compliance) of this instrument. Thus, the transducer served to measure simultaneously the volumetric flow rate of fluid through the tube and the pressure drop across the tube.

With this system, using tube diameters of 523–850 microns, the following fluids were observed to show a decay to zero pressure drop–zero flow rate within one minute: 1) water; 2) RBC in saline, 0.40 hematocrit; 3) RBC in saline plus 3% albumin, 0.40 hematocrit. However, using an RBC–plasma suspension of 0.42 hematocrit, the pressure transducer indicated that the flow had gone to zero while the pressure drop across the tube remained at a steady, nonzero value; this zero flow with finite pressure drop condition lasted for at least two to three minutes [19]. From this observed residual pressure drop and the known geometric factors of the tube, a wall shear stress was then calculated. This calculated yield shear stress was found to agree with the extrapolated value (for a portion of the same suspension) obtained using the graphical technique shown in Figure 6. Of course, the very concept of a yield shear stress implies the concept of time, and one could expect a decay to a zero pressure drop–zero flow rate state if the tube experiment had been continued for a sufficient period of time; this decay to zero conditions would be artifactual, since it would represent either RBC sedimentation in the tube or plasma movement relative to the RBC. On balance, therefore, it would appear that blood or aggregating RBC suspensions do exhibit a yield shear stress, that this yield shear stress can be measured by at least two methods, and that the magnitude of this yield shear stress, at constant hematocrit, represents a useful index to the magnitude of RBC aggregation [3, 16, 19, 20].

In addition to hematocrit and cellular aggregation, the mechanical properties of the RBC also influence the macroscopic rheological behavior of the red cell suspensions. Figure 7 compares the rheological behavior of RBC in saline, for which no aggregation is observed regardless of shear rate, with that for RBC hardened or fixed by treatment with aldehyde [21]. Microscopic observation of these aldehyde-treated RBC also indicates no cell aggregation, and since suspensions of these cells are Newtonian over a shear rate range from 0.1 to 700 sec^{-1} [1, 21], only a single curve is shown.

The normally deformable cells in isotonic saline (Fig. 7) are not Newtonian except at very low hematocrits or very high shear rates. Thus, above a hematocrit of about 0.10, these RBC–saline suspensions show a decrease in apparent viscosity with increasing shear rate; shear rates in excess of approximately 1,000 sec^{-1} do not, however, further decrease the apparent viscosity. This decrease in viscosity with shear rate is usually interpreted to indicate increased RBC deformation, with a limit on the degree of deformation being reached when the shear rate exceeds

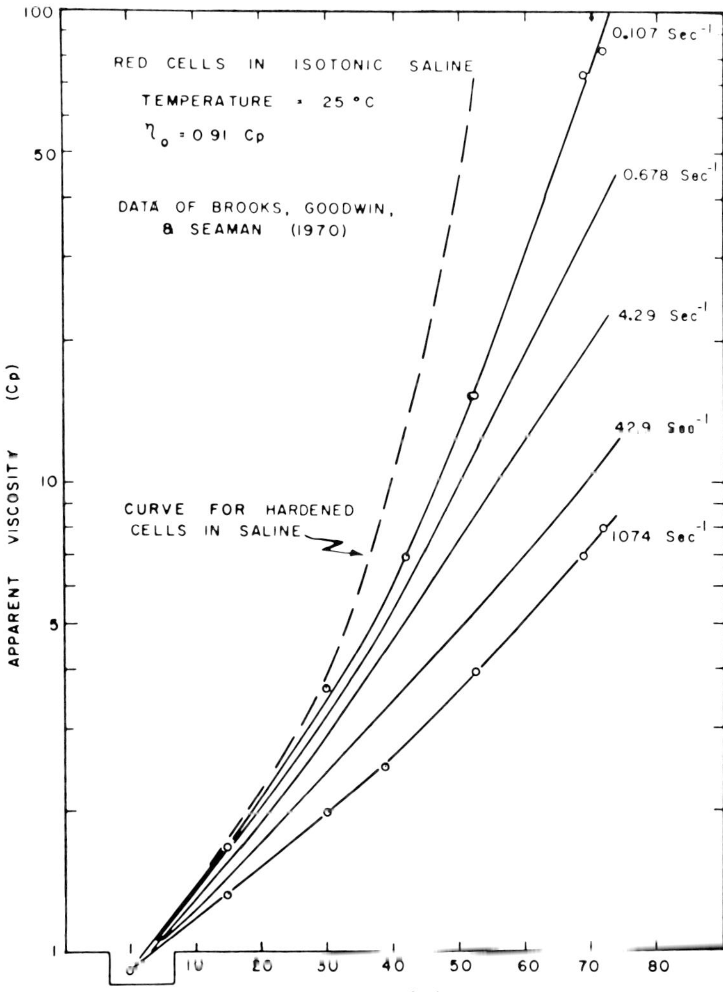

Fig. 7. The apparent viscosity–hematocrit relations for both normal and aldehyde-hardened human red cells suspended in isotonic saline. Inasmuch as the hardened cell suspensions are Newtonian, only a single curve is indicated; the normal cells are shown for five different rates of shear. Note the general agreement at low shear rates and relatively low hematocrits and the large differences between normal and hardened cells at higher hematocrits and shear rates (from [1]).

1,000 sec^{-1}. Note that the hardened RBC suspensions, for which no RBC deformation exists, are always more viscous than the deformable cells *and* that this difference in viscosity (hardened larger than normal) increases with increasing hematocrit and shear rate. Thus, as suggested in Figure 1, the ability of the normal RBC to deform is an important factor in the flow behavior of these suspensions. *Furthermore,* the difference in high shear rate apparent viscosity between suspensions of normal RBC and experimental RBC provides in index to the cellular deformability of these experimental cells.

In summary, the macroscopic rheological properties of RBC suspensions are influenced at least by hematocrit, shear rate, degree of cellular aggregation, and the mechanical properties of the cells. Figure 8 provides an overview of the shear rate and aggregation factors. Shown here, as plots of relative viscosity versus shear rate, are data for normal RBC in plasma and for RBC suspended in a Ringer solution (ie, an isotonic, balanced salt solution) to which sufficient albumin has been added to raise the suspending phase viscosity to equal that of plasma [2]. For the RBC–

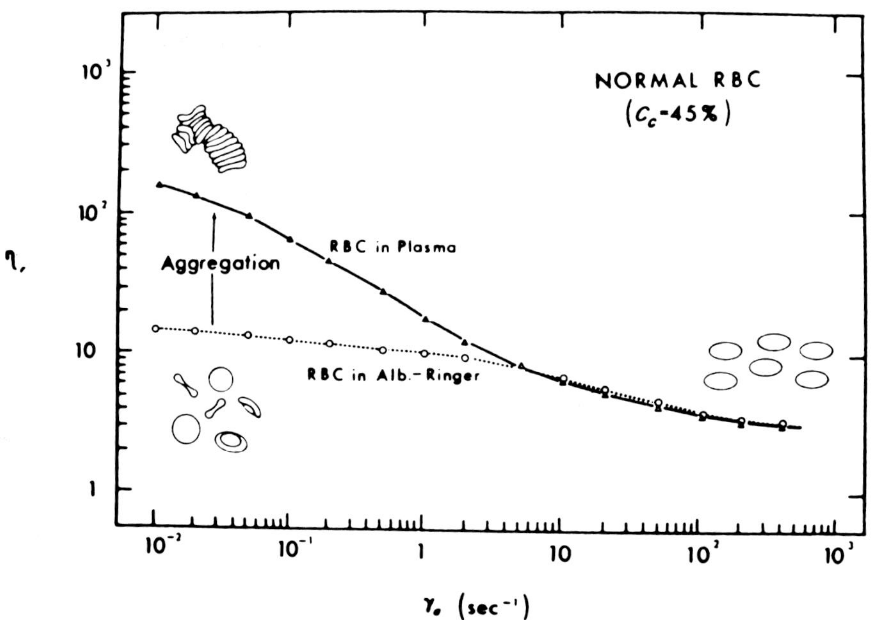

Fig. 8. Relative viscosity–shear rate data for normal human red cells in plasma and in an albumin-Ringer solution; in both suspensions the hematocrit is 0.45 and the suspending fluid viscosity is 1.2 centipoise. The drawings show the aggregation of red cells in plasma but not in the albumin-Ringer solution at low shear rates and the identical deformation of the cells, in both media, at higher rates of shear (from [2]).

plasma suspension at low shear rates, aggregation of the cells into rouleaux is known to occur, whereas the cells in the albumin—Ringer medium exist as individual particles; the physical condition of the cells is shown by the drawings adjacent to the curves at the appropriate shear rates (Fig. 8). Both the low and high shear rate regions provide interesting information: 1) at low shear rates, the difference in relative viscosity between these two suspensions can be used as an index of red cell aggregation; 2) at high shear rates, where the aggregates are dispersed and the cells are deformed, the two curves are coincident; noncoincident behavior at these higher shear rates would suggest differences in the mechanical behavior of the two RBC populations.

It should be stated that all of the above discussion has dealt with blood as a homogeneous medium. However, much work carried out in tube viscometry has been directed toward the limits of this continuum concept. The classic paper in the field is that of Fahraeus and Lindqvist [22], in which a sample of blood showed a decreasing apparent viscosity when measured in tubes of decreasing diameter. In order to deal with these observations, two basic explanations have been postulated: 1) the sigma effect, which treats blood flow in tubes as consisting of finite, cylindrical, telescoping elements which must be summed up across the tube diameter; 2) a cell-poor layer at the wall of the tube having a width greater than that predicted by simple wall exclusion (ie, greater than 1—3 microns). Overlooked in these explanations, however, is the earlier work of Fahraeus [23], which dealt with the relationship between the hematocrit of the blood in the feed reservoirs and the actual (lower) dynamic hematocrit in the tube.

Cokelet [24] has examined the Fahraeus-Lindqvist data and indicates that by assuming a uniform radial distribution at the actual tube hematocrit, the tube diameter—apparent viscosity dependence can be explained. More recent studies [25] using tube diameters from 29 to 211 microns ID have explored this effect in greater detail. These studies indicate that the ratio of tube hematocrit to feed reservoir hematocrit is a linear function of feed hematocrit for a given tube size and that the decrease of this ratio was maximal in 29 micron ID tubes (the smallest studied). In addition to these ratio measurements, it was shown that the use of the actual tube average hematocrit, rather than the feeding concentration, allowed accurate predictions of pressure—flow data. Preliminary data from Cokelet [8] indicate that this reduction in tube hematocrit continues down to 8 micron tubes; the use of the actual tube average hematocrit again appears to allow predictions of pressure—flow data in these small tubes.

Unresolved Questions

Although many features of the macroscopic flow behavior of blood have now been described in detail in the literature, there remain several unanswered (or, at least, not fully resolved) questions. The discussion that follows deals with three topics about which our curiosity exceeds our knowledge.

Pulsatile and oscillatory flow. All of the above discussion has dealt with the steady-state rheologic behavior of blood or RBC suspensions, yet the cardiovascular system operates using nonsteady, pulsatile flow with a fundamental frequency of 1–3 cycles per second. Thus, questions arise as to the applicability of these steady measurements to the living system.

Pulsatile and oscillatory flow of blood has been the subject of several investigations. Early work in medium- to large-bore tube viscometers indicates that pulsatile flow appears to be comparable to steady flow if the time average values of the driving pressure and flow rate are used [26], and a quasisteady-state approximation of the Casson equation has been developed for tube flow [27]. Neither experimental evidence for the tube diameter limits of these approximations nor the effect of a geometry more complex than a straight tube were considered. Later studies, using single rigid tubes [28], an array of rigid tubes [29, 30], or a coaxial viscometer [31] have demonstrated both the viscous and elastic components of the bulk flow properties of blood. Some caution is necessary, however, in interpreting the data from the array of rigid tube studies [29, 30] inasmuch as the amplitude of the pulsations with respect to the size of the RBC is unknown. Furthermore, none of these studies has attempted to relate the data to the shape-relaxation time (ie, viscoelastic properties) of the red cell [9]. For further details on the dynamic behavior of blood and RBC suspensions, the reader is referred to a recent review by Cokelet [32].

Limits of continuum model. The particular problem of the continuum model relates to the above discussion regarding the dynamic (lower) hematocrit of blood in small tubes and the ability to use this dynamic hematocrit to predict pressure–flow data in these tubes. That is, knowledge of the actual tube hematocrit, combined with experimental data obtained in large tubes (eg, 800–1,000 microns) at the same hematocrit, allows very accurate predictions of τ_w–$\overline{U}$ relations in tubes as small as 29 microns [25]. Furthermore, preliminary data suggest that the same procedure can be used for tubes as small as 8 microns [8].

Although experimentally valuable, these results are somewhat disconcerting. A 29 micron tube is less than four times the undeformed major dimension of a human RBC; yet data obtained in tubes 100–125 times larger than the RBC apply perfectly to this small geometry. The conceptual problem becomes worse in the case of an 8 micron tube, where the ratio of particle size to tube diameter approaches unity; yet the predictions from much larger tubes are also applicable! Collectively, these results suggest that one can tacitly ignore the particulate nature of blood and use only the macroscopic index of RBC volume fraction — ie, hematocrit. This approach must, however, have limits, since in systems with particle diameter to tube diameter ratios greater than unity, cell–wall interactions and cell deformation (as a function of cell velocity or average shear stress) should be relevant.

Relation to in vivo blood flow. A third question with broad implications relates to the applicability of the macroscopic rheologic properties of blood to flow in the cardiovascular system. The desire to relate in vitro rheological measurements to flow in the living system has led to several investigations that employed a vascular bed or organ as a biological viscometer; the results of such studies are not always in concordance [32].

The classic paper in this area is that published by Whittaker and Winton [33] in the early 1930s. In their experiments, isotonic Newtonian fluids as well as RBC suspensions were pumped through isolated hind limbs of dogs; pressure drop–flow rate data were collected for the experimental fluids. In addition, these fluids were also forced through a large-bore tube viscometer at velocities sufficient to obtain the high shear rate (ie, Newtonian) viscosities of the RBC suspensions. By taking the ratio of the Newtonian fluid flow rate to the RBC suspension flow rate at the same overall pressure drop and multiplying this ratio by the viscosity of the New-tonian fluid, Whittaker and Winton calculated the in vivo viscosity of the cell sus-pensions. Note that the flow rate ratio was determined at high flows where the cell suspension would most likely be Newtonian and, by comparing the flows of the two fluids at equal pressure drops, the vascular geometry was anticipated to be the same for both fluids.

When the RBC suspension viscosity was calculated in the above manner, it was found always to be less than that measured in the tube viscometer [33]. For ex-ample, at a hematocrit of 0.45, the in vivo viscosity was about 50% of that mea-sured in the viscometer. Whittaker and Winton attributed this difference to the Fahraeus-Lindqvist effect [22, 23]; ie, in the smaller diameter vessels in the hind limb the viscosity of the cell suspension is less than in larger vessels or in the tube viscometer, because of the lower dynamic hematocrit. Similar disagreements have been noted in experiments with the cat calf muscle [34], the isolated rabbit ear [35, 36], and the bullfrog hind limb [37].

More recently, Benis and coworkers [38–40] have suggested that the discrepancy between in vivo and in vitro rheologic data arose because of a faulty analysis of the in vivo flow data. They imply that inertial effects are significant when low-viscosity, Newtonian fluids (ie, isotonic saline) flow through the vascular system. When these inertial effects are considered, they indicate that the in vivo viscosity of an RBC suspension calculated from hind limb perfusion experiments agrees well with that obtained in the in vitro tube viscometer [40]. However, Cokelet [32] has indicated that Folkow and his associates [eg, 34] have criticized the analysis of Benis et al [40] on the grounds that the vascular geometry is not identical when the tissue is perfused with isotonic saline and with cell suspensions, unless the blood vessels are maximally dilated by the use of exercise or drugs. The issue of whether inertial effects are significant in the body has not been settled; nor has the validity of the use of the Fahraeus-Lindqvist effect in explaining the Whittaker-Winton effect been firmly established. In addition, pressure-induced geometry changes (with pressure distribution through the bed being a function of the non-Newtonian properties of the perfusate) are also unknown. Possibly, the use of hollow, rigid replicas fabricated from living vascular beds [41] may aid in resolving these questions of geometry.

MICROSCOPIC FLOW BEHAVIOR

Overview

An important conceptual innovation in hemorheology was introduced by Goldsmith and Mason [42]. Rather than studying the macroscopic flow behavior (bulk viscosity) of suspensions under defined conditions, they described in detail the microscopic behavior of particles or droplets contained in the suspension. Their microrheological approach has now been introduced into blood rheology [43, 44] in an attempt to predict both bulk properties and the flow of blood in capillaries and restricted pores. In the discussion that follows an attempt has been made to divide the material into two categories: 1) the extrinsic behavior of the RBC, usually termed RBC cellular deformability, and 2) the intrinsic behavior of the RBC membrane, which refers to various mechanical properties of the membrane independent of overall cellular topology. Clearly, this division is somewhat arbitrary, inasmuch as understanding RBC mechanical behavior usually involves factors relating to the cellular membrane. Note that the presentation is purposely brief, since other sections of this book (eg, papers by E. A. Evans, H. L. Goldsmith, R. M. Hochmuth, and R. Skalak) treat much of this material in greater detail.

Extrinsic Mechanical Behavior

The ability of the human RBC to deform is the sine qua non for effective blood flow in the cardiovascular system, since 8 micron diameter RBC must deform in order to flow through the 3 to 4 micron diameter capillaries of the microcirculation. RBC deformability is also an important determinant of blood flow in larger vessels and of the rheologic behavior of RBC suspension in various viscometers; extreme pathologic changes in RBC deformability, such as sickle cell disease, or experimental modification of RBC rigidity greatly influence the in vitro flow properties of these suspensions (eg, Figs 1 and 7). Weed [45, 46] has pointed out the in vivo importance of RBC deformability and has reviewed the various types of membrane changes that lead to in vivo red cell destruction.

Although the concept of deformability is intuitively grasped as the resistance a cell offers to a change in its shape or, in other words, the ability of the entire cell to form a new configuration, a precise definition of cellular deformability is most difficult to formulate. This difficulty is directly related to the lack of measurement units for the term. Each investigator more or less defines the mechanical forcing function on the RBC and, in turn, more or less defines the resistance to shape change offered by the cell. Exact comparisons among various literature reports are therefore extremely difficult. Thus, in reviewing the deformability literature, one is reminded of the quote from Lewis Carroll, "When I use a word, Humpty Dumpty said, in rather a scornful tone, it means just what I choose it to mean — neither more nor less" [47]. A review (more or less) of the various experimental techniques for these measurements is provided below.

Filtration methods. The flow of red cell suspensions through various filtering materials has been applied to the study of the microrheology and deformability of erythrocytes. Laboratory filter paper [48, 49], cellulosic membrane filters (ie, Millipore filters) and, more recently, precise polycarbonate micropore filters (ie, Nucleopore filters) have been employed. Figure 9 is a scanning electron microscopic picture of a polycarbonate filter with a nominal pore size of 3 microns, showing RBC in the process of passing through the pores.

Using polycarbonate filters, Gregersen and co-workers [50] have shown that all

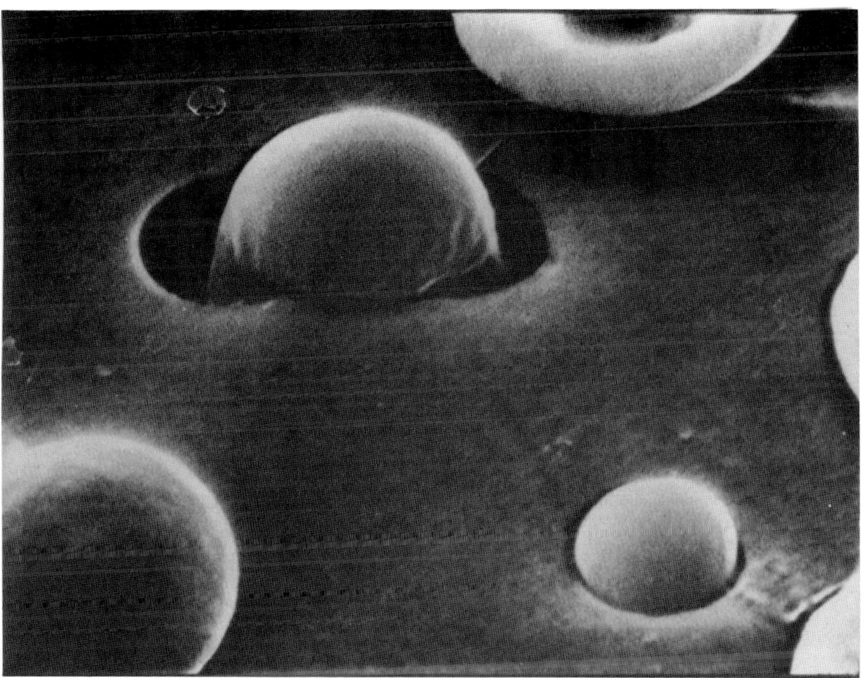

Fig. 9. Scanning electron micrograph of human red cells in various stages of flow through pores in a three-micron polycarbonate filter; the cells were fixed with glutaraldehyde under a low (about 10 mm H$_2$O) pressure head. Note that the cell in the center has entered two pores, whereas the one in the lower right corner is almost through the pore (photograph courtesy of Dr. Richard F. Baker, U.S.C. School of Medicine, Los Angeles, CA).

normal red cells will easily pass through 6.8 and 4.5 micron pores; with 2.4 micron pores, however, the filtrate contained only 70% of the original cells. These studies have been extended and pressure drop—flow rate curves have been obtained for 2.2 to 4.4 micron pores [51]. Hemoglobin and potassium leakage from red cells due to high pressure filtration has also been studied [51, 52]; a preferential release of potassium over hemoglobin was observed. Schmid-Schonbein and co-workers [53] have described an experimental system for filtration with polycarbonate filters and indicate that reduced pH as well as some disease states (eg, hereditary sphero-cytosis and pernicious anemia) lowered the flow rate through these filters; hyper-tonicity also decreased the volumetric flow rate [54]. Additional experimental data and a theoretical analysis of RBC flow through these polycarbonate filters has been presented by Lingard [55].

Micropipette suction. Small micropipettes have been used to determine the forces necessary to cause red cells to transit small-bore glass capillary tubes; a schematic drawing of this method is shown in Figure 10. A small micropipette, with a diameter between 2.8 and 3.0 microns in the case of RBC, is used to aspirate the cell. Under direct microscopic observation, the erythrocyte is brought against the tip of the pipette and the pressure within the pipette is then slowly decreased. The negative pressure required to pull the cell entirely within the micropipette (usually termed "P_t") is used as a measure of the deformability of the cell.

This micropipette suction technique has been used extensively by, for example, LaCelle [56, 57] and Leblond [58]. Normal red cells require a negative pressure of about 6.1 mm H_2O, whereas pathologic RBC (eg, thalassemia, hereditary

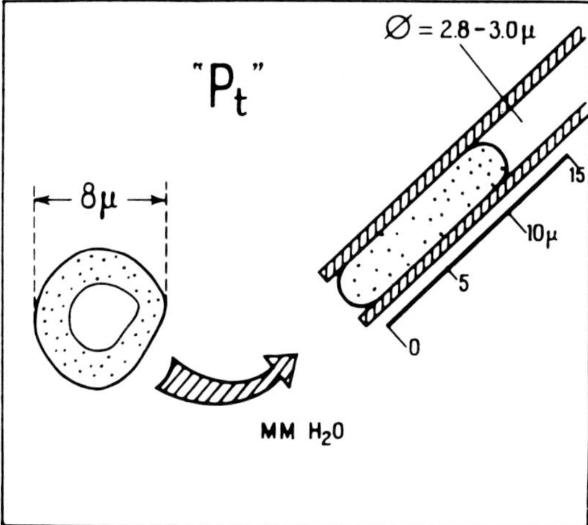

Fig. 10. Schematic drawing of the micropipette technique for measuring red cell deforma-bility. In this method, the entire cell is drawn into the pipette and the negative pressure ("P_t") required is taken as an index of this extrinsic cellular property (from [58]).

spherocytosis) require pressures from 10 to 60 mm H_2O [57]. A review of small-tube methodology has been prepared by Braasch [59], the experimental data have been reviewed by Cokelet [8], an overview of mathematical models for small-tube flow has been presented by Gross and Aroesty [60], and a finite element method for predicting RBC shape and additional pressure drop per cell has been given by Gupta and co-workers [61].

Microscopic observation in flow fields. Several techniques have been developed in which an individual RBC is observed microscopically while being subjected to fluid shear forces. Again, both the operational details and the experimental results have been described in the literature; a brief listing of these methods is provided below.

Parallel plate flow chamber. This method was developed by Hochmuth and co-workers [9, 62]. The system consists of two flat, parallel glass plates (usually a microscope slide and coverslip), separated by a liquid-tight gasket designed to provide about a 100-micron gap between the two parallel surfaces. Red cells are al-

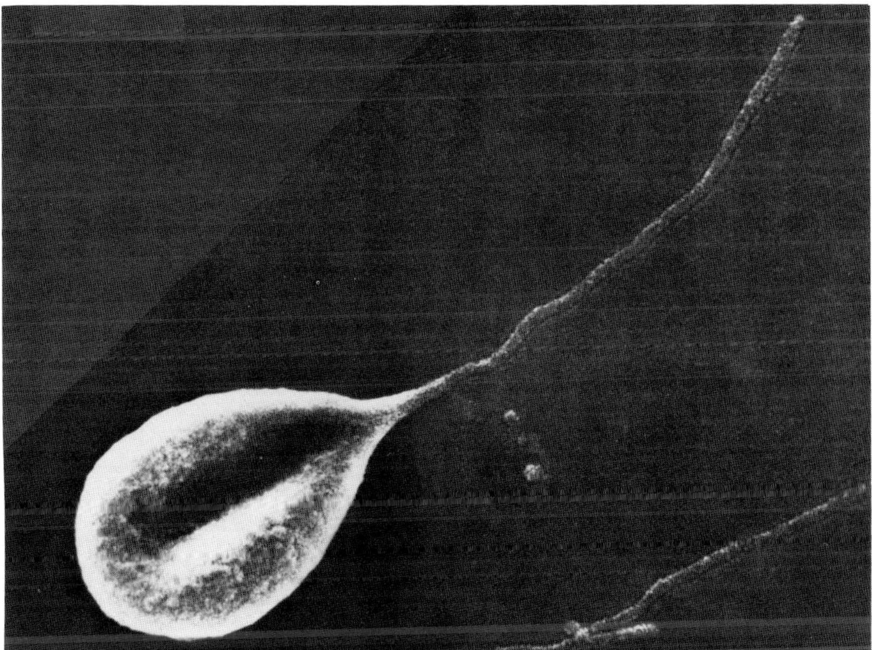

Fig. 11. Scanning electron micrograph of a normal human red cell deformed by fluid shear stress in a parallel plate flow channel. Note that the cell has a deformed shape and a long, thin tail (tether) leading from the main portion of the cell to the point of attachment to the surface. Tether formation and growth are used to measure shear viscosity in the plastic domain; lower levels of fluid shear stress result in elastic deformation of the cell (photograph courtesy of Robert M. Hochmuth, Duke University, Durham NC).

lowed to attach to one of the surfaces, following which fluid (eg, isotonic saline) is caused to flow through the gap. Knowledge of the geometry of the flow channel (ie, gap width, gap height, and channel length), the fluid viscosity, and the volumetric flow rate allows calculation of the fluid shear stress acting on the RBC.

Figure 11 provides an example of the results possible with this system. In this figure, a cell has been pulled by the flowing fluid such that it consists of a long thin tail and a residual cell body; the point of attachment of the cell to the glass surface is at the end of the tail. The deformation shown in this figure is an extreme case, for at lower fluid shear stresses, the cell reversibly elongates into a "tear drop" shape. Although data obtained with this system can be used to calculate intrinsic membrane mechanical properties [9], the extent of elongation of the tear drop shaped cell can be taken to be a measure of cellular deformability.

Poiseuille flow in tubes. The movement and deformation of liquid drops and human RBC in tube flow have been extensively studied by Goldsmith [11, 42–44, 63]. A unique aspect of his experimental system is the use of a vertical glass tube and a vertically traveling microscope, thus allowing a single particle or RBC to be followed or tracked as it moves along the tube. The behavior of an individual RBC in dilute suspension is shown in Figure 12, where it is seen to rotate like a rigid disc at local shear rates from 4 to 20 sec^{-1} (estimated local shear stress about 0.04 to 0.20 dynes/cm^2). At higher shear rates, the RBC is observed to become aligned with the direction of flow and to undergo elongation (ie, deformation), with deformation increasing as the local shear stress is increased. In addition, Goldsmith [63] has studied the deformation of RBC in a more crowded situation by us-

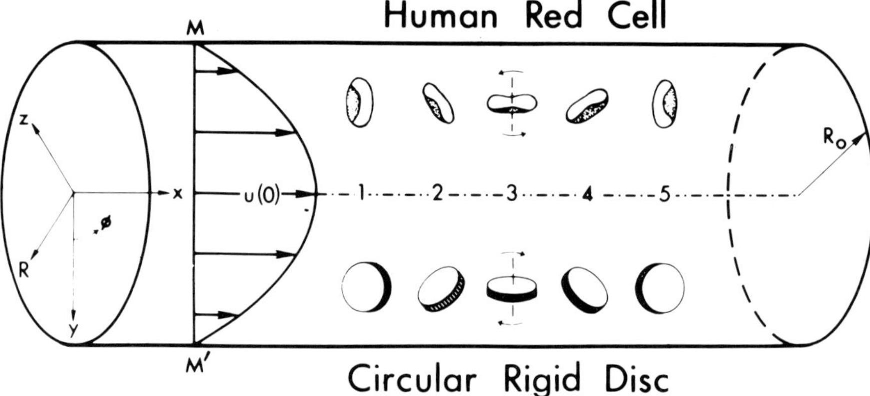

Fig. 12. Coordinate system and rotation of a rigid disc and a red cell in Poiseuille flow. The particles are viewed in the median plane of the tube, and the Cartesian coordinates constructed with their center on the median plane are shown on the left. At shear rates from 4 to 20 sec^{-1}, both the disc and the red cell are observed to rotate, whereas at higher shear rates (or shear stresses), the red cells undergo deformation (from [63]).

ing a transparent, reconstituted biconcave RBC ghost suspension containing a few intact erythrocytes. In this crowded environment, cell—cell interaction, even at low levels of local shear stress, produces significant distortion and deformation of the RBC.

Counter-rotating cone-plate system. An innovative experimental apparatus, based on the cone and plate viscometer (Fig. 3), has been developed by Schmid-Schonbein and co-workers [64, 65]. The basic design of this device is essentially that of the cone-plate instrument, and differs only in that 1) the cone and plate portions are optically transparent; 2) the cone and plate are rotated, at the same speed, in opposite directions. The suspension under study (RBC in various media) is sheared between the cone and plate and is viewed microscopically in the stationary plane halfway between these two moving surfaces. A schematic drawing of this system is shown in Figure 13. Note that, as with all cone-plate devices, the shear rate is independent of radial position and is determined by the rotational speed of the cone and plate. Using this device, individual erythrocyte behavior

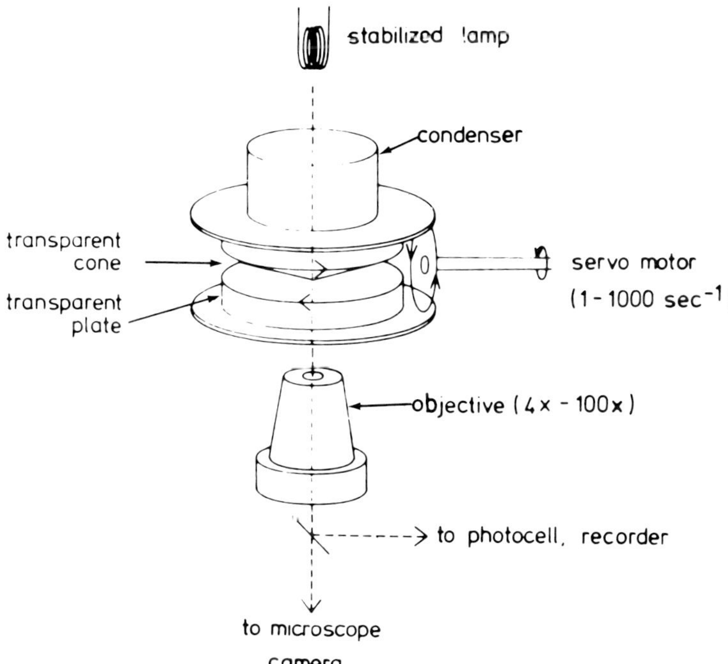

Fig. 13. Schematic representation of the counter-rotating, transparent cone and plate instrument — the rheoscope. The cell suspension under study is contained between the cone and plate and is viewed microscopically at a position halfway between the cone and plate (ie, in the stationary layer). Cell deformation results from the fluid shear stress, which is a function of the rotational speed, the medium viscosity, and the geometry of the system (from [87]).

and deformation have been described, and the "fluid drop" transition at high shear rates has been quantitated. In addition, movement of the RBC membrane around the cellular contents (the so-called tank tread motion) has been observed [65].

High-speed centrifugation. A new method for altering RBC morphology by high-speed centrifugation through a physiological buffer has recently been developed [66, 67]. In this centrifugal technique, human RBC centrifuged through isotonic buffer in relative centrifugal fields greater than 500g are reversibly deformed into a characteristic elongated shape; centrifugation of the RBC into buffer containing

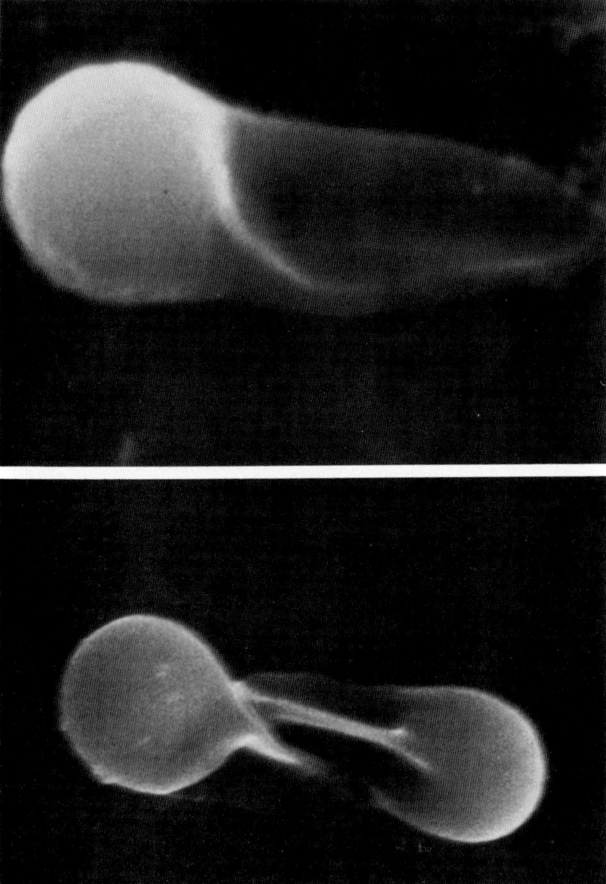

Fig. 14. Scanning electron micrographs of typical red cells deformed in a relative centrifugal field of 15,000g. The top panel shows a cell with a flat tail, whereas the bottom panel indicates cells with three-lobed tails; both tail shapes are commonly observed (from [67]).

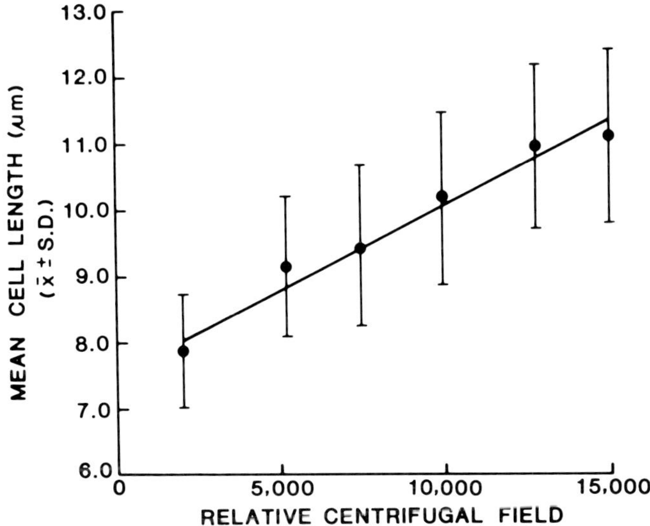

Fig. 15. Relationship between the mean length of deformed human red cells and the average relative centrifugal field used to deform the cells. The error bars represent the standard deviation of at least 90 cells (from [66]).

2% glutaraldehyde preserves the deformed shape of the cells and facilitates their examination by light and electron microscopy. Centrifugation is done in a standard 0.55 ml round-bottom microfuge tube having a Teflon insert designed to minimize mixing of the buffer with the fixative solution at the bottom of the microfuge tube. The entire assembly is spun (for a 5–7 second period) on a horizontally mounted centrifuge using a variable transformer to control the centrifuge speed. During centrifugation, the RBC pass out of their original suspending medium, down the Teflon insert through the buffer, and into the buffer plus glutaraldehyde at the bottom of the microfuge tube.

Under optimal conditions [66], normal human RBC deformed by this technique have a shape characterized by a nearly spherical head approximately 5 microns in diameter and a flat or triangular tail 4 to 10 microns in length (Fig. 14). The overall length of the cell varies with centrifugal field in a nearly linear fashion; nominal values of length are 7.9 microns at 2,000g and 10.9 microns at 15,000g (Fig. 15). Overall length of the deformed cell reflects cellular deformability, inasmuch as RBC populations treated to provide differing cellular deformability levels demonstrate decreased length under conditions where decreased cellular deformability is expected [67, 68]. Only a small volume of blood (10–30 microliters) is required for each run, and the technique provides a measure of the distribution of RBC deformability within a given population via the necessary measurement of 30—50 RBC from a given test.

Overview of RBC deformability. Although the various techniques described above differ in their experimental approach, all of them demonstrate an outstanding feature of the normal human erythrocyte: It is an extremely deformable particle, requiring relatively small forces to produce dramatic changes in its shape! Further, these studies collectively indicate that at least three factors seem to be important in determining the extent of cellular deformability:

1) The membrane surface area to cellular volume ratio — the "S/V" ratio — of the RBC (ie, a decreased S/V, produced by hypotonic media, reduces cellular deformability).

2) The mechanical properties of the cell membrane (ie, a more "rigid" membrane reduces cellular deformability).

3) The viscosity of the cellular contents (ie, an increased internal viscosity, such as that seen in sickle cell disease under low oxygen tension, reduces cellular deformability).

It should be noted, however, that while there is general agreement about the importance of these factors, the exact, quantitative manner in which they interact to determine the overall mechanical behavior of a given erythrocyte remains an enigma.

Intrinsic Membrane Mechanical Properties

Direct measurements of the mechanical behavior of the RBC membrane, independent of cellular geometry, have been the focus of several investigations within the past few years. The overall approach is to separate any influence of cell geometry from measurements of the intrinsic material properties of the cell membrane, thus defining intrinsic material behavior as the deformation and rate of deformation response of infinitesimal membrane elements to forces applied to the element [9].

In contrast to the variety of methods used to estimate cellular deformability, only two experimental approaches seem to exist for the measurement of the membrane mechanical properties, small-bore micropipettes and the parallel plate flow chamber [62]. The micropipettes used for these membrane studies are smaller than those used for cellular deformability (eg, 1.0 to 1.5 micron tip versus 2.8 to 3.0 micron tip), and the experimental protocol consists of measuring the length of the cell membrane that is pulled into the micropipette as a function of the negative pressure inside the pipette [9, 58]. A schematic representation of this micropipette method is shown in Figure 16. Note that only a portion of the cell membrane enters the pipette, whereas the cellular deformability measurement involves aspiration of the entire RBC (see Fig. 10).

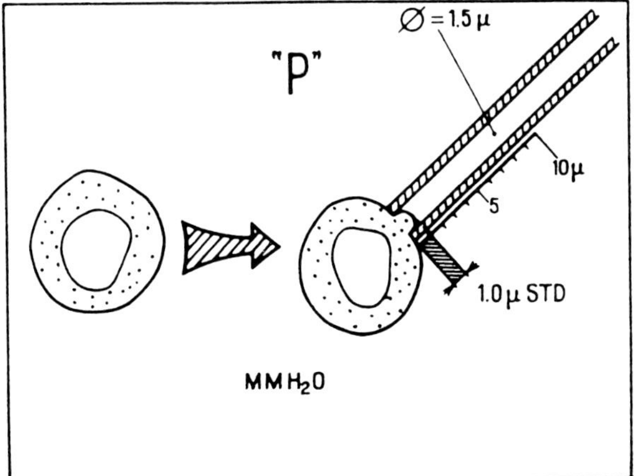

Fig. 16. Schematic representation of the micropipette method for measuring the surface shear modulus of elasticity of the red cell membrane (ie, an intrinsic material property of the membrane). A small pipette is employed, and the negative pressure required to pull a limited length of membrane into the pipette ("P") is used to compute the shear modulus. Alternatively, the negative pressure required for a hemisphere of membrane is reported as an index of membrane rigidity (from [58]).

An understanding of membrane measurements via micropipette and fluid shear stress (ie, parallel plate flow channel) methods has been materially advanced by the work of Evans and associates. These authors have applied deformation theory to micropipette and flow channel data obtained for RBC and have published extensively during the past few years [eg, 9, 62, 69–71]. It should be noted that these authors clearly separate membrane properties from extrinsic behavior (ie, cellular deformability), since the latter is related to the ability of the *entire* cell to respond to mechanical forces.

An analysis of their results [9, 62, 69–71] indicates excellent agreement between micropipette and flow channel data, that the RBC membrane exhibits both elastic and plastic deformation, and that in each domain the membrane has a characteristic shear viscosity. In the elastic domain, the intrinsic properties are defined by the shear modulus of elasticity reflecting recoverable hyperelastic response (nominal value of 10^{-2} dynes/cm) and a shear viscosity defining viscoelastic behavior with internal viscous energy dissipation (nominal value of 10^{-3} dyne · sec/cm). In the plastic domain, the intrinsic properties are defined by a yield shear stress indicating the elastic limit of the membrane material (nominal value of 2×10^{-2} dyne/cm) and a shear viscosity in the plastic domain reflecting irrecoverable extension after the elastic limit of the membrane has been exceeded (nominal value of 10^{-2} dyne · sec/cm). The area compressibility of the membrane has also been measured [72], with results indicating that uniform dilation of the membrane requires over 10^3 times more energy than for plastic flow; uniform dilation of the membrane, beyond a 1–2% change, results in membrane failure and lysis. Membrane viscoelasticity in the elastic domain is clearly demonstrated, indicating that the force required for a given deformation is dependent on the rate of deformation as well as the final deformed state [71]. An overview [9] of these experimental results thus suggests that the RBC membrane behaves as a composite material rather than a simple lipid bilayer; lipid bilayers have zero values for shear modulus and yield shear and values of surface viscosities two to four orders of magnitude less than the values reported for the RBC membrane. A spectrin-like network with a lipid "liquid-sealer" has thus been suggested as a model for the membrane [9].

Unresolved Questions

Like the macrorheological behavior of blood and RBC suspensions, the microrheological behavior suggests questions to which there are incomplete answers. The most general and long range of these, of course, is addressed to the interrelations between the various membrane mechanical properties and the overall behavior (ie, deformability) of the entire erythrocyte. That is, given accurate determinations of the relevant membrane material constants, can the mechanical response of the entire cell be predicted? To date, the answer appears to be no.

Another important, but as yet unresolved, question relates to potential RBC-suspending medium interactions associated with many of the microrheological studies [11, 43, 44, 63, 65]. In order to increase the shear stress on the RBC, at reasonable levels of shear rate (as volumetric flow rate in tube flow or rotational speed for cone-plate or coaxial instruments), the viscosity of the suspending medium is increased by adding water-soluble polymers. The most commonly used polymer is dextran, usually of 40,000 to 80,000 molecular weight. Although data are reported at various levels of shear stress (suspending medium viscosity times shear rate), erythrocyte deformation at comparable shear stress levels is not equal for all suspending media. Further, the use of a "fluid drop deformation equation" proposed by Taylor [11, 43, 63], which considers the ratio of the internal viscosity of the particle to the viscosity of the suspending medium, does *not* allow resolution of the problem. That is, a decrease of this viscosity ratio (by increasing the medium viscosity) should *decrease* the extent of deformation of the RBC at a constant level of shear stress; experimental results [11, 43, 63], however, indicate an *increase* in RBC deformation! Based on these observations, Goldsmith [11, 43] has suggested that dextran influences the RBC membrane, perhaps by lowering the effective interfacial tension at the RBC–suspending medium interface. Given the frequent use of these water-soluble polymer systems, this question deserves serious study.

Third, the importance of the viscoelastic behavior of the cell membrane and of the entire erythrocyte [73] has yet to be fully explored; note that the characteristic shape recovery time for the RBC to relax to 50% of its original shape (the half-time constant) has recently been shown to be about 0.1 seconds [73]. Two examples serve to illustrate the potential importance of this viscoelastic behavior:

1) Unpublished data generated in our laboratory indicate that the average transit time of a human RBC through a 13 micron thick, 3 micron pore diameter polycarbonate filter is on the order of 0.003 to 0.009 seconds (RBC suspended in isotonic buffer at a volume fraction of 0.0025, 10 mm H_2O average pressure drop). Thus, the time involved in deforming the RBC to allow passage through the 3 micron pore is 10–30 times shorter than the nominal 50% shape recovery time. Measurements of cell filtration at lower driving pressures (and thus slower velocities) would be of interest but are precluded by the tendency of the RBC to settle onto the face of the polycarbonate filter and thus alter the local flow conditions at the filter surface.

2) Measurements of capillary length and RBC velocity in the living microcirculation [74, 75] indicate that the average transit time through these small vessels is on the same order as the nominal 50% shape recovery time of the erythrocyte. Further, the flow is not steady in the microcirculation but rather may exhibit periods of slow or zero flow followed by periods of rapid cell motion; pulsatile RBC flow, with a frequency equal to the heart rate, has also been reported. The significance (or lack thereof) of the viscoelastic behavior of the RBC in these in vivo situations is, as yet, not clear.

RBC METABOLIC–RHEOLOGIC RELATIONS

Overview

In this final section, a brief review of the interactions between the metabolic state of the human RBC and its macro- and microrheologic behavior will be provided. It is clearly not intended to be a compendium of the voluminous biochemical and rheological literature in this area, but rather an illustration of one portion of hemorheological research now under investigation.

The relations between the metabolic state and the mechanical properties of the RBC continue to be of interest to both basic science and clinical investigators. Frequently cited in RBC rheology-metabolic state literature is the work of Weed et al [76]. Consequent to metabolic depletion via 24-hour incubation at 37°C, they indicate 1) total depletion of cellular adenosine triphosphate (ATP); 2) discocyte-echinocyte (ie, biconcave disc-crenated or spiculated) shape transformation; 3) increased apparent viscosity and more pronounced non-Newtonian flow behavior of RBC-buffer suspensions (hematocrit = 0.80; shear rate range 0.1–230 sec^{-1}); 4) 12- to 14-fold decrease in membrane deformability as judged by the negative pressure necessary to produce hemispherical deformation of the RBC membrane into a 3 micron micropipette. Nearly complete return to the control state was achieved by restoring cellular ATP via a two-hour incubation in 30 mM adenosine; ATP-calcium-dependent sol-gel changes occurring at the interface between the membrane and cell interior were proposed as the basis for the changes in membrane deformability and suspension rheology [76]. Similar ATP-related changes for RBC stored in ACD solution have also been obtained; after a 56-day storage period, negative pressure for membrane hemisphere deformation ("P") increased by nearly ninefold, and the pressure required to cause an RBC to entirely enter a micropipette ("P_t") increased by a factor of 20 [77].

More recent studies of the membrane mechanical properties of depleted RBC are, however, not consistent with the above-mentioned results [76, 77]. Employing 1.5 micron micropipettes, Leblond [58] reports an identical twofold increase in "P" for fresh echinocytic RBC produced by sodium oleate, alkaline pH, or aged plasma *and* for echinocytes produced by incubation at 37°C for 20 hours. Using experimental apparatus similar to that of Weed and co-workers [76], but with an improved pressure measuring system, Heusinkveld [78] indicates *no* significant difference in "P" between fresh RBC and RBC incubated at 37°C for up to 45 hours. Cells partially fixed with low concentrations of glutaraldehyde did, however, demonstrate increased membrane rigidity, thus validating his experimental technique. A complete reconciliation between these later studies [58, 78] and the earlier data [76, 77] is not yet possible.

Macrorheologic Behavior

ATP depletion results. Figure 17 indicates typical rheologic behavior, as measured in a cone-plate viscometer [79], for various RBC suspensions in buffer at a hematocrit of 0.60: 1) fresh RBC; 2) ATP-depleted RBC produced by incubation at 37°C for 24 hours in a calcium-containing buffer; 3) initially depleted RBC in which the cellular ATP had been repleted via incubation with 30 mM adenosine for two hours at 37°C. These data, which are consistent with earlier reports [76], indicate the expected increase in low shear rate viscosity following ATP depletion and the nearly complete restoration to control following the incubation with 30 mM adenosine. Note that the major changes occur at the lower rate of shear, whereas the data for all three suspensions are coincident at the highest shear rate.

The morphology of the RBC used for Figure 17, as observed by both light and scanning electron microscopy, is also in agreement with previous reports. As indicated in Figure 18, ATP depletion via incubation results in the discocyte—echinocyte shape transformation, with the majority of RBC after incubation being either echinocytes III or spheroechinocytes I [79, 80]. Following ATP repletion (Fig. 18, right panel), the morphology of the cells is seen to be somewhat variable, but with a general tendency toward the discocytic form.

Although the data in Figure 17 are suggestive of RBC aggregation (ie, elevated viscosity at lower rates of shear; see Fig. 8), it is important to note that *no* microscopic evidence of RBC aggregation was ever observed in either fresh or incubated RBC—buffer suspensions [79].

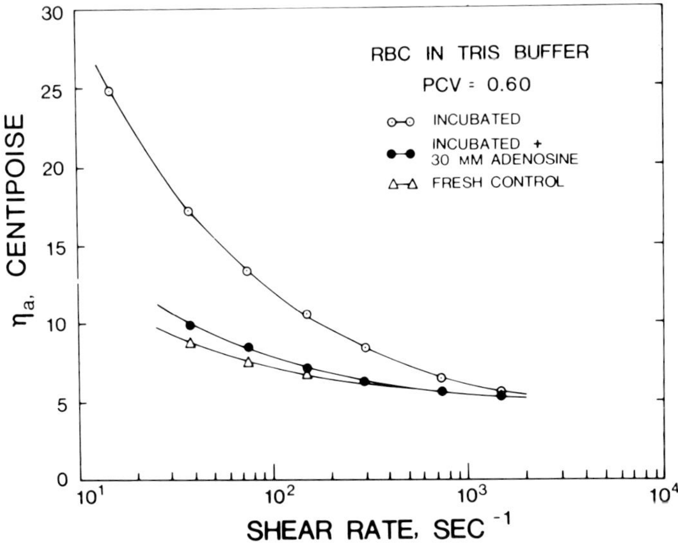

Fig. 17. Rheological comparison of fresh, ATP-depleted, and ATP-repleted red cells suspended in an isotonic buffer at a hematocrit of 0.60. The ATP-repleted cells (solid circles) were depleted by incubation at 37°C for 22 hours, then incubated for an additional two hours in the presence of 30 mM adenosine. The morphology of the cells used in these suspensions is shown in Figure 18 (from [79]).

Shape-transformed fresh RBC. Although in agreement with other studies, the rheologic data presented in Figure 17 give rise to a question: Why, in the absence of RBC aggregation, is the lower shear rate viscosity of ATP-depleted cells elevated above that for the fresh, control cell suspension? Arguing that this change in rheologic behavior may be related to alterations in cellular morphology as shown in Figure 18, a series of experiments have been carried out using fresh, shape-transformed (ie, echinocytic) human erythrocytes [79, 81].

Various amphiphilic agents can be employed to transform the normal biconcave shape (discocyte) into either the crenated (echinocytic) or cupped (stomatocytic) form [82]. These agents appear to act as true antagonists, although at high concentrations there is an irreversible process of smooth sphere to hemolysis. Sheetz and Singer [83] have proposed a theory for these shape transformations based on an asymmetric bilayer model of the RBC membrane: Echinocytic agents intercalate preferentially into the exterior half of the bilayer, whereas stomatocytic agents are suggested to act mainly on the interior half.

Figure 19 shows typical RBC shapes obtained by exposing fresh erythrocytes to increasing concentrations of the crenating agent 2,4-dinitrophenol (DNP). Note the progressive, dose-related transformation toward the echinocyte form. Light microscopy of dilute wet mounts confirmed these scanning microscopy results

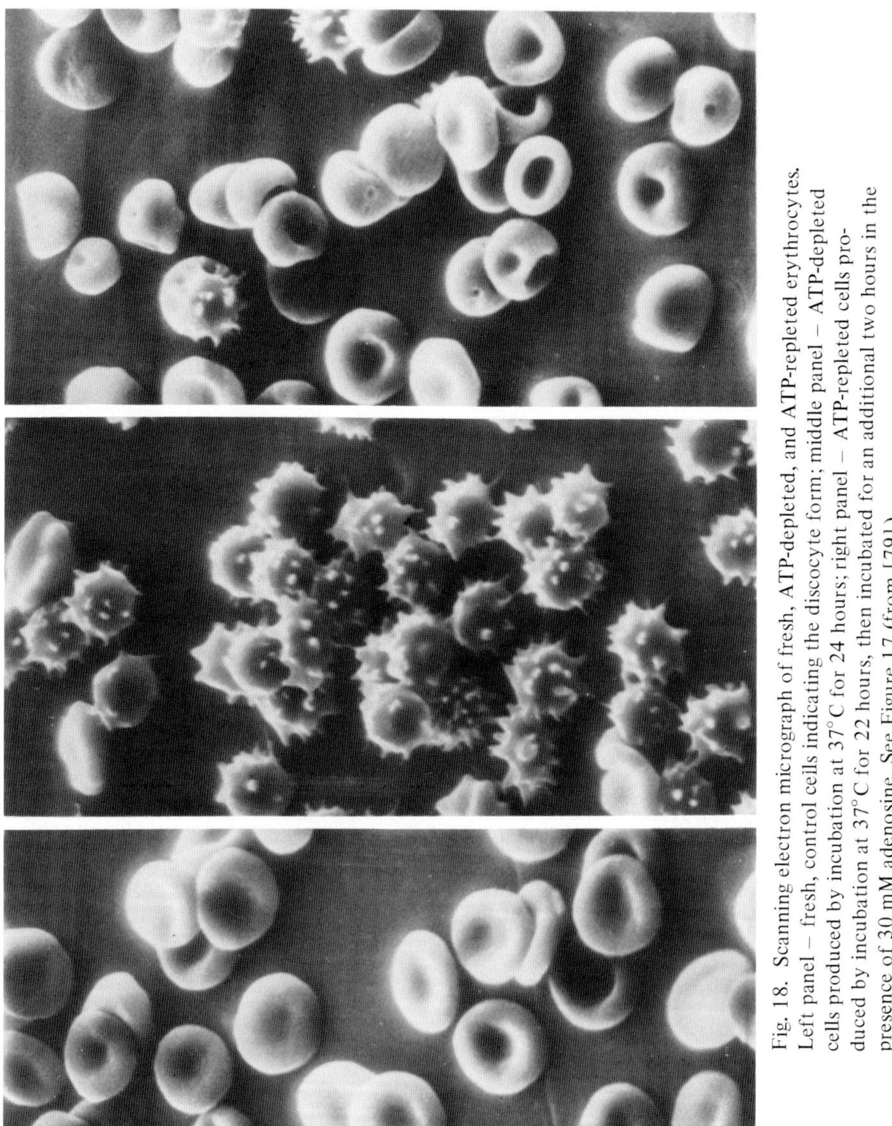

Fig. 18. Scanning electron micrograph of fresh, ATP-depleted, and ATP-repleted erythrocytes. Left panel — fresh, control cells indicating the discocyte form; middle panel — ATP-depleted cells produced by incubation at 37°C for 24 hours; right panel — ATP-repleted cells produced by incubation at 37°C for 22 hours, then incubated for an additional two hours in the presence of 30 mM adenosine. See Figure 17 (from [79]).

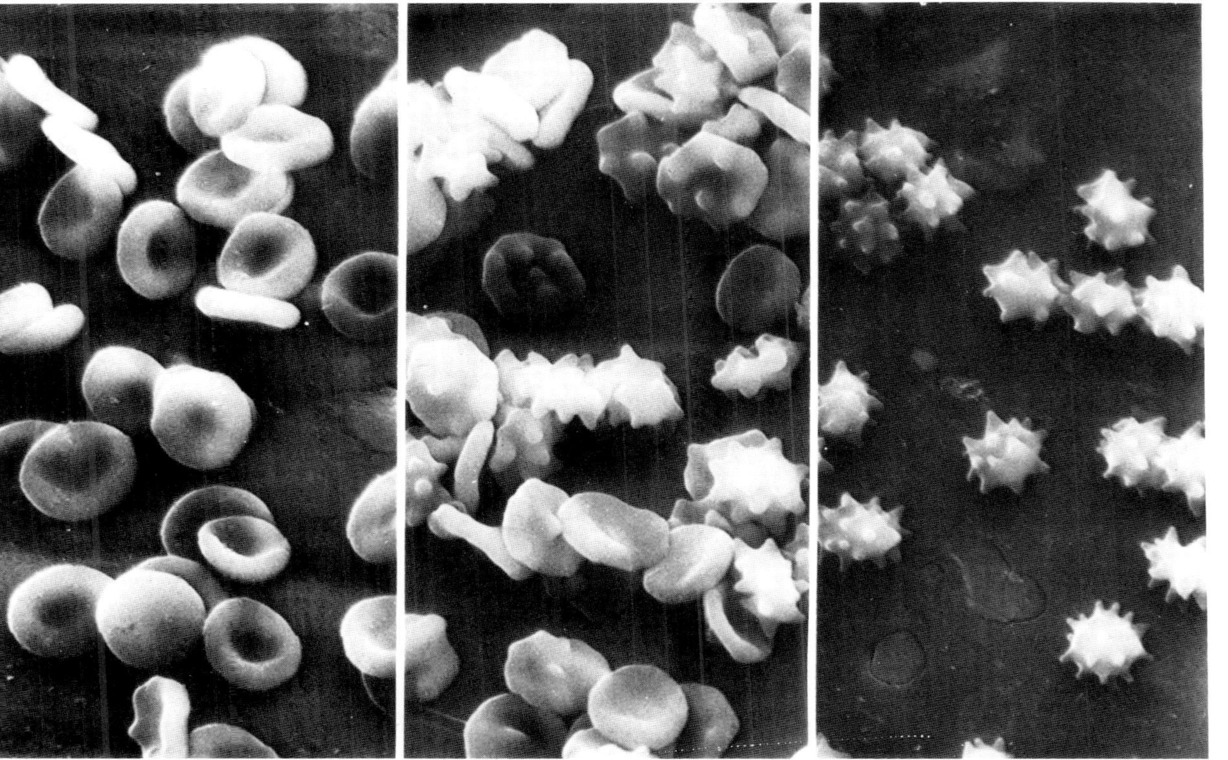

Fig. 19. Scanning electron micrograph of fresh red cells in buffer containing various concentrations of 2,4-dinitrophenol (DNP). Left to right: 0, 1, 5 mM DNP. Note the dose-related discocyte—echinocyte shape transformation (from [79]).

[79, 81] and, in addition, demonstrated that 1) at a given DNP concentration, the shape transformation was stable and rapid; 2) there was no evidence of RBC aggregation regardless of DNP concentration; and 3) the RBC could be restored to the original biconcave shape by washing in a DNP-free buffer. This dose-related shape change can also be induced by other nitrophenols or a substituted benzoate and, within the concentrations studied, results in no measureable change in the volume of the cell [79, 81].

The rheologic effects of increasing DNP concentration, and thus enhanced echinocyte formation, are presented in Figure 20 as plots of apparent viscosity versus shear rate for RBC in buffer at a constant hematocrit of 0.60; the inset shows the same data plotted as shear stress–shear rate, where the dashed line of unity slope represents the behavior of a Newtonian fluid. Note the dose-related increase in apparent viscosity at the lower shear rates and the coincidence of the data, for all DNP concentrations, at the highest shear rate. The shear rate at which the rheologic behavior becomes identical to that of the control suspension is, however, a function of DNP concentration; the 1 mM DNP suspension coincides above about 600 sec^{-1}, whereas shear rates in excess of about 1,000 sec^{-1} are necessary for the 5 mM DNP suspension. Similar data have been obtained using 2,4,6-trinitrophenol and sodium salicylate as the crenating agents [81].

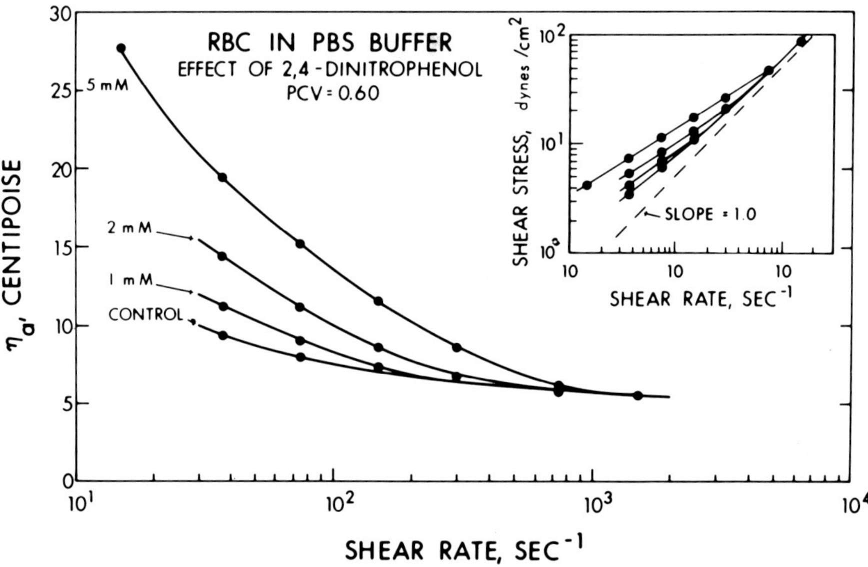

Fig. 20. Rheologic behavior of red cells in buffer containing various concentrations of DNP. The inset shows the same data, plotted using a shear stress–shear rate coordinate system. The dashed line of unity slope shown in the inset represents the behavior of a Newtonian fluid (from [81]).

A comparison between the rheological behavior of fresh, DNP-treated RBC suspensions and the results obtained for incubated RBC indicates a substantial degree of agreement. Both types of RBC suspensions show an elevated apparent viscosity at the lower shear rates and viscosity values essentially identical to fresh, control RBC at the highest shear rates. By comparing Figures 17 and 20, it can be seen that the nominal behavior of ATP-depleted RBC corresponds well with that for fresh RBC at a DNP concentration of about 5 mM. Further, by examining these rheologic data vis-à-vis the morphologic information given in Figures 18 and 19, it is obvious that this correspondence occurs at comparable degrees of discocyte–echinocyte shape transformation. Thus, the common factor appears to be the shape of the cell, not the means employed to induce the shape change (ie, metabolic depletion or addition of a shape-altering agent).

Shape-restored, ATP-depleted RBC. The above results, suggesting that the shape of the ATP-depleted RBC rather than its ATP content per se determines its macro-rheological behavior, have led to a series of experiments in which the biconcave shape was restored without restoring the cellular ATP level [79]. Based upon both literature reports and preliminary experiments, the stomatocytic agent chlorpromazine hydrochloride (CPZ) was chosen as an "antagonist" to the echinocytic shape produced by ATP depletion. Viscometric and morphologic studies using fresh, echinocytic RBC treated with various CPZ concentrations indicated that restoration of the biconcave shape (in the presence of the echinocytic agent) was possible *and* that when the biconcave shape was restored the flow behavior returned nearly to control [79, 81].

Figure 21 is a scanning electron micrograph of ATP-depleted RBC treated with various concentrations of CPZ; all RBC in this figure had been incubated, as before, at 37°C for 24 hours. CPZ produced a dose-related echinocyte–discocyte shape transformation, with an optimal concentration for biconcave shape restoration of about 1.0×10^{-4} M. Concentrations of CPZ above this optimal level produced an excess of the stomatocytic shape, whereas concentrations below this level yielded a majority of echinocytic forms.

The rheologic behavior of ATP-depleted RBC at various CPZ concentrations is shown in Figure 22. Inasmuch as several measurements on fresh, nonincubated RBC had been made on RBC from the donor, the rheologic behavior of these fresh RBC suspensions is indicated by a continuous band representing the mean + 1 SD at each shear rate. Inspection of these data indicates that 1) increasing concentrations of CPZ results in decreasing apparent viscosities at the lower shear rates; and 2) at the CPZ concentration optimal for restoration of the biconcave shape (1.0×10^{-4} M; see Fig. 21), the rheologic behavior of the ATP-depleted RBC suspension is indistinguishable from that for fresh RBC over the entire range of shear rates.

These morphologic and viscometric data [79, 81] thus offer an alternative explanation for the rheologic results obtained in prior studies [76], in that they do

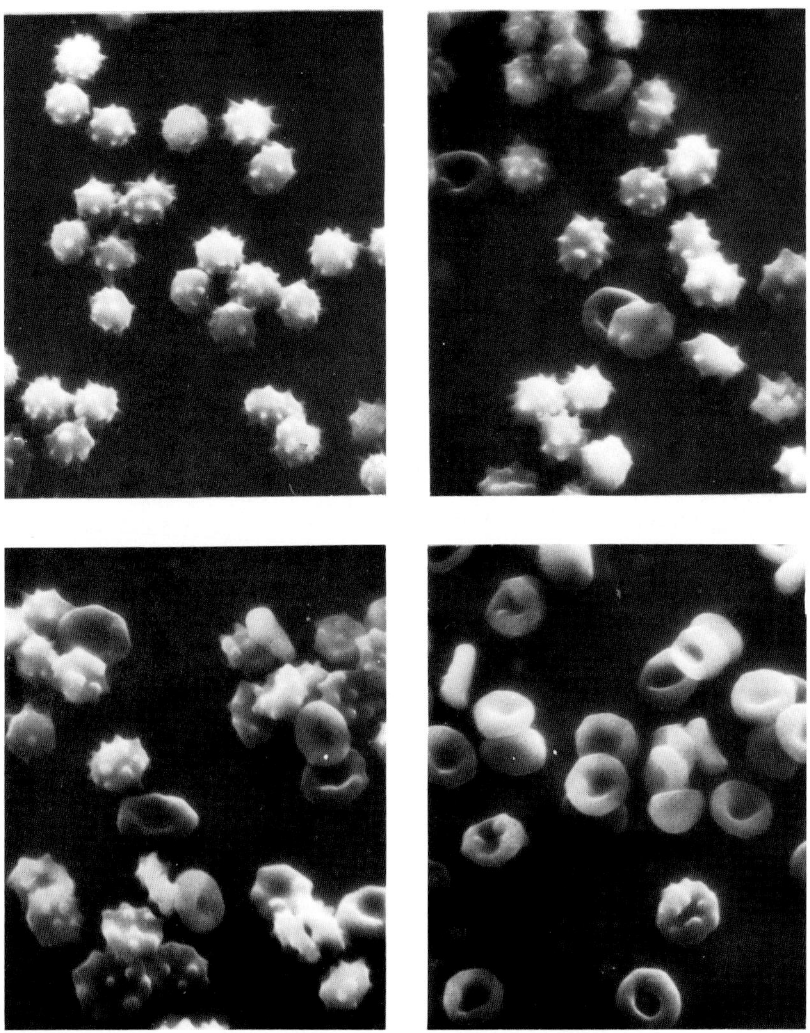

Fig. 21. Scanning electron micrograph of ATP-depleted red cells treated with various concentrations of CPZ; all cells in this figure had been incubated at 37°C for 24 hours prior to the addition of CPZ. Top left – control, no CPZ; top right – 2.5×10^{-5} M CPZ; bottom left – 5×10^{-5} M CPZ; bottom right – 1.0×10^{-4} M CPZ. Note the dose-related echinocyte–discocyte shape transformation (from [79]).

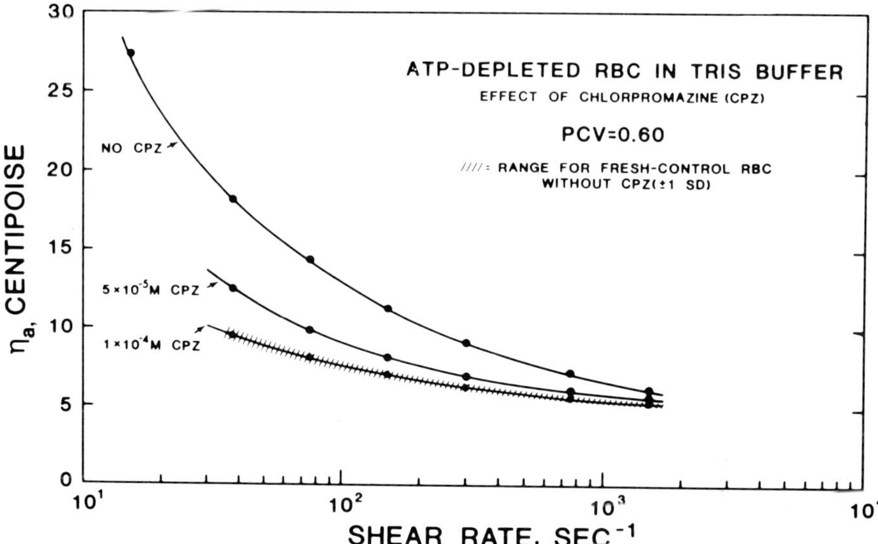

Fig. 22. Rheological behavior of ATP-depleted human erythrocytes in buffer with various concentrations of CPZ added after metabolic depletion. The rheologic behavior of fresh, control cells in buffer is shown by the continuous band (mean ± 1 SD). Note that the ATP-depleted cells with 1.0×10^{-4} M CPZ show flow behavior identical to the fresh, control suspension (from [79]).

not directly involve intrinsic (ie, membrane) factors. Rather, the results suggest that cellular morphology has an overriding influence on the flow behavior of ATP-depleted RBC suspensions, with the echinocytic form causing enhanced cell–cell interactions at the lower shear rates. At the higher shear rates, deformation of the cell to an ellipsoid-like form should occur [43, 63], thus greatly reducing echinocytic-type cell–cell interactions. Note that at the highest shear rates, where the ATP-depleted and DNP-treated RBC suspensions are coincident with the fresh control data (Figs. 17, 20, and 22), the shapes of the cells, regardless of their initial resting morphology, should be identical.

Microrheologic Behavior

As mentioned above in the Overview, literature reports detailing the microrheologic behavior of RBC with varying ATP levels are not consistent. In particular, measurements of changes in the membrane mechanical properties consequent to metabolic depletion yield estimates of from no change to a 14-fold increase in rigidity [58, 76–78]. A summary of the results from these earlier studies is shown in Figure 23, where the negative pressure required to pull a hemisphere of RBC membrane into a micropipette ("P") is plotted versus the time the cells were incubated at 37°C to reduce their intracellular ATP level. Note the wide range of results, especially at the 24-hour time period.

The values of several intrinsic material constants of human RBC membranes both before and after metabolic depletion via incubation at 37°C for 24 hours have recently been reported [84]. Using both micropipette and parallel plate flow channel techniques, three properties were measured: 1) Surface shear modulus of elasticity; 2) elastic area compressibility modulus; and 3) shear viscosity in the plastic domain. Note that the surface shear modulus of elasticity is qualitatively analogous to the earlier measures of negative pressure for hemisphere formation in micropipettes [56, 58, 76, 78] in that similar basic data are obtained; calculation of the modulus from these data only involves the application of a first-order tension deformation law [69].

The results of this study [84] indicate *no* significant differences in these three

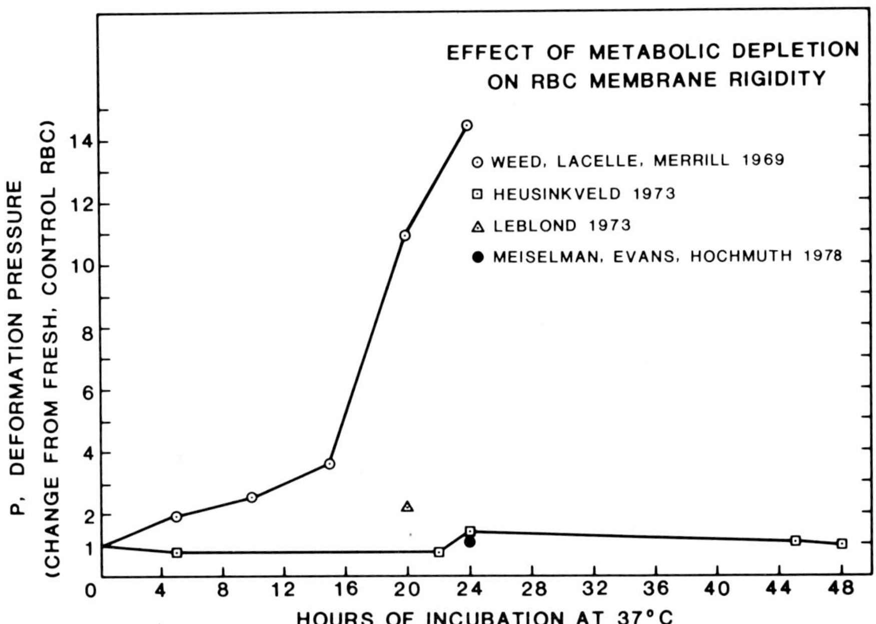

Fig. 23. Summary of several literature studies describing the change in red cell membrane rigidity as a function of time of incubation at 37°C. The vertical axis represents the change in negative pressure required to produce a hemisphere of cell membrane in a micropipette (see Fig. 16). Note that a value of unity on this vertical axis indicates no change from the fresh, control red cell membrane.

material constants between fresh and ATP-depleted human RBC membrane. In particular, the surface shear modulus of elasticity was found to be 0.018 ± 0.004 dynes/cm (mean ± 1 SD, n = 14) for fresh cells and 0.021 ± 0.004 dynes/cm (mean ± 1 SD, n = 38) for cells depleted of ATP. These results are thus clearly in disagreement with the earlier studies [76]. The basis for this discrepancy appears to be the methods of analysis and instrumentation used for these earlier studies: 1) The original design of the pressure-measuring manometer and the use of mouth suction to draw the hemisphere of RBC membrane into the micropipettes produces overestimates of the actual negative pressures involved [58], particularly for the shape-transformed ATP-depleted RBC. 2) The use of large-bore (2.5–3.5 micron) micropipettes in these earlier studies [76] may have produced overestimates of this negative pressure, because the magnitude of the pressure associated with the application of large pipettes depends on the intrinsic membrane shear modulus convolved with the extrinsic geometry of the whole cell. Regardless of these particular explanations, however, the current literature does appear to be in agreement in indicating a lack of measurable differences in this mechanical property between fresh and ATP-depleted erythrocytes [84, 85].

Unresolved Questions

Perhaps of paramount importance to our understanding of these metabolic–rheologic data is the relationship between the membrane mechanical properties and the overall deformability of the entire erythrocyte. Note that there is general agreement regarding the three factors that determine cellular deformability (ie, membrane surface area to cellular volume ratio, membrane mechanical properties, viscosity of cellular contents). Further, note that neither cell membrane surface area, cellular volume [79], membrane mechanical properties [84, 85] nor, most probably, the viscosity of the cell contents is modified by metabolic depletion via incubation at 37°C for 24 hours. How, then, to explain the significant decrease in cellular deformability of ATP-depleted cells that has been reported [58, 76] using a variety of microrheologic techniques? An easy, but not totally rewarding answer would probably include mention of yet unmeasured factors (perhaps membrane in nature), which could resolve this uncertainty; energy storage via the echinocytic form of the membrane has been proposed as one possibility. Deformation of these echinocytic cells might then be viewed as a two-stage process, in which the first step would involve smoothing the membrane, thereby removing the echinocytic spicules. This smoothing process would require work (for example, as a larger negative pressure for RBC entrance into a micropipette), thus leading to a measured decrease in cellular deformability. Alternative solutions to this dilemma also appear tenable, therefore suggesting the need for additional experimental studies dealing with the microrheological behavior of nondiscocytic erythrocytes.

ACKNOWLEDGMENTS

This work was supported by NIH research grants HL15722, HL15162, and by AHA-GLAA Investigative Group Fellowship Award 537IG3.

REFERENCES

1. Cokelet GR: The rheology of human blood. In Fung YC, Perrone N, Anliker M (eds): "Biomechanics, Its Foundations and Objectives." Englewood Cliffs, NJ: Prentice-Hall, 1972, pp 63–103.
2. Chien S: Biophysical behavior of red cells in suspension. In Surgenor DM (ed): "The Red Blood Cell." New York: Academic Press, 1975, pp 1031–1133.
3. Merrill EW: Rheology of blood. Physiol Rev 49:863–888, 1969.
4. Whitmore RL: "Rheology of the Circulation." Oxford, England: Pergamon Press, 1968.
5. Schmid-Schonbein H, Wells RE: Rheological properties of human erythrocytes and their influence upon the anomalous viscosity of blood. Rev Physiol 63:147–219, 1971.
6. Dintenfass L: "Blood Microrheology – Viscosity Factors in Blood Flow, Ischemia and Thrombosis." New York: Appleton-Century-Crofts, 1971.
7. Gabelnick HL, Litt M: "Rheology of Biological Systems." Springfield, Illinois: Charles C Thomas, 1973.
8. Cokelet GR: Microscopic rheology and tube flow of human blood. In Grayson J, Zingg W (eds): "Microcirculation," Vol 1. New York: Plenum Press, 1976.
9. Evans EA, Hochmuth RM: A solid-liquid composite model of the red cell membrane. J Membr Biol 30:351–362, 1977.
10. Meiselman HJ, Cokelet GR: Blood rheology – Instrumentation and techniques. In Harders H (ed): "Advances in Microcirculation." Basel: Karger, 1973, 5:32–61.
11. Goldsmith HL, Marlow J: Flow behavior of erythrocytes. Proc R Soc Lond B 182:351–384, 1972.
12. Vassar PS, Hards JM, Brooks DE, Hagenberger B, Seaman GVF: Physicochemical effects of aldehydes on the human erythrocyte. J Cell Biol 53:809–818, 1972.
13. Gilinson PJ, Dauwalter CR, Merrill EW: A rotational viscometer using an A.C. torque to balance loop and air bearing. Trans Soc Rheol 7:319–331, 1963.
14. Krieger IM: Shear rates in the Couette viscometer. Trans Soc Rheol 12:5–11, 1968.
15. Wells RE, Merrill EW: Shear rate dependence of the viscosity of whole blood and plasma. Science 133:763–764, 1961.
16. Merrill EW: Rheology of human blood and some speculations on its role in vascular homeostatis. In Sawyer PN (ed): "Biophysical Mechanisms in Vascular Homeostatis and Intravascular Thrombosis." New York: Appleton-Century-Crofts, 1965, pp 121–137.
17. Evans A, Weaver JPA, Walder DN: A viscometer for the study of blood. Biorheology 4:169–174, 1967.
18. Meiselman HJ: The influence of dextran on the sedimentation behavior of human red cells. Biol Anat 10:20–31, 1969.
19. Merrill EW, Benis AM, Gilliland EW, Sherwood TK, Salzman EW: Pressure-flow relations of human blood in hollow fibers at low flow rates. J Appl Physiol 20:954–967, 1965.
20. Meiselman HJ, Merrill EW, Salzman E, Gilliland ER, Pelletier GA: The effect of dextran on the rheology of human blood. J Appl Physiol 22:480–486, 1967.
21. Brooks DE, Goodwin JW, Seaman GVF: Interactions among erythrocytes under shear. J Appl Physiol 28:172–177, 1970.

22. Fahraeus R, Lindqvist T: The viscosity of blood in narrow capillary tubes. Am J Physiol 96:562–568, 1931.
23. Fahraeus R: The suspension stability of blood. Physiol Rev 9:241–274, 1929.
24. Cokelet GR: Comments on the Fahraeus-Lindqvist effect. Biorheology 4:123–126, 1967.
25. Barbee JH, Cokelet GR: Prediction of blood flow in tubes with diameters as small as 29 microns I.D. Microvasc Res 3:17–21, 1971.
26. McComis WT, Charm SE, Kurland G: Pulsing of blood flow in capillary tubes. Am J Physiol 212:1–12, 1967.
27. Aroesty J, Gross JF: The mathematics of pulsatile flow in small vessels. Microvasc Res 4:1–12, 1972.
28. Krishnakumar CK, Rovick AA, Cavan Z: The effect of pressure pulsations on time mean flow rate. Microvasc Res 11:41–49, 1976.
29. Thurston GB: Elastic effects in pulsatile blood flow. Microvasc Res 9:145–157, 1975.
30. Thurston GB: The viscosity and viscoelasticity of blood in small diameter tubes. Microvasc Res 11:133–146, 1976.
31. Chien S, King RG, Skalak R, Usami S, Copley AL: Viscoelastic properties of human blood and red cell suspensions. Biorheology 12:341–346, 1975.
32. Cokelet GR: Hemodynamics. In Johnson PC (ed): "Peripheral Circulation." New York: John Wiley, 1978, pp 81–110.
33. Whittaker SRF, Winton FR: The apparent viscosity of blood flowing in the isolated hindlimb of the dog and its variation with corpuscular concentration. J Physiol (Lond) 78:339–369, 1933.
34. Djojosugito AM, Folkow B, Oberg B, White S: A comparison of blood viscosity measured in vitro and in a vascular bed. Acta Physiol Scand 78:70–84, 1970.
35. Skovberg F, Nielsen AV, Schlicktkrull J: Blood viscosity and vascular flow rate. Scand J Clin Lab Invest 21:83–88, 1968.
36. Rasmussen SN: Influence of plasma hypertonicity on blood studied in vitro and in an isolated vascular bed. Acta Physiol Scand 84:472, 1972.
37. Inouye A, Kamino K, Ogawa M, Uyesaka N: Pressure-flow relation of erythrocyte to suspension in perfusion of bullfrog hind limb and marginal zone theory. Biorheology 13:251–256, 1976.
38. Benis AM, Usami S, Chien S: Effect of hematocrit and inertial losses on pressure flow relations in the isolated hindpaw of the dog. Circ Res 20:1047–1068, 1970.
39. Benis AM, Usami S, Chien S: Evaluation of viscous and inertial pressure losses in isolated tissue with a simple mathematical model. Microvasc Res 4:81–93, 1972.
40. Benis AM, Chien S, Usami S, Jan KM: A reappraisal of Whittaker and Winton's results on the basis of inertial losses. Biorheology 11:153–161, 1974.
41. Meiselman HJ, Cokelet GR: Fabrication of hollow vascular replicas using a gallium injection technique. Microvasc Res 9:182–189, 1975.
42. Goldsmith HL, Mason SG: The microrheology of dispersions. In Eirich FR (ed): "Rheology – Theory and Applications," Vol 5. New York: Academic Press, 1967, pp 85–250.
43. Goldsmith HL: The microrheology of human erythrocyte suspensions. In Becker E, Mikhailov GK (eds): "Proceedings of the 13th International Congress of Theoretical and Applied Mechanics." New York: Springer-Verlag, 1973, pp 85–103.
44. Goldsmith HL, Skalak R: Hemodynamics. Annu Rev Fluid Mech 7:213–247, 1975.
45. Weed RI: The importance of erythrocyte deformability. Am J Med 49:147–150, 1970.
46. Weed RI: Disorders of red cell membranes – History and perspectives. Semin Hematol 7:249, 1970.

47. Carroll L: Through the Looking Glass and What Alice Found There. New York: Random House, 1946.

48. Teitel P, Galecki G, Xanakis A, Marcu I: Clinical investigations with an improved filtration test on red blood cell rheology in hemolytic anemias. Biorheology 9:164–165, 1972.

49. Teitel P: Basic principles of the filterability test and analysis of erythrocyte flow behavior. Blood Cells 3:55–70, 1977.

50. Gregersen MI, Bryant CA, Hammerle WE, Usami S, Chien S: Flow characteristics of human erythrocytes through polycarbonate filters. Science 157:825–827, 1967.

51. Chien S, Luse SA, Bryant CA: Hemolysis during filtration through micropores. Microvasc Res 3:183–203, 1971.

52. Blackshear PL, Anderson RJ: Hemolysis thresholds in microporous structures. Blood Cells 3:377–395, 1977.

53. Schmid-Schonbein H, Weiss J, Ludwig H: A simple method for measuring red cell deformability in models of the microcirculation. Blut 26:369–379, 1973.

54. Schmid-Schonbein H, Wells RE, Goldstone J: Effect of ultrafiltration and plasma osmolality upon the flow properties of blood. Pflugers Arch 338:93–114, 1973.

55. Lingard PS: Capillary pore rheology of erythrocytes. Microvasc Res 13:29–77, 1977.

56. LaCelle PL: Alteration of deformability of the erythrocyte membrane in stored blood. Transfusion 9:238–245, 1969.

57. LaCelle PL: Pathologic erythrocytes in the capillary microcirculation. Blood Cells 1: 269–284, 1975.

58. Leblond P: The discocyte-echinocyte transformation of the human red cell. In Bessis M, Weed RI, Leblond PF (eds): "Red Cell Shape." New York: Springer-Verlag, 1973, pp 95–104.

59. Braasch D: Red cell deformability and capillary blood flow. Physiol Rev 51:679–701, 1971.

60. Gross JF, Aroesty J: Mathematical models of capillary flow. Biorheology 9:225–264, 1972.

61. Gupta BB, Natarajan R, Seshadri V: Analysis of blood flow in capillaries by finite element method. Microvasc Res 12:91–100, 1976.

62. Hochmuth RM, Mohandas N: Uniaxial loading of the red cell membrane. J Biomech 5: 501–509, 1972.

63. Goldsmith HL: Red cell motions and wall interactions in tube flow. Fed Proc 30:1578–1588, 1971.

64. Schmid-Schonbein H: Erythrocyte rheology and optimization of mass transport in the microcirculation. Blood Cells 1:285–306, 1975.

65. Fischer T, Schmid-Schonbein H: Tank tread motion of red cell membranes in viscometric flow. Blood Cells 3:351–362, 1977.

66. Corry WD, Meiselman HJ: Deformation of human erythrocytes in a centrifugal field. Biophys J 21:19–34, 1978.

67. Corry WD, Meiselman HJ: Centrifugal method for determining red cell deformability. Blood 51:693–701, 1978.

68. Corry WD, Meiselman HJ: Modification of erythrocyte physiochemical properties by milli-molar concentrations of glutaraldehyde. Blood Cells 4:465–480, 1978.

69. Evans EA: New membrane concept applied to the analysis of fluid shear and micropipette deformed red blood cells. Biophys J 13:941–954, 1973.

70. Evans EA, LaCelle PL: Intrinsic material properties of the erythrocyte membrane indicated by mechanical analysis of deformation. Blood 45:29–43, 1975.

71. LaCelle PL, Evans EA, Hochmuth RM: Erythrocyte membrane elasticity, fragmentation and lysis. Blood Cells 3:335–350, 1977.

72. Evans EA, Waugh R, Melnik L: Elastic area compressibility modulus of red cell membranes. Biophys J 16:585–595, 1976.

73. Hochmuth RM, Worthy PR, Evans EA: Red cell extensional recovery and the determination of membrane viscosity. Biophys J 26:101–114, 1979.

74. Lipowsky HH, Zweifach BW: Methods for the simultaneous measurement of pressure, differentials and flow in single unbranched vessels of the microcirculation for rheological studies. Microvasc Res 14:345–361, 1977.

75. Fronek K, Zweifach BW: Microvascular blood flow in cat tenuissimus muscle. Microvasc Res 14:181–189, 1977.

76. Weed RI, LaCelle PI, Merrill EW: Metabolic dependence of red cell deformability. J Clin Invest 48:795–809, 1969.

77. LaCelle PI: Alteration in deformability of the erythrocyte membrane in stored blood. Transfusion 9:238–245, 1969.

78. Heusinkveld R: Studies of erythrocyte membrane rigidity. Ph.D. Thesis, University of Rochester, Rochester, New York, 1973 (Ann Arbor, Michigan: University Microfilms, 73-26544).

79. Meiselman HJ, Baker RF: Flow behavior of ATP depleted human erythrocytes. Biorheology 14:111–126, 1977.

80. Bessis M: Red cell shape. In Bessis M, Weed RI, Leblond PF (eds): "Red Cell Shape." New York: Springer-Verlag, 1973, pp 1–24.

81. Meiselman HJ: Rheology of shape-transformed human red cells. Biorheology 15:225–237, 1978.

82. Deuticke B: Transformation and restoration of biconcave shape of human erythrocytes. Biochim Biophys Acta 163:494–500, 1968.

83. Sheetz MP, Singer SJ: Equilibrium and kinetic effects of drugs on the shapes of human erythrocytes. J Cell Biol 70:247–251, 1976.

84. Meiselman HJ, Evans EA, Hochmuth RM: Membrane mechanical properties of ATP-depleted human erythrocytes. Blood 52:499–504, 1978.

85. Heusinkveld RS, Goldstein DA, Weed RI, LaCelle PL: Effect of protein modification on erythrocyte membrane mechanical properties. Blood Cells 3:175–182, 1977.

86. Meiselman HJ, Merrill EW: Hemorheology, hemodilution and the effect of low molecular weight dextran. Bibl Anat 9:288–295, 1967.

87. Schmid-Schonbein H, Gallasch G, Volger E, Klose HJ: Microrheology and protein chemistry of pathological red cell aggregation. Biorheology 10:213–227, 1973.

6

Mechanisms of Erythrocyte Aggregation

Donald E. Brooks, Russell G. Greig, and Johan Janzen

Interactions among cellular elements of the blood in the form of cellular aggregation or adhesion are observed in a variety of normal and pathological conditions. Rouleaux formation in static blood and the intravascular red cell aggregation (sludging) associated with shock, which has also been observed subsequent to acute myocardial infarction and tissue trauma, are examples in which erythrocytes are involved. Platelet adhesion and aggregation in both thrombosis and hemostatic plug formation, granulocyte adhesion to endothelial surfaces in the initial stages of inflammation, immune agglutination of cells exposed to appropriate antibodies, and the adhesion of circulating tumor cells to vessel walls during metastasis are other documented examples of such interactions involving the blood elements. Although the general features of some of these reactions are known (eg, rouleaux formation [1–4] and immune agglutination [5]), none of the phenomena are understood in detail at the biochemical level. With the molecular mechanisms largely undefined, there seems to be little possibility of designing rational programs of management for those clinical conditions in which it would be desirable to manipulate or eliminate such interactions. The formation and maintenance of close contacts between cells is a fundamental phenomenon in biology and medicine [6–8], yet the mechanisms involved are not well understood for any in vivo system [9–11].

Analogies drawn between cellular aggregation and the aggregation of nonliving particles of similar dimension (colloids) have provided neither convincing arguments nor accurate predictions with respect to biological systems. We have shown that colloid stability theory [12], which has achieved reasonable success in explaining the behavior of simple nonliving dispersions, is of little help in rationalizing the aggregation reactions known to occur among the blood elements [13, 14]. Calculations carried out with parameters representative of blood cells predict that under physiological conditions all cells should be strongly aggregated in the absence of shear forces [15], a result that is at marked variance with observation.

Erythrocyte Mechanics and Blood Flow, pages 119–140
© 1980 Alan R. Liss, Inc., 150 Fifth Avenue, New York, NY 10011

This discrepancy is probably due to incorrect modeling of the interacting cell surfaces, since the physical theories involved have been reasonably well tested in simpler systems [16, 17]. In any event, simple addition of electrostatic repulsion and van der Waals attraction — the basis of colloid stability theory — is not capable of explaining blood cellular interactions. An additional force would seem to be required for an adequate description of the system.

As macromolecular dimensions are comparable in magnitude to the separation distance over which electrostatic repulsion is significant, a simple way to produce aggregation in the face of such repulsion would be to introduce bridging macromolecules between cells that could simultaneously bind to adjacent cell surfaces. Whether aggregation occurred would then be determined by the relative magnitudes of the binding energy involved in the cell—macromolecule bond and the electrostatic repulsive energy operating at that intercellular separation.

This aggregation mechanism by macromolecular bridging can in fact be demonstrated both in colloidal [18] and cellular [19] suspensions when appropriate polymers are added to these systems. A physiological example of the bridging mechanism is found in immunological agglutination, where immunoglobulin molecules act as the intercellular links [20]. Lectin-induced agglutination occurs analogously [21]. Erythrocyte rouleaux formation, the loose association of red cells into roughly linear aggregates, which occurs in static whole blood, is probably due to bridging, in this case primarily by fibrinogen molecules [22]. We originally showed that red cell aggregation induced by dextran, widely used as a model for rouleaux formation [23] and blood sludging [24], was probably due to this mechanism [14], a concept that was strongly supported by results from Chien's group [23, 25]. Reversible platelet aggregation has also been discussed in terms of fibrinogen bridging [26], and platelet adhesion to foreign surfaces always occurs via an adsorbed protein layer [27]. Fibrinogen uptake by platelets is enhanced when aggregation is induced by addition of ADP [28, 29]. Macrophages and peripheral blood monocytes apparently contain subpopulations of cells that bind fibrinogen or fibrin avidly [30]; the adherence reactions these cells undergo may involve fibrin bridging [31]. Therefore, although conclusive in only a few cases, there is evidence that macromolecular bridging may be the operative aggregation mechanism in a variety of interactions undergone by the formed elements of the blood.

The model described above is particularly appealing for blood cell contacts because of the concentration and variety of macromolecules with which the cells are in equilibrium in vivo. A wide range of both specific and nonspecific interactions is therefore possible, depending on the plasma constituents present and their ability to bind in an appropriate configuration to cell membrane sites. For several years our laboratory has been involved in the study of blood cell interactions induced by macromolecules that interact nonspecifically with cell surfaces [22]. That is, systems where macromolecular adsorption and bridge formation occur as the result of relatively weak binding of the molecule to the membrane, probably via a large number of individual low-energy contacts such as hydrogen bonds or electrostatic interactions, which do not require a specific molecular configuration to be effective [32]. Rouleaux formation and dextran-induced aggregation are believed to occur via such nonspecific interactions [14]. More recently, in order to develop a model for aggregation mediated by specific bridging agents, we have begun studies on the aggregation of red cells by lectins, a group of multivalent proteins that bind to particular carbohydrate residues and sequences of cell surface glycoproteins and glycolipids. Our results to date in these areas, all of which were obtained with human erythrocytes, are summarized in the pages that follow.

AGGREGATION ASSOCIATED WITH NONSPECIFIC ADSORPTION OF MACROMOLECULES

The Bridging Model

It has been known for some time that colloidal suspensions can be aggregated by the nonspecific adsorption of macromolecules to two or more particles simultaneously [18] There is therefore reason to suppose that a similar mechanism is operative when washed red cells suspended in physiological saline aggregate upon the addition of a neutral polymer species such as dextran. Typical results are shown in Figure 1, which correlates the degree of aggregation, as judged by the microscopic appearance of the cell suspension, with the presence of increasing concentrations of dextrans of different molecular weights. The left-hand boundary of the shaded region in Figure 1 defines the dependence of the critical concentration required to induce aggregation on polymer molecular weight. It demonstrates that a minimum molecular weight is required to produce aggregation at physiological ionic strength (mol wt $\simeq$ 50,000), and as molecular weights greater than 50,000 are used, aggregation appears at progressively lower concentrations. Both these characteristics are consistent with the bridging model. In the presence of electrostatic repulsion very low molecular weight fractions would not be of sufficient size to bridge the intercellular gap. The observed molecular weight dependence might be expected since progressively higher molecular weight polymers, which form thicker adsorbed layers [33, 34], ought to form more stable bridges (more bonds per molecule) assuming the amount adsorbed does not vary drastically with molecular weight.

More direct support for the bridging model derives from study of the adsorption behavior of dextran in erythrocyte suspensions. The measurement of polymer adsorption isotherms is straightforward in nonbiological systems, since generally the polymer is tightly bound to the substrate and is easily detected. The measurement of dextran adsorption to red cells proves to be much more difficult, however, because the polymer is relatively weakly bound at the cell surface. This weak binding is seen in the behavior of the aggregated systems when washed in dextran-free media: A single wash at a supernatant:cell volume ratio greater than ten is generally sufficient to reverse completely any aggregation present. Technical problems there-

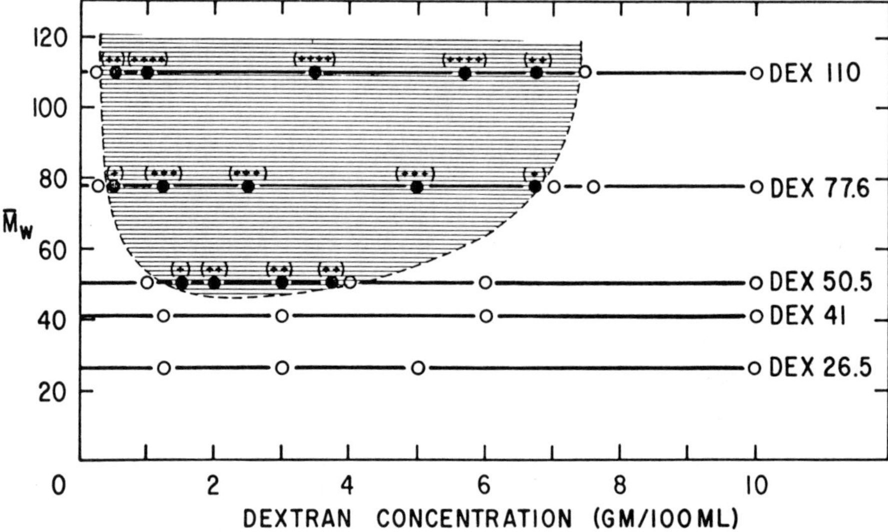

Fig. 1. The degree of human erythrocyte aggregation induced by dextran fractions of the molecular weight ($\overline{M}_W$) shown as a function of concentration, determined by microscopic examination of static suspensions. The degree of aggregation was estimated subjectively at the points shown on a scale of 0 (no aggregation) to ++++ (heavy aggregation). Aggregation of some degree was present within the shaded region. Fresh washed erythrocytes were suspended in bicarbonate-buffered physiological saline, pH 7.2, T = 21°C, in which various concentrations of sharply cut dextran fractions were included.

fore arise in attempting to estimate the equilibrium binding of the polymer due to the difficulty in distinguishing material bound to the cell surface from material trapped in the pellet. Examining the distribution of a "nonadsorbing" solute provides ambiguous results (probably due to adsorption or uptake by the cells), so we have resorted to more elaborate kinetic procedures to characterize the binding.

In spite of the difficulties in interpretation, it is of interest to examine the apparent equilibrium adsorption isotherm for dextran, as estimated from the total pellet radioactivity following high-speed centrifugation of washed erythrocytes suspended in ^{3}H-dextran 78 (mol wt = fraction number $\times$ 10^3). The results given in Figure 2 [14] demonstrate that the isotherm is linear up to about 7% w/v, above which concentration the specific activity of the pellet increases strongly. Assuming the trapped (nonadsorbed) activity in the pellet is small, the sharp increase in the amount bound per cell is interpreted as supporting the bridging model because the concentration at which it is found coincides with that at which disaggregation is observed in the suspensions (right-hand boundary in Figure 1). The disaggregation process, which is probably mediated through an electrostatic mechanism, will not be reviewed here but is discussed in detail elsewhere [14, 22, 35]. If, when the cells

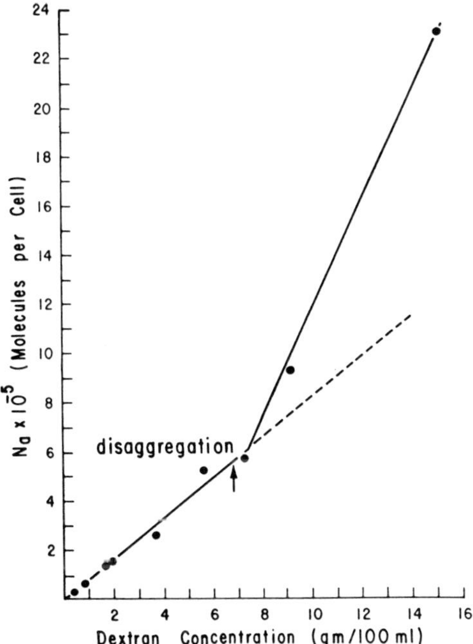

Fig. 2. The apparent number of molecules of dextran 78 bound per cell as a function of equilibrium dextran concentration, T = 21°C. Measurements based on total counts in a cell pellet equilibrated with ^{3}H-dextran 78 then ultracentrifuged at 165,000g for 60 minutes.

are in the aggregated state, dextran adsorption occurs in a bridging configuration, disaggregation should free some of the binding sites previously occupied by bridging polymer, and the amount bound per cell ought to increase more rapidly with concentration, as observed. If aggregation were due to the interaction of two adsorbed layers [36, 37] in which each bound molecule interacted with only one of the two cell surfaces, no increase would be expected.

Further Studies on Dextran Adsorption

In our original work, only the apparent equilibrium binding of a single molecular weight fraction of dextran was examined [33]. We have recently begun looking in more detail at the dextran/erythrocyte system with regard to the molecular weight dependence of adsorption and desorption behavior [4]. The results derived from the apparent equilibrium binding in the cell pellet, measured as described [33], for washed human erythrocytes and dextran 36.5 and dextran 150 are given in Figure 3. The data are analogous to those in Figure 2 and again assume that negligible trapping occurs in the pellets. The curve for dextran 150 is qualitatively similar to

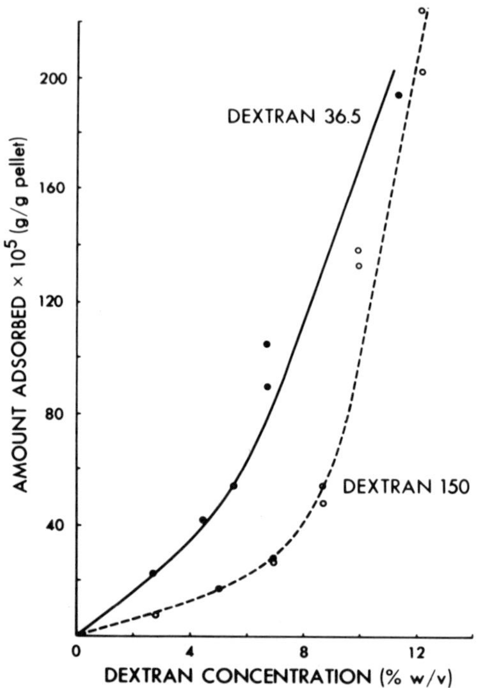

Fig. 3. The apparent adsorption of the dextran fractions indicated as a function of equilibrium dextran concentration, $T = 21°C$, measured as in Figure 2.

that for dextran 78 in that a sharp increase in apparent binding is seen at concentrations where disaggregation occurs. A similar, though less dramatic, increase is also seen with the dextran 36.5 fraction. This result was unexpected, based on the above reasoning, because dextran 36.5 produces no aggregation at any concentration due to its low molecular weight. A more extensive analysis of the systems in Figure 3 was therefore carried out.

Figure 4 represents the results of an experiment in which washed erythrocytes were equilibrated with ^{3}H-dextran 36.5 at the concentrations, C_o, indicated, then washed seven times in dextran-free saline (16 minutes/wash; 6:1 supernatant:cell volume ratio). The data are expressed in terms of the logarithm of $W(i)$, the amount of dextran released into the i-th wash supernatant per gram of cells, corrected for carry-over from the previous wash, as a function of the number of washes. Linearity of such a plot implies that a constant fraction of the total amount of polymer in the pellet is released into each supernatant. Physically such linearity could occur either if polymer were being diluted out of a volume of solution trapped in the equilibrium pellet or if dextran were desorbing by first-order kinetics from a uni-

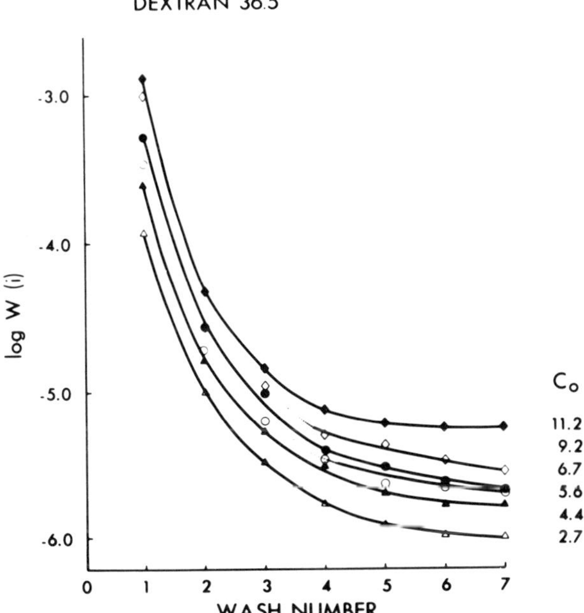

Fig. 4. The weight of dextran released into the supernatant per gram of packed red cells, W(i), as a function of the number of washes, i, of the pellet in dextran-free saline, T = 21°C. W(i) is corrected for carry-over of counts from the previous wash and represents only the amount of material released from the pellet during the i-th exposure to saline. Cell pellets were suspended in six volumes of saline for 16 minutes per wash. C_0 = equilibrium dextran concentration, gm/100 ml, before washing commenced.

form set of adsorption sites. The nonlinearity shown in Figure 4 implies that more than one type of binding is present, that both desorption and dilution of trapped material are occurring simultaneously, or that dextran desorption does not obey first-order kinetics.

The data in Figure 4 were analyzed by extrapolating back to i = 1, the linear portion of the curves for i = 5, 6, and 7 (representing release of the most slowly desorbing material), substracting this line from the original data and plotting the differences for i = 1–4 as in Figure 4. These derived curves for the more rapidly released components were again linear only for i = 2–4, so the extrapolation process was repeated. The total amounts of material associated with the slow and more rapid release processes were calculated by integrating the two extrapolated lines from i = 1 to ∞. The excess material in the equilibrium pellet not accounted for by these two desorption processes was identified with nonadsorbed polymer trapped in the pellet and expressed as TSF, the volume fraction of pellet occupied by dextran solution of the equilibrium concentration.

Analyses of this type were carried out for each of the curves of Figure 4 and for the data, which were qualitatively similar, obtained with dextran 150. The results are given in Figures 5–7. The first two of these are adsorption isotherms for the more rapidly (fast site) and more slowly (slow site) released portions of the total amount bound for each fraction. For dextran 36.5, which does not cause aggregation, two linear isotherms are obtained, with the bulk of the adsorption being associated with the fast site (Fig. 5). More complex results were obtained for the aggregating fraction dextran 150, shown in Figure 6. Adsorption of the slowly desorbing component, linear in concentration, is indistinguishable from that for dextran 36.5. The isotherm for the more rapidly desorbing component is linear up to about the disaggregation concentration, above which it appears to increase. This conclusion cannot be firmly stated at present, as the errors involved in deriving this curve can be considerable, but the data available to date indicate that the increase is present.

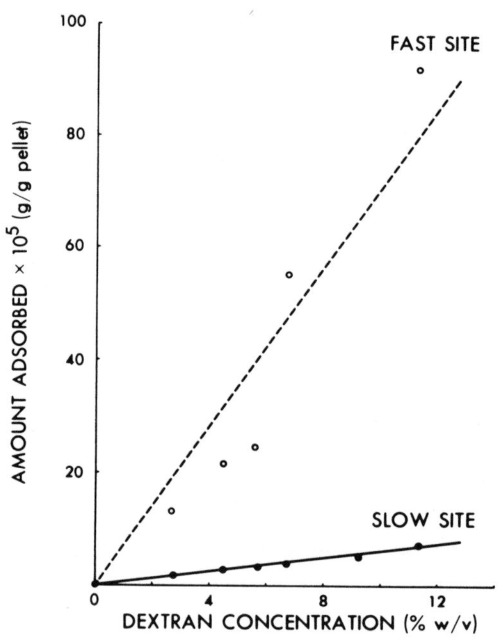

Fig. 5. Adsorption isotherms calculated as the total amount of dextran 36.5 per gram of packed red cells desorbed via the rapid (fast site) and slower (slow site) desorption processes after equilibration at each dextran concentration noted, calculated as described in the text.

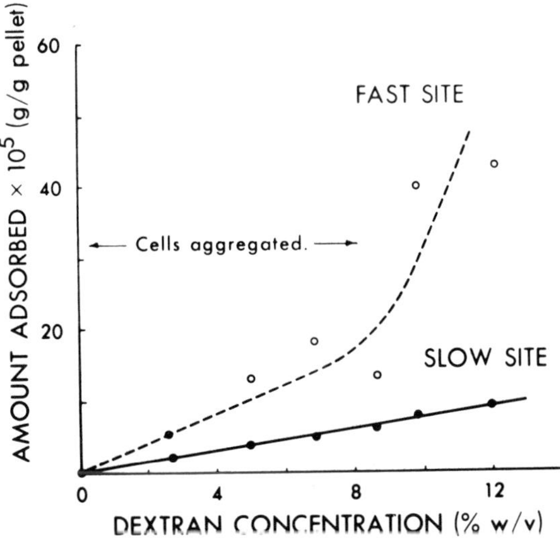

Fig. 6. Adsorption isotherms calculated as the total amount of dextran 150 per gram of packed red cells desorbed via the rapid and slower desorption processes after equilibration at each dextran concentration noted, calculated as described in the text.

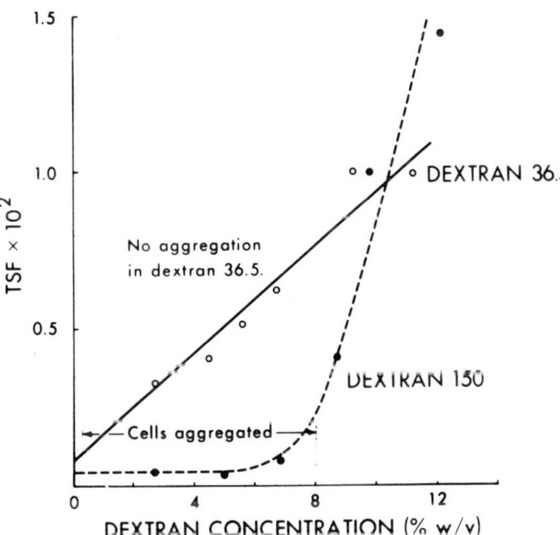

Fig. 7. The fraction of the equilibrium pellet, TSF, occupied by nonadsorbed dextran at the equilibrium concentrations indicated, calculated as described in the text.

The concentration dependence of the trapped solution fraction (TSF) for the two molecular weights also appears to differ significantly (Fig. 7). For dextran 36.5 the TSF after high-speed centrifugation (165,000g, 60 minutes) increases approximately linearly with the equilibrium dextran concentration. It is this increase in the fraction of total pellet counts associated with unbound polymer that is responsible for the concavity in the apparent dextran 36.5 isotherm (Fig. 3). Presumably the increased trapping is due to increased repulsion between the cells associated with adsorption of polymer.

Trapping of dextran 150 differs from that of the lower molecular weight fraction. Over the concentration range in which aggregation occurs the TSF remains roughly constant, but a sharp increase is seen near the disaggregation concentration. This behavior is reasonable, since aggregated cells should show little change in packing as dextran concentration is increased. At higher concentrations, where no aggregation is possible, intercellular repulsion is believed to be markedly enhanced, more so than for equivalent concentrations of lower molecular weight fractions [14], and presumably causes the observed loosening of the pellets.

It is interesting that the "fast site" isotherm for dextran 150 has a considerably lower slope than that for dextran 36.5. Qualitatively similar results were obtained by Chien et al by a method that did not distinguish among types of binding [38]. This behavior is the opposite of that found in most studies of homolog adsorption to solid surfaces where higher molecular weight polymers bind more avidly than lower molecular weight fractions. Apparently the lower molecular weight fraction has access to regions of the cell coat from which higher molecular weights are excluded. The layer of glycoprotein and glycolipid external to the membrane lipid bilayer has been described as a porous, nonrigid polyanion [39]. Its properties in situ are not well understood. Based on our dextran adsorption results it would seem that portions of this region can exhibit exclusion properties similar to those of the cross-linked polymers used in gel chromatography [40].

Assuming that our decomposition of the wash curves into two components is correct, it seems clear that the more rapidly desorbing component is the one responsible for aggregation. This is suggested by the break in the dextran 150 "fast site" isotherm, by the linearity and molecular weight dependence of the "slow site" isotherms, and by the desorption rate of this component. Aggregation induced by dextran 150 is reversed after one wash at all concentrations. Very little is lost from the "slow site" after one wash, however. Given the concentration dependence of the wash curves (Fig. 4) the observed reversal would not be expected to occur were aggregation caused by adsorption of this sort. Analysis of the data shows that the amount remaining on the "slow site" after one wash, at which point the cells are disaggregated, can be greater than that present before washing is carried out in a lower concentration system in which the cells are aggregated. Thus it is unlikely that the "slow site" is directly involved in the aggregation mechanism.

Little is known about the membrane features responsible for the observed adsorption behavior. Dextran is believed to interact with lipid bilayers [41], but severalfold higher concentrations are required to induce aggregation in liposome suspensions than in erythrocyte suspensions, suggesting that adsorption is weaker to lipid surfaces than to the sites responsible for cell aggregation. Since the adsorption isotherm characterizing the "slow site" binding has a lower slope than that associated with the "fast site," the more slowly desorbing component must be more weakly bound under equilibrium conditions. It is therefore conceivable that the "slow site" represents adsorption to lipid, but the evidence is singularly circumstantial. Neuraminidase treatment, which removes all of the sialic acid from cell surface oligosaccharides, has been claimed to affect dextran binding [38], but it is not yet known whether both or only one of the types of adsorption is affected.

To summarize, dextran adsorption to erythrocytes appears to be of two types, which are distinguishable by their desorption kinetics. The more rapidly desorbing component appears to be that which adsorbs in the bridging configuration and causes aggregation. The structures to which adsorption occurs are not yet known, but the molecular weight dependence suggests that the "fast site" binding occurs to glycoprotein or glycolipid structures which are capable of size-dependent exclusion. Such exclusion suggests that the "fast site" adsorption of high molecular weight fractions occurs at the periphery of the cell coat, a location consistent with rapid desorption and one that would encourage a bridging configuration.

Fibrinogen Adsorption

It has been known for a long time that fibrinogen is intimately involved in normal rouleaux formation [1, 3], and it is generally believed that macromolecular bridging is the relevant mechanism [3]. There is surprisingly little information available regarding the details of the reaction, however. In particular, no detailed studies of the interaction of pure fibrinogen with red cell surfaces have been carried out, although a few measurements (which appear to be inaccurate) have been published [42, 43]. We have therefore undertaken an examination of this problem [4], and our preliminary results are summarized here.

Human fibrinogen is prepared from fresh anticoagulated plasma by a modified version of Johnson's procedure [44], based on the poly(ethylene glycol) (PEG) precipitation technique described by Polson [45]. Using barium citrate adsorption, pH-dependent PEG precipitation, and affinity chromatography on lysine-Sepharose in the presence of proteolytic inhibitors, a produce is obtained in about 30% yield that is 98–100% thrombin clottable. Radioiodination is carried out via the lactoperoxidase procedure, and the labeled product is used immediately to study the adsorption of fibrinogen to red cells. A washing technique similar to that developed to follow dextran–cell interactions is then employed.

When fresh washed human erythrocytes are equilibrated with solutions of ^{125}I-labeled fibrinogen and then washed in protein-free buffer, a desorption curve such as that shown in Figure 8 is obtained. It is clear that, as in the dextran experiments, the radioactivity desorbs via an initially rapid process that appears to be superimposed on a second, much slower release. Rather than indicating two distinct types of binding for fibrinogen, however, these results seem to describe the adsorption and desorption of two distinct molecular entities.

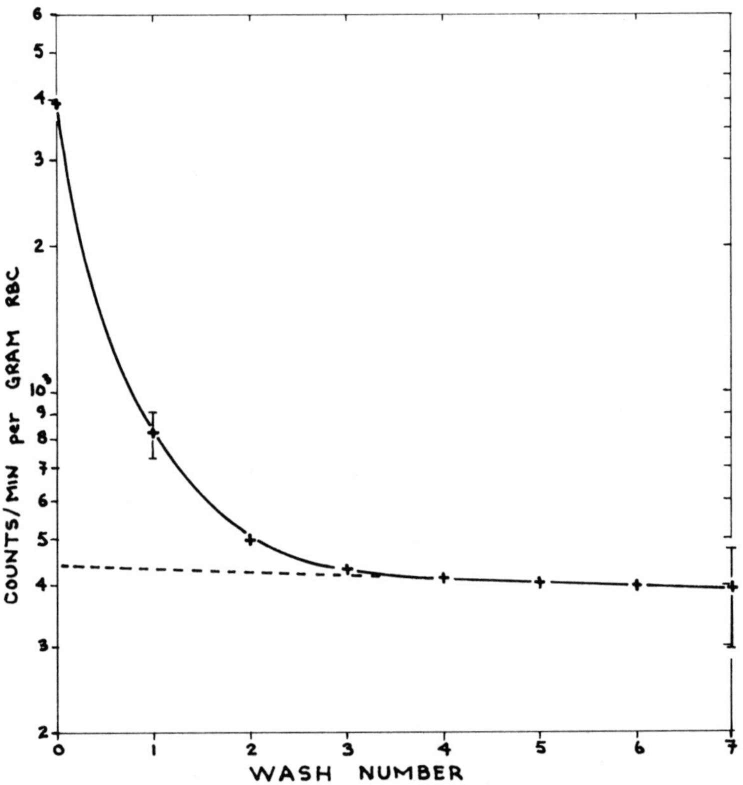

Fig. 8. The fibrinogen-associated radioactivity per gram of packed red cells remaining associated with the cell pellet as a function of the number of times the pellet is washed in protein-free buffer, T = 21°C, corrected for carry-over and tube wall adsorption. Cell pellets were suspended in ten volumes of phosphate-buffered saline, pH 7.2. Equilibrium fibrinogen concentration = 1.5 mg/ml before washing commenced; specific activity = 2.87 × 10⁵ dpm/gm.

Although our fibrinogen is virtually pure, as judged by its ability to sediment as
a clot when treated with thrombin, it does not migrate as a single species on sodium
dodecyl sulfate polyacrylamide gel electrophoresis (SDS-PAGE), a procedure that
separates mainly on the basis of molecular weight [46]. Figure 9 shows an optical
density tracing of an unreduced fibrinogen sample analyzed on a 4% gel stained
with Coomassie Brilliant Blue. Superimposed on this tracing is the distribution of
total radioactivity when the same gel was sliced and counted for [125]I. Native fibrino-

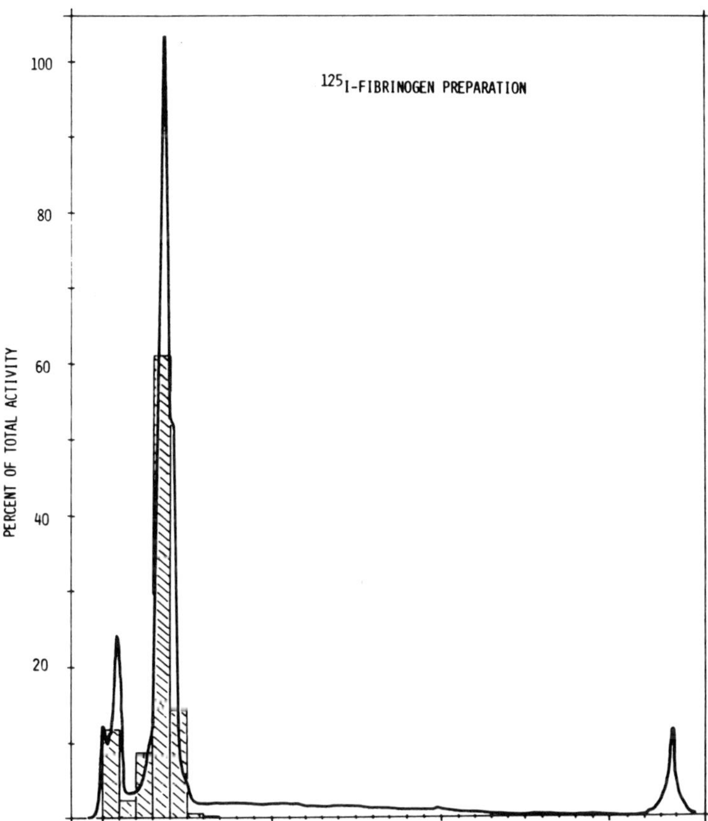

Fig. 9. Optical density scan at 595 nm (solid line) of purified ^{125}I-labeled human fibrinogen
analyzed on a 4% SDS-PAGE gel. Hatched bars represent the percentage of recovered radioactivity located in each 2 mm gel slice. Protein stained with Coomassie Brilliant Blue. Optical
density peaks at gel extremities are artifacts due to ends of the gel.

gen of molecular weight 340,000 is the major peak, containing about 85% of the recovered activity, while the remaining 15% resides in a higher molecular weight fraction. However, further analysis on SDS-PAGE showed that, upon reduction, the higher molecular weight product comigrated with reduced fibrinogen. It was also incorporated into thrombin catalyzed clots. Together these observations suggest that, while the high molecular weight product and native fibrinogen may share a common subunit structure, under nonreducing conditions they behave as distinct molecular weight species. High molecular weight fibrinogen-derived species have been recognized for some time; they are believed to be indicators of fibrin production in vivo [47].

It is possible to follow the desorption of native fibrinogen (NF) and the higher molecular weight fibrinogen-related material (HMWF) from erythrocytes by analyzing on SDS-PAGE the cell pellets obtained at different points in the washing procedure. Figure 10 shows the optical density trace and radioactivity distribution of a mixture of red cell membranes (ghosts [48]) and the labeled fibrinogen preparation. This pattern represents the components in the equilibrium unwashed pellet (in the absence, for clarity, of hemoglobin). The fibrinogen peaks are clearly seen superimposed on the normal red cell ghost protein pattern. After three washes in buffer the pellet of Figure 8 is analyzed by SDS-PAGE, either with no further manipulations (not shown) or after making ghosts by hypotonic lysis (Fig. 11). In both cases the distribution of radioactivity is the same. Essentially all of the counts migrate in the HMWF region and virtually no NF can be detected. Reference to Figure 8 shows that after three washes all of the bound material is associated with the slowly desorbing fibrinogen component. This slowly desorbing material can therefore be identified as HMWF.

We have not completed our studies on this system, so we cannot yet relate adsorption isotherms for NF and the HMWF to the degree of cell aggregation present. Again, however, it seems unlikely that the slowly desorbing HMWF is responsible for rouleaux formation, based on the ease with which such aggregation is reversed upon washing in protein-free media. The more intriguing possibility is that the binding of HMWF, believed to be associated with fibrin formation in vivo, might be related to red cell—fibrin interactions. Demonstration of binding of polymerized fibrin to erythrocyte surfaces would be of both physiological and pathological interest. Work now in progress should clarify most of the above issues.

Because of the apparent similarity between fibrinogen- and dextran-induced red cell aggregation, it is relevant to ask whether the two species interact with one another on the cell surface. Representative results of a mixed adsorption/desorption experiment are given in Figure 12. In this experiment aliquots of a washed erythrocyte pellet were incubated either with 1.5 mg/ml ^{125}I-fibrinogen, 3% w/v ^{3}H-dextran 150, or a combination of the two. The pellets were then washed and analyzed as in the earlier experiments. It can be seen from Figure 12 that neither species significantly affected the desorption behavior of the other. The only effect

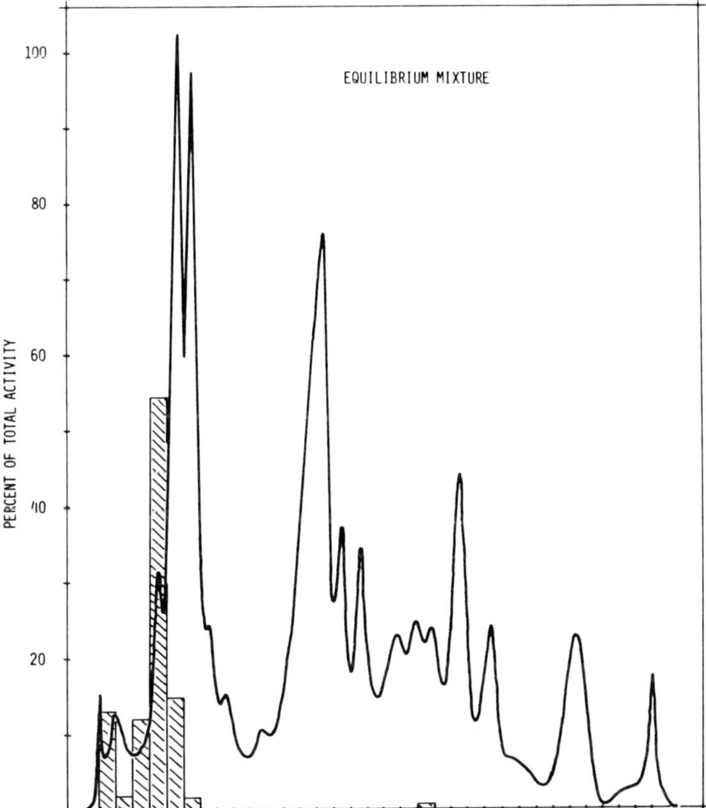

Fig. 10. Optical density trace (solid line) of equal volumes of human erythrocyte ghosts and 1.5 mg/ml ^{125}I-fibrinogen analyzed on a 4% SDS-PAGE gel. This is an artificial mixture made up to show the relative locations of HMWF, NF, and erythrocyte ghost protein bands. Hatched bars give percentage of recovered radioactivity in each gel slice.

of combining the two was to enhance the degree of aggregation relative to that present when only one of the species was included in the suspending medium. Again, in all cases the aggregation was reversed after one wash. These results indicate that dextran and fibrinogen act independently at the concentrations examined and do not compete for binding sites on the cell surface. Such behavior is completely consistent with the weak binding observed for both species. The aggregating effects of the two molecules appear to be roughly additive, as would be anticipated from the mixed binding results. Apparently the density of bridging macromolecules is rather low under the conditions used in this study.

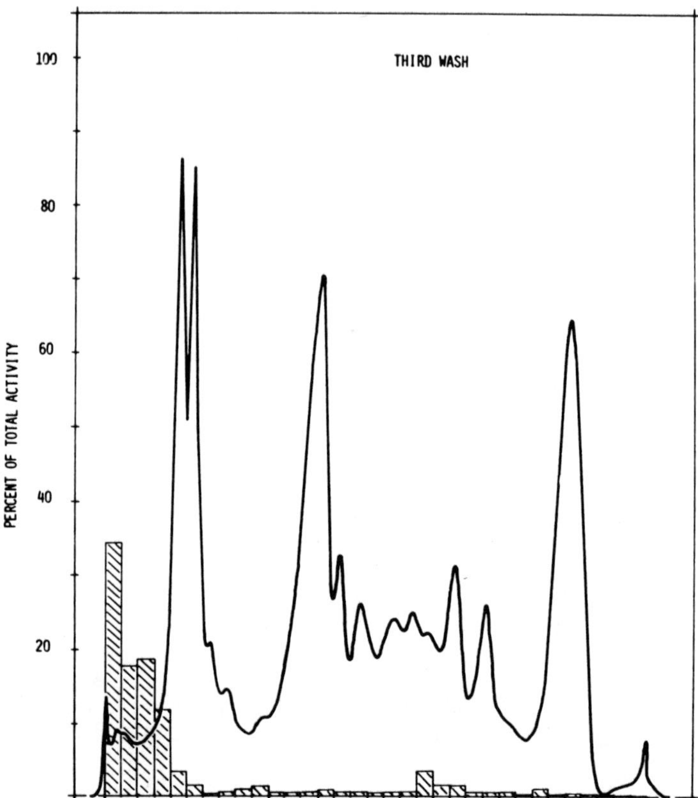

Fig. 11. Optical density trace of ghosts made by hypotonic lysis from the pellet of Figure 8 after three washes analyzed on a 4% SDS-PAGE gel. Hatched bars give percentage of recovered radioactivity in each gel slice. Large peak on extreme right is residual hemoglobin retained in the ghost preparation.

Considering the qualitative similarities between the dextran and fibrinogen desorption behavior, one might expect that similar mechanisms were operative. The molecular weight distributions of the ^{3}H-dextran fractions show no evidence of high or low molecular weight contaminants on gel filtration, however. We therefore have no reason to believe that the observed dextran wash-out kinetics represent desorption of different components.

AGGREGATION INDUCED BY AN AGENT THAT BINDS TO SPECIFIC CELL SURFACE STRUCTURES: CONCANAVALIN A

As was pointed out earlier, aggregation induced by macromolecular bridging can be classified into two types according to the degree of specificity associated with binding of the bridging agent to the cell surface. Our current studies on

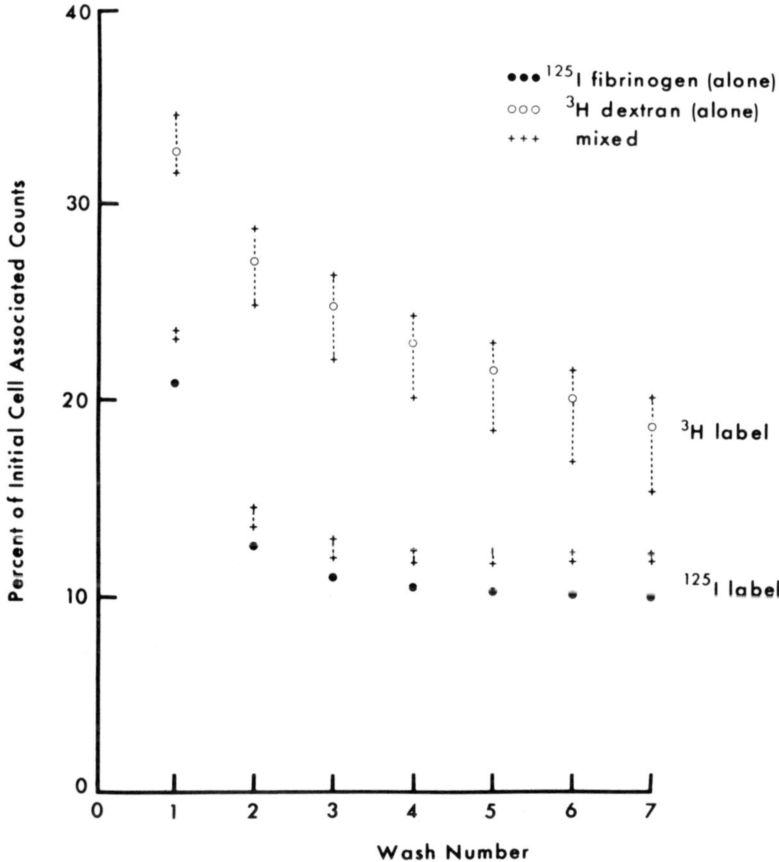

Fig. 12. The percentage of equilibrium pellet counts remaining with the pellet when red cells were equilibrated with 1.5 mg/ml ^{125}I-fibrinogen, 3% ^{3}H-dextran 150, or 1.5 mg/ml ^{126}I-fibrinogen, plus 3% ^{3}H-dextran 150 in phosphate-buffered saline pH 7.3, then washed the indicated number of times.

dextran- and fibrinogen-induced aggregation are providing a body of information on what appear to be nonspecific bridging reactions. We class them as nonspecific because we have, as yet, been unable to find low molecular weight agents that will competitively inhibit their adsorption and also because they bind extremely weakly to cell membranes. A variety of agents that do interact with specific cell surface structures, such as lectins and antibodies, are known to cause strong agglutination, which can be inhibited by small molecules. The study of aggregation induced by such agents is of particular relevance to aggregation among blood cellular elements because there is accumulating evidence that in some systems, particularly platelets, receptors for macromolecules appear when the cell is stimulated to aggregate [29]. Rapid aggregation occurs subsequent to stimulation, a response that is necessary if aggregation/adhesion is to occur locally in flowing blood.

It seems likely a priori that if macromolecular bridging is responsible for aggregation of this sort the relevant binding reactions would be reasonably strong and of the specific type. Strong aggregation would be required to maintain the patency of a platelet plug, for instance, in the face of fluid shear stresses tending to disperse it. To model reactions of this sort we have begun studying erythrocyte agglutination induced by the lectin concanavalin A (con A) in the presence of shear. Con A is a tetravalent protein that binds strongly to cell surface glycoproteins and glycolipids containing glucose- or mannose-like residues. The con A–red cell system demonstrates some unique characteristics — ie, that under some conditions shearing has been found to *induce* strong aggregation rather than break it down.

The composite results of a class of experiments we have performed on this system are illustrated in Figure 13. The experiment is carried out in the cup of a variable shear rate couette viscometer, the Contraves LS-2. Washed red cells suspended in glucose-free Hanks balanced salt solution are introduced at a hematocrit of 45–60% and sheared steadily to obtain a baseline shear stress. Concentrated con A in a negligible volume is then introduced, the sample mixed by raising and lowering the bob several times while shearing is maintained, and the effects on shear stress recorded as a function of time. The results depend on con A concentration, hematocrit, shear rate, and temperature, but the data presented in Figure 13 — obtained at $37°C$ with an equilibrium con A concentration of $120\ \mu g/ml$, hematocrit of 47%, and shear rate of about $49\ s^{-1}$ — are representative. Upon addition of the lectin an immediate rise in shear stress is recorded. We have shown previously that this increase provides a quantitative index of the degree of aggregation present in the suspension [49, 50]. The initial rise tends to level off within a few minutes, the slowly changing shear stress plateau defining what we call type I aggregation. The same plateau value occurs if con A is added to static suspensions and the aggregation allowed to develop for up to at least 60 minutes before shearing is commenced. If α-methyl mannoside (αmm), which is a competitive inhibitor of con A binding, is added while type I aggregation is present, the shear stress returns to the level characteristic of the disaggregated suspension. Microscopic examination of the system at this point shows the type I aggregation to be completely reversed. This reversal is expected for lectin agglutination in the presence of the appropriate specific saccharide and is due to occupation of the sugar binding sites by αmm. The bridging macromolecules then desorb from the cell surface.

If no αmm is added and type I aggregation is allowed to continue, a second marked increase in shear stress is observed, commencing in the case illustrated 12 minutes after the addition of con A (Fig. 13). This increase marks the onset of a second, stronger type of aggregation, type II, which is induced by shearing. That this second increase in shear stress represents a stronger type of aggregation and not sedimentation has been demonstrated in a number of ways. If the sample is mixed in the type II region, the shear stress recovers rapidly to its pre-mix value. If the suspension is sampled at different times after adding con A and the

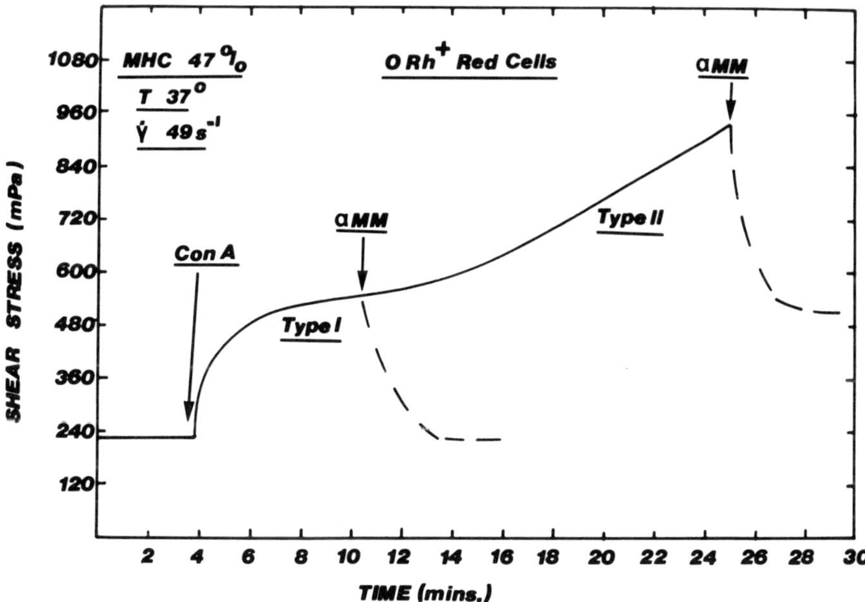

Fig. 13. The shear stress recorded as a function of time when washed human erythrocytes suspended at a hematocrit of 47% in a total volume of 0.9 ml in glucose-free, HEPES-buffered Hanks balanced salt solution, pH 7.4, are sheared at a shear rate of approximately 49 S^{-1}. At the first arrow sufficient con A is added in a volume of 23 μl to give an equilibrium solution concentration of 120 μg/ml. The dotted line at ten minutes indicates the shear stress behavior if the suspension is made 10 mM in α-methyl mannoside by the addition of 10 μl of αmm at the ten-minute arrow. If instead the αmm is added after 25 minutes, the behavior indicated by the second broken line is observed.

erythrocyte sedimentation rate is measured, the sample in type II sediments much more rapidly than that in type I. Both sedimentation rates exceed that for red cell/con A suspensions not exposed to shear. The microscopic appearance of the sheared and static suspensions also shows the former to be more strongly aggregated. It seems clear that shearing can induce a unique strong agglutination reaction among red cells exposed to con A.

One of the most interesting features of type II aggregation is that complete reversal is not affected by αmm. Figure 13 shows that if αmm is added after the type II stress increase has developed the trace does not return to its pre-con A level but remains somewhat elevated. Microscopic examination confirms that aggregation is still present. The shear-induced type II aggregation therefore appears to be associated partly with αmm-insensitive con A interactions.

Given the number of variables in this system it is perhaps not surprising that a complex dependence of the time development of shear stress has been found when the shear rate, hematocrit, con A concentration, and temperature are varied. Detailed analyses will be published elsewhere, but some of the observed features follow.

1) Type II aggregation is always preceded by type I.

2) The type I plateau saturates rapidly with con A concentration; further increases decrease the lag time for the appearance of type II and increase its slope. Type II aggregation eventually plateaus to a final shear stress, which is independent of con A concentration.

3) A minimum shear rate, dependent on hematocrit and con A concentration, is required for the type II response to be observed.

4) Fixation of the cells with glutaraldehyde before exposure to con A reduces the degree of type I aggregation and eliminates type II.

5) Incubating the cells with con A before shearing prolongs the lag period preceding the appearance of type II aggregation.

6) At a given concentration the total amount of con A adsorbed to the cells is not measurably different in static, type I, or type II aggregation.

We do not know the detailed mechanisms responsible for the observed behavior. Taken together, however, the above characteristics suggest that con A adsorbs initially to glucose/mannose-like cell surface sugars, causing static or type I aggregation. Once type I aggregation has been established, shearing allows a change in the bridging reaction to take place, which produces a more stable cell–cell bond. The change presumably results from stressing the cell membranes in such a way that a finite number of new binding configurations for the lectin are attainable. Either the number of molecules involved in this change is small or the new configurations must still block initially occupied sugar residues. The shear-induced bridging reactions appear to involve, at least in part, con A binding either to saccharide structures which interact much more strongly with con A than does αmm or to structures on the cell surface which react with regions on the con A molecule other than the carbohydrate binding sites. Work in progress in our laboratory should clarify the above issues somewhat and provide a more definite picture of this shear-induced aggregation reaction.

CONCLUSION

The experiments outlined above suggest that red cell aggregation can be induced by both specifically and nonspecifically bound macromolecules that adsorb simultaneously to two cells in a bridging configuration. The evidence available indicates that dextran and fibrinogen cause aggregation by the latter mechanism; con A, by the former. We have shown that in con A- (but not dextran- or fibrinogen-) induced agglutination, shearing can actually enhance the degree of aggregation present, presumably through its effect on the erythrocyte membrane. This is a provocative observation considering the variety of cell—cell interactions that are known to take place in the bloodstream under shear. It remains to be seen whether such effects prove to have physiological or pathological counterparts in vivo.

ACKNOWLEDGMENTS

Portions of this work were supported by grants from the Medical Research Council of Canada and the B. C. Health Care Research Foundation. One of us (J.J.) was a recipient of a research fellowship from the B. C. Heart Foundation while working on this project. We gratefully acknowledge the assistance of all the above agencies. We also wish to thank Mr. Wai-Pan Chan and Mr. Joe Charalambous for expert technical assistance.

REFERENCES

1. Fahreus R: Physiol Rev 9:241, 1929.
2. Thorsen G, Hint H: Acta Chir Scand (Suppl) 154:1, 1950.
3. Merrill EW, Gilliland ER, Lee TS, Salzman EW: Circ Res 18:437, 1966.
4. Brooks DE, Charalambous J, Janzen J: Biophys J 21:194a, 1978.
5. Kabat EA: "Structural Concepts in Immunology and Immunochemistry." New York: Holt, Rinehart and Winston, 1968.
6. Curtis ASG: "The Cell Surface: Its Molecular Role in Morphogenesis." London: Logos Press, 1967.
7. Weiss L: "The Cell Periphery, Metastasis and Other Contact Phenomena." Amsterdam: North-Holland Publishing Co., 1967.
8. Auerback R: Contacts and communications between cells in their relationship to morphogenesis and differentiation. In Wallach DFH, Fischer H (eds): "The Dynamic Structure of Cell Membranes." Berlin: Springer-Verlag, 1971, p 37.
9. Edwards J: Intercellular adhesion. In Paine R (ed): "New Techniques in Cell Biology and Biophysics." New York: Marcel Dekker, 1973.
10. Curtis ASG: Cell adhesion. In Butler JAV, Noble D (eds): "Progress in Biophysics and Molecular Biology," Vol 26. Oxford: Pergamon Press, 1973, p 317.
11. Greig RG, Jones MN: Biosystems 9:43, 1977
12. Verwey EJW, Overbeek J: "Theory of the Stability of Lyophobic Colloids." Amsterdam: Elsevier Publishing Co., 1948.
13. Seaman GVF, Brooks DE: Thromb Diath Haemorrh (Suppl) 42:93, 1970.
14. Brooks DE: J Coll Interface Sci 43:714, 1973.
15. Parsegian VA, Gingell D: J Adhes 4:283, 1972.
16. Brooks DE, Levine YK, Requena J, Haydon DA: Proc R Soc Lond Ser A 347:179, 1975,
17. Israelachvili JN, Ninham BW. J Coll Interface Sci 58:14, 1977
18. La Mer VK: J Coll Sci 19:291, 1964.
19. Katchalsky A, Danon D, Nevo A, DeVries A: Biochim Biophys Acta 33:120, 1959.
20. Prokop O, Uhlenbruck G: "Human Blood and Serum Groups." London: Maclaren, 1969.

21. Goldstein IJ, Hayes CE: Adv Carbohydr Chem Biochem 35:127, 1978.

22. Brooks DE: Red cell interactions in low flow states. In Grayson J, Zingg W (eds): "Microcirculation," Vol 1. New York: Plenum Press, 1976, p 33.

23. Chien S, Jan K-M: Microvasc Res 5:155, 1973.

24. Gelin LE, Shoemaker WC: Surgery 49:713, 1961.

25. Jan K-M, Chien S: J Gen Physiol 61:638, 1973.

26. Born GVR: In Deutsch E, Gerlach E, Moser K (eds): "Metabolism and Membrane Permeability of Erythrocytes and Thrombocytes." First International Symposium, Vienna. Stuttgart: Schattauer, 1969, p 294.

27. Salzman EW: Fed Proc 30:1503, 1971.

28. Mustard JF, Regoeczi E, Perry DW, Kinlough-Rathbone RL: Blood 52:453, 1978.

29. Marguerie GA, Plow EF, Edgington TS: Fed Proc 38:1207, 1979.

30. Linnemans WAM, Spies F, De Ruyter De Wildt ThM, Elbers PF: Exp Cell Res 101:191, 1976.

31. Colvin RB, Dvorak HF: J Exp Med 142:1377, 1975.

32. Silberberg A: J Polymer Sci Part C Polymer Symposia 30:393, 1970.

33. Brooks DE: J Coll Interface Sci 43:700, 1973.

34. Eirich FR: J Colloid Interface Sci 58:423, 1977.

35. Brooks DE: J Colloid Interface Sci 43:687, 1973.

36. von Hummel K, Szczepanski LV: Blut 9:145, 1963.

37. von Hummel K: Blut 9:215, 1963.

38. Chien S, Simchon S, Abbott RE, Jan KM: J Colloid Interface Sci 62:461, 1977.

39. Seaman GVF, Heard DH: J Gen Physiol 44:251, 1960.

40. Fischer L: "An Introduction to Gel Chromatography." Amsterdam: North-Holland Publishing Co., 1975.

41. Schachter D: Biochem Biophys Res Commun 84:840, 1978.

42. Traber DL, Kolmen SN: Tex Rep Biol Med 23:782, 1963.

43. Muller HE, Gramlich F: Acta Haematol 34:239, 1965.

44. Johnson AJ: personal communication.

45. Polson A, Potgeiter GM, Largier JF, Mears GEF, Joubert FJ: Biochim Biophys Acta 82:253, 1963.

46. Weber K, Osborn M: J Biol Chem 244:4406, 1969.

47. Fletcher A, Alkjaersig N, O'Brien J, Tulevski V: Trans Assoc Am Physicians 83:159, 1970.

48. Dodge JT, Mitchell C, Hanahan DJ: Arch Biochem Biophys 100:119, 1963.

49. Brooks DE, Goodwin JW, Seaman GVF: Biorheology 11:69, 1974.

50. Knox RJ, Nordt FJ, Seaman GVF, Brooks DE: Biorheology 14:75, 1977.

7

Dynamics of Red Blood Cell Deformation and Aggregation, and In Vivo Flow

Giles R. Cokelet

Most of the blood flow processes in the body are unsteady in nature, whereas most of the commonly used in vitro processes used to characterize the flow properties of blood and its constituent cells are steady or, if unsteady, they are likely to have time constants that are longer than the time constants for the in vivo flow processes.

One can take the in vitro data and attempt to predict in vivo performance if the blood and the blood cells are of such a nature that they respond very rapidly to changes in conditions. The words "very rapidly" are relative and must be defined for the case in hand: if the basic flow processes of blood and cells have characteristic time constants, say a tenth or less than the time constants of in vivo flow processes, then the blood and the cells can be considered to respond very rapidly. Under such conditions, the quasi-steady-state approximation is satisfactory: at any instant in time, the blood and the cells are in the state in which they would be under steady flow conditions, with the same values for the independent variables. If this approximation cannot be made, then we must be sure that we have performed suitable in vitro tests so that the time-dependent properties of the blood and the cells are uncovered. Two kinds of information are presented here: 1) data from which time constants for in vivo processes can be estimated, and 2) data from which material property time constants for blood and blood cells can be deduced.

COMMENTS ON IN VIVO PHENOMENON

The fundamental time constant of the circulatory system is the one imposed by the heart; in the physiological range this corresponds to 0.3 to 3 seconds per beat. This time constant persists even into the capillaries of the microvasculature. In terms of pressure, this was first shown in detail by Wiederhielm et al [1]. The fact that the velocity also is pulsatile in nature, even in the capillaries in the microvasculature, has been demonstrated (for example, by Intaglietta [2]).

Erythrocyte Mechanics and Blood Flow, pages 141–148

As blood flows along a vessel, it is subjected to continually changing stresses, partly because the vessel lumen changes its dimensions and shape with axial position and (usually) with time. The residence times between vessel junctions are an upper limit on this type of characteristic time. These times reflect gross changes in flow conditions caused by vessel diameter changes, especially when the blood flows from larger supply vessels into smaller branch vessels (or vice versa), but they don't indicate the local changes in lumen diameter and cross-sectional shape that occur along the length of a vessel. Table I gives some typical in vivo values for the residence time of blood between junctions in a single vessel. The first set of data was obtained by Mayrovitz et al [3] for blood flow in the wing of the bat. The characteristic times recorded here range from about 0.4 seconds up to 12 seconds, with the shorter times being in the smaller vessels. The other set of data in this table was obtained by Gaehtgens et al [4] and was obtained from a study of the circulation in the mesentery of the cat. These latter data indicate very short residence times, from less than 0.02 seconds up to over 2 seconds, with the residence times increasing as one goes to the smaller vessels. The diversity of these data indicates that each microcirculatory bed must be considered separately if one is to obtain reliable residence time values. However, data such as these indicate interjunctional vessel residence times that range from less than 0.02 second up to several seconds.

When one considers the flow of blood in a vascular bed from the viewpoint of continuum fluid mechanics, one must recognize the effects of the nonuniform flow in the entrance length of each vessel. In the case of steady flow of single-phase fluids, this entrance length is the region over which the fluid entering the vessel establishes the uniform velocity profile it will maintain further down-

TABLE I. In Vivo Residence Times

Experimental material	Vessel	Diameter (μ)	Mean RBC velocity (mm/sec)	Mean residence time (seconds)
Batwing [3]	Main artery	76	3.3	12.0
	Artery	42	3.0	5.7
	Small artery	22	2.0	1.7
	Arteriole	8.0	1.2	0.79
	Terminal arteriole	5.7	0.50	0.40
Cat mesentery [4]	Arteriole	60	20	< 0.02
		23	18	0.09
		20	13	0.13
		12	6	0.06
	Arteriole capillary	7.5	4	0.11
	Capillary	7.5	1.0	2.1
	Venule	14	1.6	0.16
		19	10	0.12
	Venule	60	10	< 0.02

stream in that vessel. For the flow of a Newtonian fluid in a circular cylindrical tube, the zero Reynolds number flow entrance length is given by

$$L_e = 1.3 \, R_t \tag{1}$$

[5], where L_e is the entrance length and R_t is the vessel diameter. For flows with Reynolds number less than one, $L_e \leqslant 1.4 \, R_t$. This length is short. For particulate systems, the entrance length may not be given by the above equations because of an additional effect due to the radial migration of particles in the entrance region. This particle radial migration establishes a concentration profile that differs from that found in the suspension as it enters the vessel. Since the rheological properties of suspensions are strongly influenced by particle concentration, and the velocity profile is influenced by the fluid's local rheological properties, one would expect the entrance length to significantly exceed the length prediction from Equation 1 if the rate of particle radial migration is slow compared to the time required to establish the uniform velocity profile in the absence of particle radial migration. This question of whether L_e is much larger than the value predicted from the equations above, for blood flowing in a vessel, must be answered from experimental data. At this time, the only piece of experimental work bearing on this matter is that of Palmer and Betts [6].

In their experiments, blood and red cell suspensions were forced to flow from a reservoir into and through a channel formed by two parallel plates. The distance between the two parallel plates was 25 microns. By using a multi-wedge arrangement at the exit of the parallel plate channel, they were able to separate the suspension flowing down the wall region of the parallel plate channel from that flowing down the center and intermediate parts of the channel. By varying the length of the parallel plate channel, they were able to get a measure of the hematocrit profile as a function of distance from the entrance of the parallel plate channel. The hematocrits were measured for blood samples collected at the exit of the system and so do not directly represent the hematocrits in the parallel plate channel. However, changes in hematocrit within the parallel plate channel will be reflected by changes in the measured hematocrits. In short, their results showed that, for suspensions of either fresh red blood cells in mixtures of plasma and acid citrate dextrose solution or hardened red blood cells in physiological saline with feed reservoir hematocrits of 10, 20, and 40 percent, the suspension had to travel five millimeters from the entrance of the parallel plate channel before the hematocrit profile was fully established. In the process of arriving at this profile, the peripheral layers of suspension reached the equilibrium concentrations faster than the central areas. For the lower feed reservoir hematocrit, hardened cells tended to move slower than fresh cells in establishing the hematocrit profile, whereas at the highest feed reservoir hematocrit, hardened cells tended to move faster than the fresh red blood cells. The suspension average velocities used in this study were in the range of 1—2 cm/sec, which corresponds to a Reynolds number of about 1 if the parallel plate separation is used as the characteristic dimension of the system. The conclusion from this is that, because of particle radial migration, the entrance length is much longer than the distance predicted from the equations for Newtonian fluids. Indeed, if this single set of data is any indication, the blood vessels in the microcirculation may be so short that their entire length is, in effect, an entrance length. If true, this would make any analytical analysis of blood flow in the microcirculation that much more difficult. Obviously, a much more extensive set of experimental data is needed before this question can be answered.

DIRECT OBSERVATION OF RED CELL DEFORMATION AND RED CELL AGGREGATION

Two flow processes that can greatly affect blood flow are red blood cell aggregation and red blood cell deformation.

Characteristic time constants for the process of red cell deformation can be obtained from the literature. Hochmuth et al [7] used a micropipette with a diameter of the order of one micron to aspirate into the pipette a small hemispherical tongue of the red cell membrane. This tongue of red cell was then expelled from the micropipette, and the time required for the tongue to relax to half its initial amplitude was measured. For unnucleated red blood cells, this half time was about 0.06 seconds (opossum and human red cells). This characteristic time should be a material property of the red cell membrane, since Waugh and Evans [8] showed that the viscosity of the red blood cell contents has a negligible effect on the deformation of red cells, and, because of geometric factors, the viscosity of the suspending media would be expected to exert little if any viscous effect on this experiment.

Skalak and Branemark [9] observed the flow of blood in the capillaries and other microvascular vessels of the armfold skin of the human. They observed large deformation of the red blood cell as it flowed through bifurcations. Their observations are summed in the statement ". . . large deformations and recovery can occur in very short times (of the order of 0.06 seconds), particularly at branchings of blood vessels and at tapered sections." These deformations took place under the influence of external forces that induced red cell shape changes. It is a bit surprising that forced cell shape changes took roughly the same time to occur as unforced, or self-relaxing, deformations. In any case, the time required for a cell to respond to a deforming stress is extremely short.

The kinetics of red cell aggregation have been studied extensively by Schmid-Schoenbein and his co-workers [10]. A "half time of aggregate reformation" has been determined for bloods that undergo a step change in shear rate from $460 \sec^{-1}$ to zero $\sec^{-1}$ (or a very low shear rate). At the higher shear rate, the cells are dispersed, whereas at steady state at the lower shear rate the cells are aggregated into rouleaux and aggregates of rouleaux. At hematocrits of 45% and a temperature of 37°C, the half time for aggregate formation in normal blood is 3–5 seconds, whereas the corresponding time in pathological bloods is of the order of 0.5–1.5 seconds. (For normal blood, a shear rate of about 55 sec^{-1}, equivalent to a shear stress of 3.0 dynes per cm^2, is required for complete hydrodynamic disaggregation of the red cells; in pathological bloods this critical shear stress is much higher.) If a blood containing aggregated cells is suddenly subjected to a shear stress above the critical shear stress for aggregation, one would expect the aggregates of cells to disperse very rapidly. The aggregation process involves collision of red cells and rearrangement of red cells into rouleaux; these are time-consuming steps that do not occur when rouleaux are dispersed by high stress. Therefore, we would expect the aggregation time to be much longer than the disaggregation time. Consequently, it appears that the process of red cell aggregation is normally very slow, of the order of several seconds, whereas the disaggregation process is probably very rapid, with a time constant comparable to that of cell deformation.

INDIRECT DETERMINATION OF TIME CONSTANTS FOR BLOOD FLOW MECHANISMS

The characteristic times of the blood flow processes have been estimated from both steady and unsteady rheological data. In making these estimates, a hypothetical model of the fluid is often envisioned, and the stress-rate of deformation equation for the model is derived. Parameters in this constitutive equation can then be related to response times of the mechanisms in the model fluid.

Such equations can be viewed in two ways: 1) as being faithful reflections of the actual flow mechanisms, in which case the time constants embedded in the constitutive equation are descriptive of the flow mechanisms, and 2) as being an analytical form that can be used to represent the continuum rheological properties of the fluid (by some curve-fitting procedure), in which case the time constants reflect response times for the fluid as a whole but may or may not be the time constants for the actual flow processes. In the latter case the time constants are useful when considering flows where the blood can be treated as a continuum and only continuum behavior is of interest, but these time constants may or may not be useful when considering flows where the continuum model is invalid, such as when the interest is in the flow behavior on a cellular scale.

Many of the equations used to analyze unsteady viscometric blood flows have been taken from the vast reservoir of work done with polymer systems. Such polymeric fluids are demonstrably viscoelastic; consequently, the constitutive equations have been for viscoelastic fluids. Such viscoelastic relationships have been adopted by many workers to describe unsteady blood flow, although the practical significance of elastic behavior in the flow of normal blood under physiological conditions has not been unequivocally demonstrated. Nevertheless, it is instructive to look at some examples of work that has been conducted using viscoelastic fluid approaches, since the uncovered time contants are of interest to us here.

The first example is the report of Lessner et al [11]. From data obtained for oscillatory flow of blood in a concentric cylinder viscometer, in which the outer surface is stationary and the inner surface is driven in a sinusoidal manner, they calculated a relaxation time spectrum, which is interpreted to be $H(\tau)$, the number of relaxation mechanisms per unit volume of fluid that have a given time constant, τ, expressed as a function of the value of the time constant. Since their measurements were made under conditions in which the amplitude of the fluid motion was very small, their calculated relaxation spectrum is thought to reflect relaxation mechanisms between and within rouleaux. Their conclusion was that over the relaxation time range of 0.3–0.8 seconds, $H(\tau)$ was relatively constant but began to drop rather rapidly with increasing τ, for τ near or greater than 0.8 second. This can be interpreted to mean that, under limiting low rates of deformation, the flow mechanisms involving making, breaking, and rearranging rouleaux and structures of rouleaux have characteristic time constants of about 1 second or less.

Another approach is shown in a second example, the work of Deutsch and Phillips [12]. A linear viscoelastic expression is assumed to be the constitutive equation of blood. As a first approximation, an equation with four constants is tested; one constant is a zero shear rate viscosity, and the other three constants are time constants. From steady viscometric data, the zero shear rate viscosity is obtained, as well as two of the time constants expressed as a function of the third time constant. By experimentally determining the critical conditions under which Taylor vortices are first formed in Couette flow, the time constants can be evaluated. Although this procedure can be criticized on both theoretical and experimental grounds (eg, while polymer systems show a low shear rate range Newtonian viscosity as represented in the chosen constitutive equation, blood does not show such behavior; the use of outdated blood), it does give a rough qualitative representation for several of the features of blood rheological data. On this basis, the stress relaxation time is about 3 seconds and the rate of deformation relaxation is about 1 second. These times are, of course, composite times, suitably averaged over all the flow mechanisms in the blood, and so cannot be directly related to particular flow mechanisms. Nevertheless, they are continuum characteristic times and, interestingly, are close to the longest characteristic time for the various flow processes (ie, for the red cell aggregation process).

DISCUSSION

In summary, the following statements can be made:

1) Under normal physiological conditions in human blood, erythrocyte deformation is a much faster process than red cell aggregation (characteristic time of several seconds). Disaggregation probably is much faster than aggregation. The effects of pathological conditions on these characteristic times is not established.

2) Relative to in vivo flow processes. a) For the erythrocyte deformation process: Normally the time required for deformational shape changes may be comparable to red cell interbifurcational residence times in some microcirculatory vessels, but generally it is shorter than the fundamental frequency time of the heart and the residence times in larger vessels. b) For the aggregation process: usually the required time is longer than characteristic times of the in vivo flow processes, especially in smaller vessels. If the flow becomes very slow, say in venous vessels, then aggregation times may become comparable to or shorter than the flow processes times. c) For the disaggregation process: Probably the characteristic times would be comparable to those for erythrocyte deformability. d) For pathologic conditions: One would expect aggregation kinetics to become faster and disaggregation kinetics slower. The nature of changes in erythrocyte deformation kinetics is not apparent; changes in cell deformability and shape would also influence aggregation.

3) The influence of changes of cell aggregation and/or deformability on vessel entrance lengths is not known, although the results of Palmer and Betts' steady flow experiments imply that little effect would occur.

One implication of these features of blood flow in the circulation is that often, when the characteristic time for erythrocyte aggregation is long compared to the characteristic times for the unsteady flow conditions, red cell aggregation will be less significant than expected from a knowledge of local shear rates. That is to say, the quasi-static assumption will not be valid. The instantaneous rheological properties of the blood will be somewhere between the steady flow rheological properties of red cells in plasma and red cells in serum, even if the blood should be a purely viscous fluid with time (history)-dependent properties.

There is also an implication for experimental rheological studies in which oscillatory flow in tubes is studied. Unlike concentric cylinder and cone-and-plate devices, tube instruments do not subject all of the test fluid to (essentially) the same time history. At the ends of the tube, fluid is either entering the tube from the reservoir or leaving the tube. If the volume of "new" blood drawn into the tube from the reservoir during a cycle is small compared to the total tube volume, the influence of the difference in properties of this "new" blood and the blood that always remains in the tube will be negligible. However, if the "new" blood volume is an appreciable fraction of the tube total volume, a difference in properties of the "new" blood and the "tube" blood (due to different stress histories, or other different histories) can result in a flow performance that depends on how much "new" blood enters the tube per cycle and the history of this blood. These considerations carry over to the study of pulsatile flow in tubes.

ACKNOWLEDGMENTS

The author's work has been partially supported by a research grant (HL 23355) from the National Heart, Lung and Blood Institute, NIH, and partially performed under contract with the U.S. Department of Energy at the University of Rochester, Department of Radiation Biology and Biophysics and has been assigned report number UR-3490-1681.

REFERENCES

1. Wiederhielm CA, Woodbury JW, Kirk S, Rushmer RF: Pulsatile pressure in the microcirculation of the frog's mesentery. Am J Physiol 207:173, 1964.
2. Intaglietta M: Pressure flow relationships in the in vivo microcirculation. In Grayson J, Zingg W (eds): "Microcirculation I. New York: Plenum Press, 1976, pp 71–76.
3. Mayrovitz HN, Wiedeman MP, Noordegraaf A: Microvascular hemodynamic variations accompanying microvessel dimensional changes. Microvasc Res 10:322, 1975.
4. Gaehtgens P, Meiselman HJ, Wayland H: Erythrocyte flow velocities in mesenteric microvessels of the cat. Microvasc Res 2:151, 1970.
5. Lew HS, Fung YC: Entry flow into blood vessels at arbitrary Reynolds number. J Biomech 3:23, 1970.
6. Palmer AA, Betts WH: The axial drift of fresh and acetaldehyde-hardened erythrocytes in 25 μm capillary slits of various lengths. Biorheology 12:283, 1975.
7. Hochmuth RM, Worthy PR, Evans EA: Red cell extensional recovery and the determination of membrane viscosity. Biophys J 26:101, 1979.
8. Waugh R, Evans EA: Viscoelastic properties of erythrocyte membranes of different vertebrate animals. Microvasc Res 12:291, 1976.
9. Skalak R, Branemark P-J: Deformation of red blood cells in capillaries. Science 164:717, 1969.
10. Schmid-Schoenbein H, Volger E, Klose HJ: Microrheology and light transmission of blood. II. The photometric quantification of red cell aggregation formation and dispersion in flow. Pfluegers Arch 333:140, 1972.
11. Lessner A, Zahavi J, Silberberg A, Frei EH, Dreyfus F: The viscoelastic properties of whole blood. In Hartert HH, Copley AL (eds): "Theoretical and Clinical Hemorheology." New York: Springer-Verlag, 1971, pp 194–205.
12. Deutsch S, Phillips WM: The use of the Taylor-Couette stability problem to validate a constitutive equation for blood. Biorheology 14:253, 1977.

8

Physical and Mathematical Models of Blood Flow: Theoretical Analysis

Richard Skalak

INTRODUCTION

Theoretical analysis of blood flow from first principles has so far derived extensive results for only a few cases of capillary flow. Most of our knowledge of the flow properties of blood comes from experiments [see the chapter by H. L. Goldsmith in this volume]. The main reasons for limited theoretical progress are that the red blood cells are very flexible, and the motion of a concentrated suspension produces very complicated flow fields. A multiple interaction between neighboring particles takes place in a shear flow. Such interactions have been analyzed in detail for only two particles at a time [Batchelor and Green, 1972]. The bulk flow of a suspension with a particle concentration (hematocrit) in the range of normal blood (35–40%) has not been rigorously treated by an existing theory, even for rigid particles.

On the other hand, in capillary flow, where the red blood cells are of the same order of diameter as the capillary blood vessels, the situation may be simplified by considering only one cell at a time or a line of identical cells. Such flows are amenable to analytic and numerical solutions for a variety of cases for rigid particles and various flexible models. This is the class of flows that is treated in this chapter.

Although capillary blood flow is far from the situation in larger blood vessels, it is nevertheless suggestive of the mechanism by which the deformability reduces the apparent viscosity. The basic mechanism is the generation of fluid pressures between particles on a collision course. These pressures may be computed by applying lubrication theory, and they can produce very large deformations of flexible particles.

Erythrocyte Mechanics and Blood Flow, pages 149–164
© 1980 Alan R. Liss, Inc., 150 Fifth Avenue, New York, NY 10011

The properties of red blood cells are sufficiently well known [see chapters by R. M. Hochmuth and E. A. Evans in this volume] to allow good models of the red blood cells to be postulated. For the purposes of flow mechanics up to the point of hemolysis, the red blood cell membrane may be assumed to have constant area, even though the deformations are severe. For a steady capillary flow, the viscous properties of the red cell membrane and of the hemoglobin are of no consequence, since the cell moves as a rigid body. Its shape does not change with time in a steady flow. Due to the incompressibility of the cell contents, the volume of the cell is constant. It is important to take this behavior into account. One consequence of the incompressibility of the cell is that the mean pressure of the surrounding plasma does not have any effect on the shape of the cell. A uniform pressure rise throughout the fluid does not produce any deformation.

The important elastic properties of the red blood cell for purposes of modeling capillary blood flow are the shear modulus and the bending stiffness of the membrane. The strains in the surface of the cell membrane are moderate, but the shear modulus is important as an elastic restoring force. The bending stiffness is necessary to the production of the smooth curved surfaces that the red blood cells exhibit under moderate stresses. At very high stresses, the elastic properties of the membrane are less important. The membrane can then be modeled as a flexible shell of constant area, but without elastic stiffness, like a two-dimensional sheet of fluid.

In the present chapter, the capillary flows with rigid particles and elastic spheres are considered first. For the idealized geometry of a sphere it is possible to have sufficiently complete results to illustrate the significant effects in capillary flow. Spheres are good first approximations of white blood cell geometry. In addition, red blood cell models are analyzed, taking into account the elasticity of the membranes. The computations based on rigid cells allow demonstration of the effects of shape and spacing of cells in a capillary. Computations for flexible cells are difficult and limited, but they give realistic representations of red blood cell shapes observed experimentally. The apparent viscosities computed from these models are remarkably low and are in reasonable agreement with experimental results.

RIGID PARTICLES

The mechanics of capillary blood flow can be understood to a considerable extent by consideration of the flow of rigid particles suspended in a Newtonian, incompressible fluid, as shown in Figure 1. This motion of the suspending fluid is governed by the Stokes equation

$$0 = -\nabla p + \mu \nabla^2 U \qquad [1]$$

and the continuity equation

$$\nabla \cdot U = 0 \qquad [2]$$

where U is the velocity vector, p is the pressure, and μ is the viscosity of the fluid. It is assumed that the particles are neutrally buoyant and that there are no body forces.

The particles shown in Figure 1 are assumed to move at a uniform axial direction. It is convenient to view the flow from coordinates fixed to the solid particles. In these coordinates the wall of the capillary moves at velocity U to the left. The velocity field in the gap between the particle and the tube wall may be broken up into two parts, as shown in Figure 2. The first part, shown in Figure 2A, is a Couette flow having a linear velocity profile of axial velocity μ_z vs radius R. This velocity distribution cannot be the full solution, because the discharge of fluid to the left would not be the same at each section. Since the boundaries are the same at any time, this discharge, Q (called the leakback), must be a constant with respect to axial position z. To achieve the constancy of Q, a velocity field such as shown in Figure 2B must be added. When the gap is thin, there must be a falling pressure gradient $[(\delta p/\delta z) < 0]$ at sections 2 and 3 and a rising pressure $[(\delta p/\delta z) > 0]$ in section 2 to produce the flows shown in Figure 2B. The typical form of the resultant pressure curve vs z is shown in Figure 2D.

Another aspect of capillary flow that may be seen from Figure 2 concerns the force balance. The particles moving at uniform velocity must be under zero net force. This means force due to the pressure drop across the particle must be balanced by the shear stresses and the pressures on the particle shown in Figure 2D. The pressure contributions can be quite important, because the positive pressure

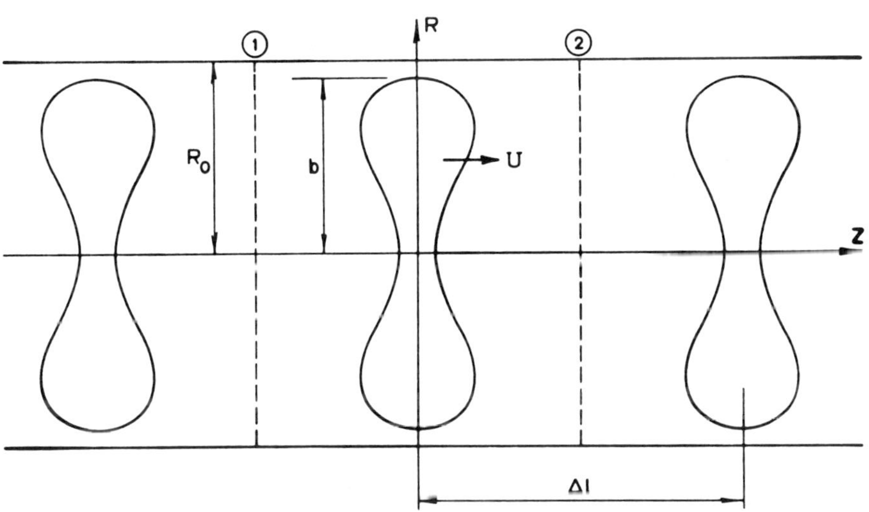

Fig. 1. Line of red cells axisymmetric in a capillary of circular cross-section. Diameter ratio λ is equal to b/R_0.

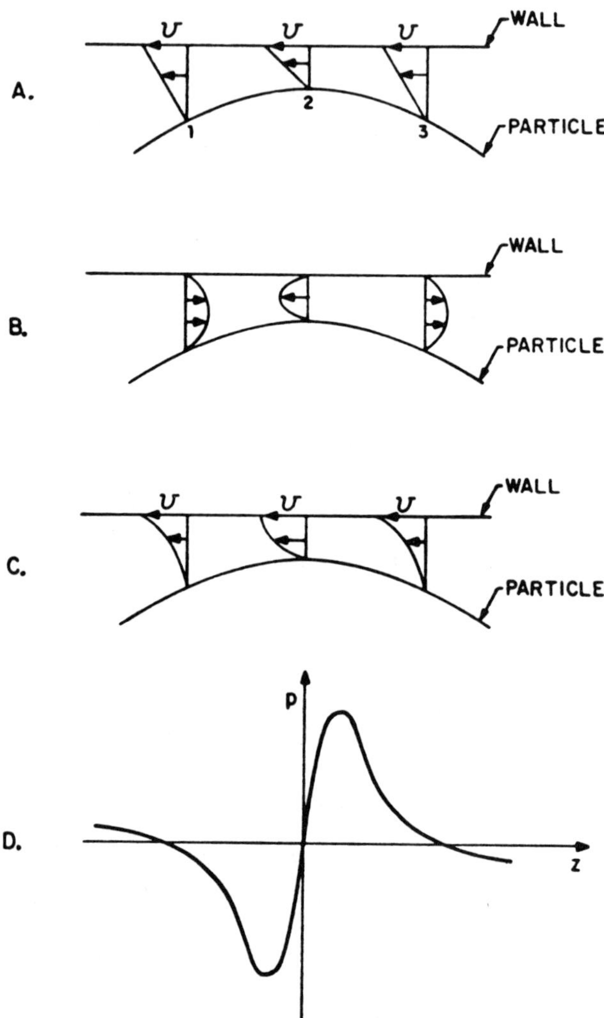

Fig. 2. A. Linear velocity profiles between particle and capillary wall, associated with wall motion at velocity U (relative to particle). B. Parabolic velocity distributions associated with pressure gradient. C. Total velocity profiles (relative to particle) equal to sum of A plus B. D. Pressure distribution associated with the velocity profiles shown in C.

on the right-hand side of Figure 2D acts on a sloping region so that it tends to retard the particle. The negative pressure on the left-hand side also has a resultant that retards the particle.

The zero drag condition is simpler to apply by considering the entire unit between sections 1 and 2 in Figure 1, containing both fluid and particle. Here the force due to the pressure drop is balanced by the shear stress at the tube wall only because the pressures on the tube wall are normal to the tube axis.

The pressure distribution in Figure 2D also suggests how the particle will tend to deform if it is flexible. The high pressures on the right tend to compress the leading end of the particle. The low pressures on the left tend to increase the diameter of the particle at the upstream end. This results in a typical "fish tail" shape observed at the trailing end of very flexible particles.

For rigid particles having the shape of red blood cells, as shown in Figure 1, the relative apparent viscosities computed for various hematocrits and capillary diameters are shown in Figure 3. [Skalak et al, 1972]. The capillary diameter is indicated by the parameters λ defined as the ratio of the maximum particle diameter to the capillary diameter, Figure 1, $\lambda = b/R_0$. It may be seen in Figure 3 that unless λ is greater than 0.9, the relative apparent viscosity is less than about 2.0 for normal hematocrits. This means the pressure drop of the capillary flow is less than twice the pressure drop for the flow of plasma alone at the same mean velocity when $\lambda \leq 0.90$.

The various curves in Figure 3 show the effect of rouleaux formation in which the cells are again assumed to be axisymmetric but are bunched in groups of three or five instead of single cells as in Figure 1. There is a reduced pressure drop for rouleaux as compared to single cells at the same hematocrit.

Another interesting feature of Figure 3 is that the initial portion of each curve is a straight line, so that the increase in apparent viscosity is proportional to the hematocrit. In this region there is no interaction between particles. There is another interesting region at higher hematocrits (for example, from H = 25% to H = 50% with $\lambda = 0.80$) where there is very little change of apparent viscosity with increasing hematocrit. In this region the fluid is more or less trapped between cells and moving as a rigid body. Hence the addition of more cells in this region has little effect.

ELASTIC SPHERES

The steady flow of a line of elastic, incompressible spheres equally spaced along the axis of a capillary tube has been investigated by Tözeren and Skalak [1978]. The computations for this geometry can be carried out more easily, and the results show the qualitative features of the flow of elastic particles in suspension in capillaries. The spheres are assumed to be closely fitting; ie, the cases of the sphere diameter slightly smaller, equal to, and slightly greater than the tube diameter are all considered. Because the gap between the sphere and the tube wall is always very thin in the region of interest near the tube wall, the reduced equations of lubrication theory are adopted for the fluid.

The elastic spheres are assumed to be incompressible, so that a uniform change in pressure does not produce any deformation of the spheres. But they are elastic in shear, and any nonuniform pressure over the surface of such a sphere produces an elastic change of shape at constant volume. The incompressibility of the particles is realistic for the physiological situation and distinguishes this theory from the earlier work of Lighthill [1968] and Fitz-Gerald [1969], who assumed that the particles could be compressed by a uniform pressure. Another difference is that these earlier theories assumed that the force due to the pressure drop across the particle is approximately equal to the resultant of shear stresses acting along the surface of the particle. As discussed earlier, this neglects the effects of the lubrication pressures in the narrow gap near the tube wall. In the results quoted below, the zero drag condition has been corrected.

Pressure distributions computed for rigid spheres are shown first for comparison purposes in Figure 4. The parameter λ is the ratio of the sphere radius to the tube radius, $\lambda = a/R_O$. The curves are quantitative examples of the qualitative pressure curve sketched in Figure 2D. The examples in Figure 4 show that the lubrication

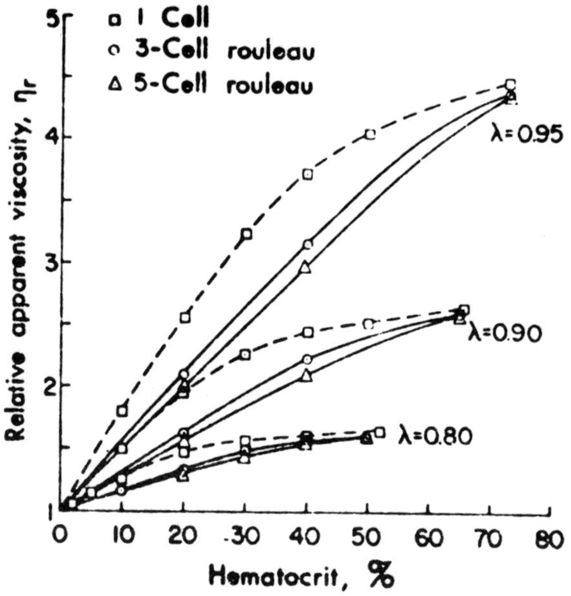

Fig. 3. Relative apparent viscosity η_r for a train of equally spaced rigid biconcave cells such as shown in Figure 1 (dashed curve). Solid curves are η_r for equally spaced rouleaux of three and five cells at the same hematocrits and diameter ratios λ. $\eta_r = \eta_a/\eta_O$, where η_a = apparent viscosity of the suspension and η_O = viscosity of the Newtonian suspending fluid [Skalak et al, 1972].

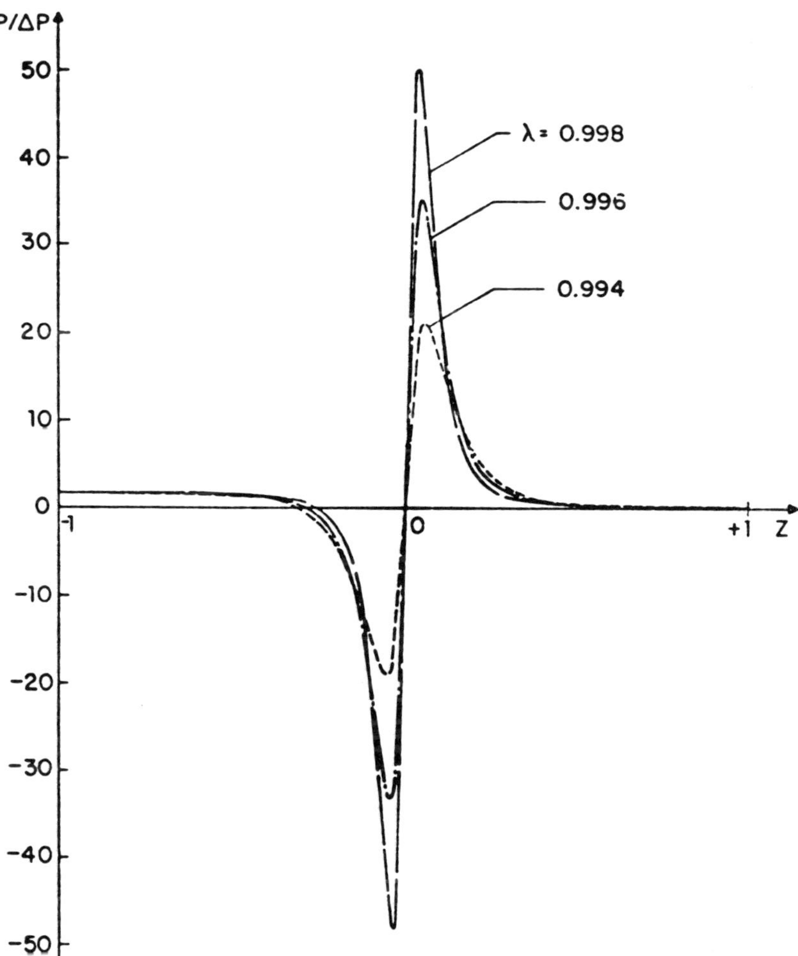

Fig. 4. Pressure distribution along the surface of rigid spherical particles flowing in a rigid circular capillary computed by lubrication theory. The pressure is normalized with $\triangle p$, the pressure drop across the particle. The diameter ratio $\lambda = a/R_O$, where a = radius of the sphere and R_O = capillary radius [Tözeren and Skalak, 1978]

pressures can be many times the applied pressure drop, which is normalized to unity in Figure 4.

The relative apparent viscosity for a line of rigid spheres is shown by the uppermost curve in Figure 5. As the diameter ratio λ_i approaches unity, the apparent viscosity tends to infinity. The apparent viscosities shown in Figure 5 are for a line of spheres that are just touching. For wider spacing of the spheres (ie, for lower hematocrits) the apparent viscosity may be found by a weighted average. The values given in Figure 5 apply for the length occupied by spheres. In the spaces between spheres the relative apparent viscosity is taken to be unity (Poiseuille flow). This is a reasonable approximation for tightly fitting particles, and it becomes increasingly accurate as the particle spacing is increased.

Results for elastic spheres are also shown in Figure 5 for various values of a velocity parameter, A, defined by

$$A = \mu Ua/GR_O{}^2 \tag{3}$$

where

    μ = viscosity of suspending fluid
    U = velocity of the spheres
    a   = initial radius of the spheres
    G = shear modulus of the elastic spheres
    R_O = tube radius

A small value of A corresponds to a very small velocity or a very stiff particle (large value of G). As may be seen in Figure 5, results for small values of A approach the curve for rigid spheres, which is labeled A = 0. As the flexibility of the particles increases, the values of A increase, the particles become more deformed, and the apparent viscosities decrease. For elastic spheres, the initial diameter ratio λ_i can be greater than unity, as shown on the abscissa in Figure 5. For such cases, the elastic deformation will bring the final diameter ratio λ_f below unity, so that the particle can fit into the tube. The final (deformed) diameter ratio, λ_f, is shown as a function of initial diameter ratio λ_i and velocity parameter A in Figure 6.

Some pressure distributions computed for flexible spheres are shown in Figure 7. Comparing Figure 7 to Figure 4 shows that the positive pressure region increases in extent and the negative pressure region decreases. This is the result of the particle deformation, which tends to make the gap more uniform.

RED BLOOD CELL MODELS

Some computations have been carried out for the capillary flow of flexible red blood cell models, starting from a configuration such as shown in Figure 1 [Zarda et al, 1977]. The membrane of the red blood cell is assumed to be elastic in shear and in bending, but its area is assumed to remain constant during any deformation.

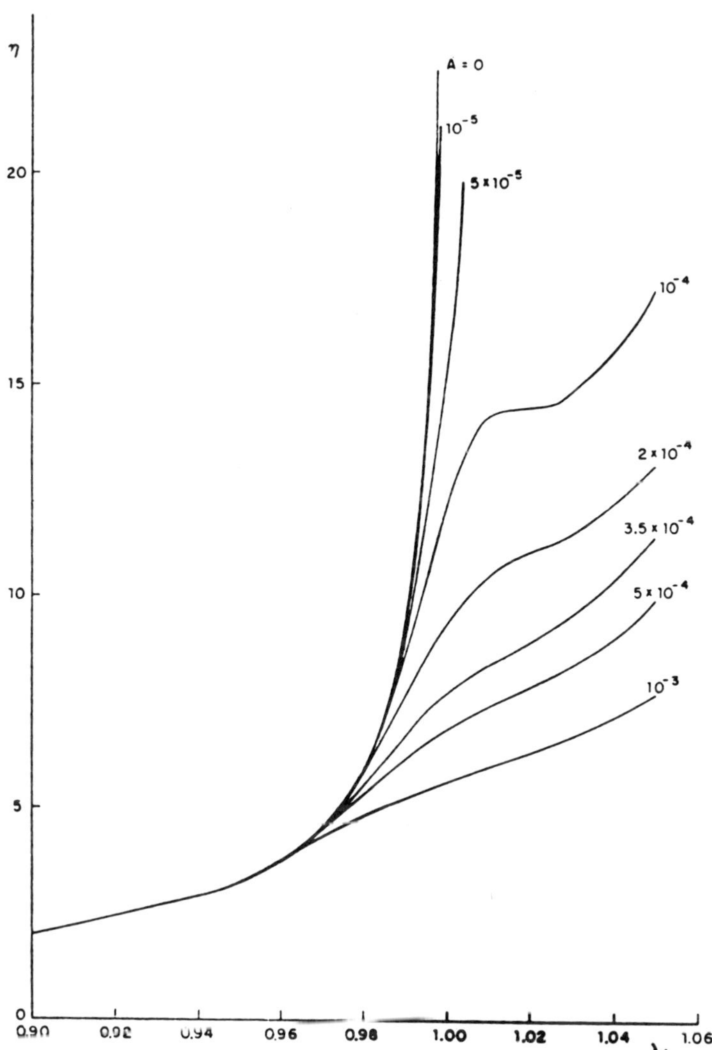

Fig. 5. Relative apparent viscosity, η, for a line of elastic spheres in a circular capillary as a function of the initial diameter ratio $\lambda = a/R_0$ for various values of the velocity parameter $A = \mu Ua/GR_0^2$. The center-to-center distance of the spheres is 2a, so that each sphere touches its two neighbors [Tözeren and Skalak, 1978].

The tensile stress resultants T_1, T_2 (dyne/cm) in principal axis are adopted from Skalak et al [1973] in the form:

$$T_1 = B (\lambda_1^2 - 1) \, \lambda_1/2\lambda_2 + T_O \qquad [4]$$

$$T_2 = B (\lambda_2^2 - 1) \, \lambda_2/2\lambda_1 + T_O \qquad [5]$$

where λ_1 and λ_2 are the principal extension ratios and T_O is an isotropic stress, which is the analog of pressure in an incompressible fluid. The coefficient B is an elastic modulus which is taken to be 0.005 dyne/cm in the computations.

The principal bending moments, M_1, M_2, are assumed to depend on the changes of curvature K_1, K_2 according to the formulas

$$M_1 = D (K_1 + \nu K_2) / \lambda_2 \qquad [6]$$

$$M_2 = D (K_2 + \nu K_1) / \lambda_1 \qquad [7]$$

where D is a bending stiffness coefficient taken equal to 10^{-12} dynes/cm in the computations.

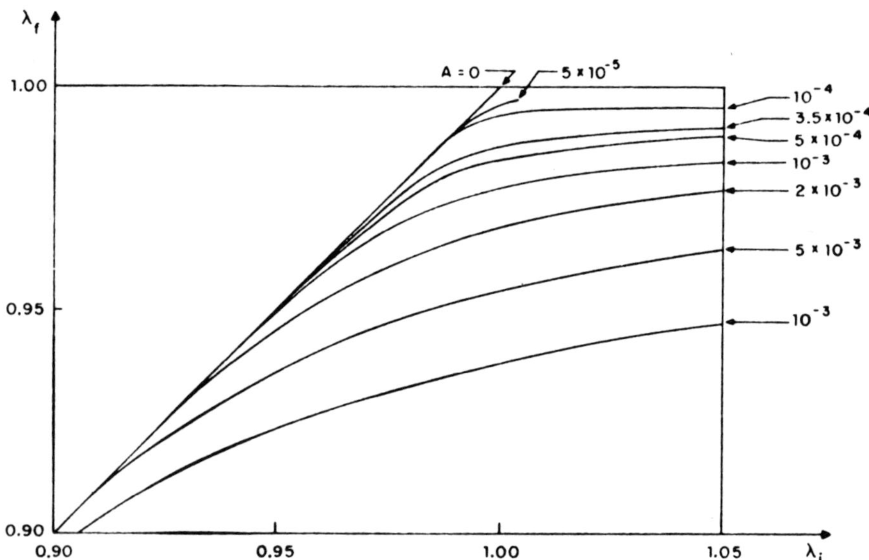

Fig. 6. Final diameter ratio λ_f vs initial diameter ratio λ_i for several values of the velocity parameter $A = \mu Ua/GR_O^2$ for a line of elastic spheres in a circular capillary [Tözeren and Skalak, 1978]. A = 0 corresponds to rigid spheres that lie on the line $\lambda_f = \lambda_i$ terminating at $\lambda_i = \lambda_f = 1$ [Tözeren and Skalak, 1978].

The interior of the red blood cell is assumed to be an incompressible Newtonian fluid with a viscosity of 6.2 cp. This viscosity plays no role in the solutions discussed below, because the shape of the red blood cells is assumed to be constant. The interior of each cell is then at a constant pressure whose value is one of the unknowns of the problem. The exterior fluid is the blood plasma, which is also assumed to be a Newtonian fluid with a viscosity of 1.2 cp. The pressure drop over a typical length of the capillary containing one blood cell is assumed to be given, and the computations seek the shape and velocity of the red blood cells.

The numerical procedures used to compute the results given below are given in more detail by Zarda et al [1977]. A finite element method is used in conjunction with a variational principle. The deformation of the red blood cell and the flow of the external fluid are computed simultaneously in seeking the stationary point of

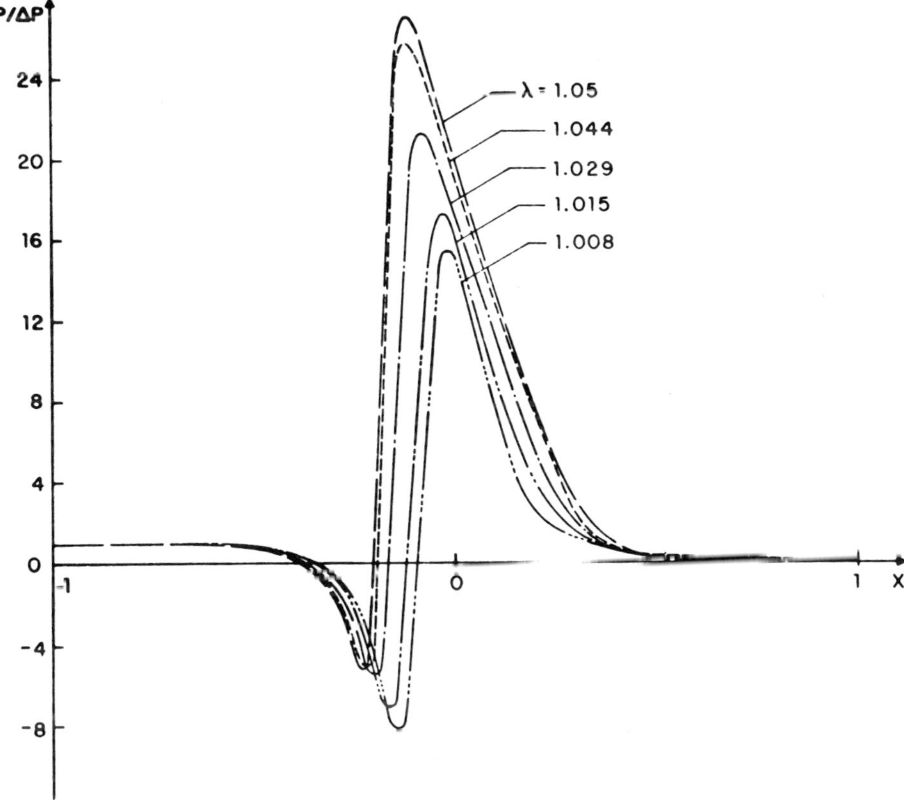

Fig. 7. Pressure distribution along the surface of elastic spheres flowing in a rigid circular capillary for parameter $A = 10^{-4}$. The pressure is normalized with Δp, the pressure drop across the particle, and the abscissa is axial distance normalized with the initial sphere radius, a. Compare to Figure 4 for rigid particles [Tözeren and Skalak, 1978].

the variational function. A Newton-Raphson method is used to arrive at a converged solution. The problem is highly nonlinear because the geometry of the red cell is not known in advance and it undergoes large deformations and large strains. Moreover, the cell membrane elastic properties are nonlinear in the strains, so there is material nonlinearity as well as geometric nonlinearity. The geometric nonlinearity is handled by moving the grid points of the finite element mesh so that nodal points on the membrane are also nodal boundary points of the fluid.

Some computed results are shown in Figures 9, 10, and 11. These results are for a line of cells such as shown in Figure 1, with a hematocrit close to 26% in each case. The diameter of the capillary is assumed to have different values to produce the initial diameter ratios of 0.95, 1.00, and 1.05. The spacing of the particles is varied for each diameter ratio to keep the hematocrit constant. The assumed unstressed shape of the cell in shown in Figure 8.

Figure 9 shows converged shapes computed for red blood cells moving at constant velocity under several different pressure gradients. Only one cell is shown, but the computations are carried out for a hematocrit of 26%. In Figure 9A, the various shapes are drawn with the leading node on the axis coincident for all cases. In Figure 9B, the same shapes are shown with the rear node on the axis coincident for all cases. It happens that the curvature of the rear face of the cell is nearly the same for all the cases computed. No particular significance is attached to this circumstance. The pressure inside the cell is roughly half-way between the upstream and downstream pressures, so that both the upstream and downstream faces are under tension by Laplace's law, like a sphere with an internal pressure greater than the outside pressure. Both faces are also under some bending stress, but this is

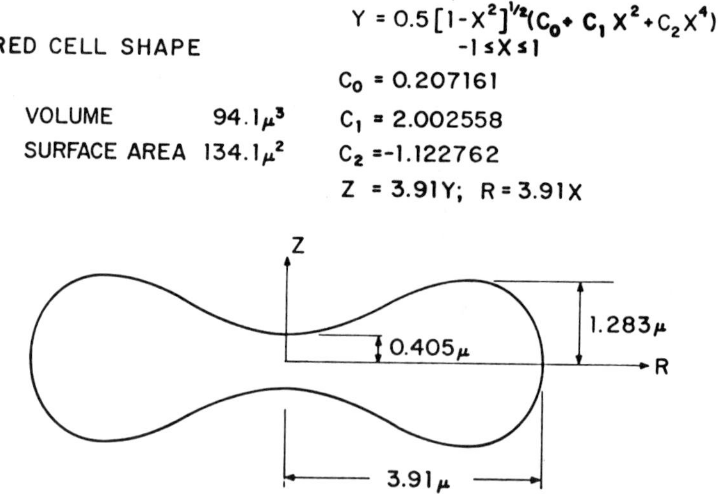

Fig. 8. Unstressed shape assumed for elastic model of human red blood cells. [Zarda et al, 1977].

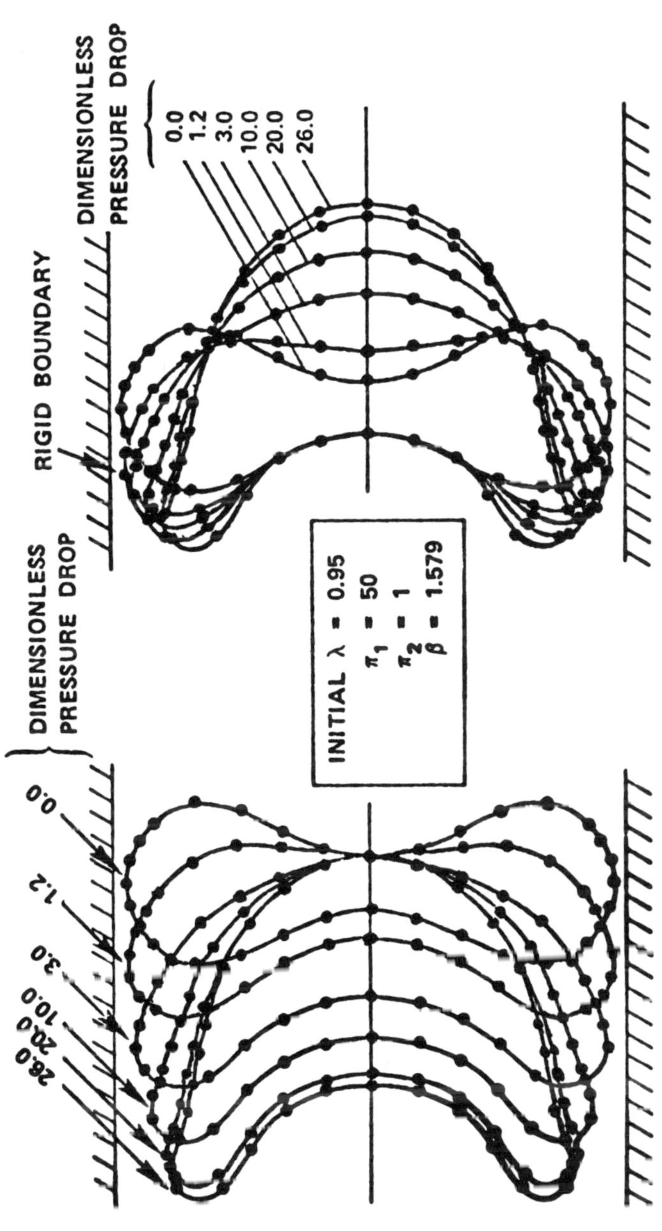

Fig. 9. Computed shapes of elastic red blood cells flowing in circular capillary. All cases shown have the same area and volume. The applied pressure gradient is different for each case as indicated. The two sets of curves are the same cases, drawn with leading nodes or trailing nodes coincident [Zarda et al, 1977].

very small on the rear face and is greater on the front face, which also carries the larger portion of the total pressure drop.

Some observations of interest regarding the strains can be made from the geometry of the results shown in Figure 9. It can be seen from the initial and final node locations (indicated by the black dots) that the radial position of most nodes changes very little. The maximum change is for the outermost rim node, where the change is about 15%. The circumferential stretch ratio, λ_2, is proportional to the radius and the azimuthal stretch ratio, λ_1, is $\lambda_1 = 1/\lambda_2$. Thus, the elastic part of the tensions (the first terms in Equations 4 and 5 will be comparatively small. Most of stress, particularly in the early stages of deformation, will be carried by bending stresses and the isotropic tension, T_O. The shear elasticity will also provide some restoring force. To define the relative roles of bending stiffness D and membrane shear modulus B it would be necessary to vary their ratio and compute a variety of cases. This has not yet been done. However, the bending stiffness is essential to the production of smooth curved shapes such as shown in Figure 9. If there were no bending stiffness, the rear edge of the cell would probably show a sharp cusp. This shape should be approached as the pressure drop is increased to very large values. The computational scheme used was found to be very difficult to converge at high pressure drops, and no such results are available.

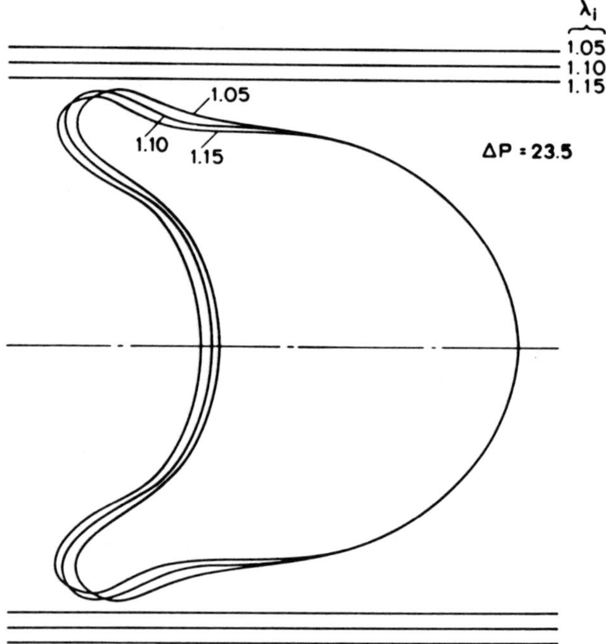

Fig. 10. Computed shapes of elastic red blood cells for initial diameter ratio λ_i = 1.05, 1.10, and 1.15 for the same dimensionless pressure drop ΔP = 23.5.

Figure 10 shows three cases of different diameter ratios under approximately the same pressure drop and hematocrit (26%). In the case of the initial diameter ratio $\lambda_i = 1.05$ the cell has a diameter 5% greater than that of the capillary when the cell is unstressed. The results shown were not computed by starting from the rest position, but the walls of the tube were successively moved inward while maintaining the pressure drop. This corresponds physically to the flow of the red blood cell down a capillary, which has a slight and gradual taper.

The results in Figure 10 show that the shape of the cell does not vary much for the small range of tube diameters shown, but the cell deforms easily under the given pressure gradient to maintain a nearly constant minimum gap. This has significance for the apparent viscosities developed, as shown in Figure 11. As the pressure drop is increased, the cells deform and the apparent viscosities decrease in all cases. But

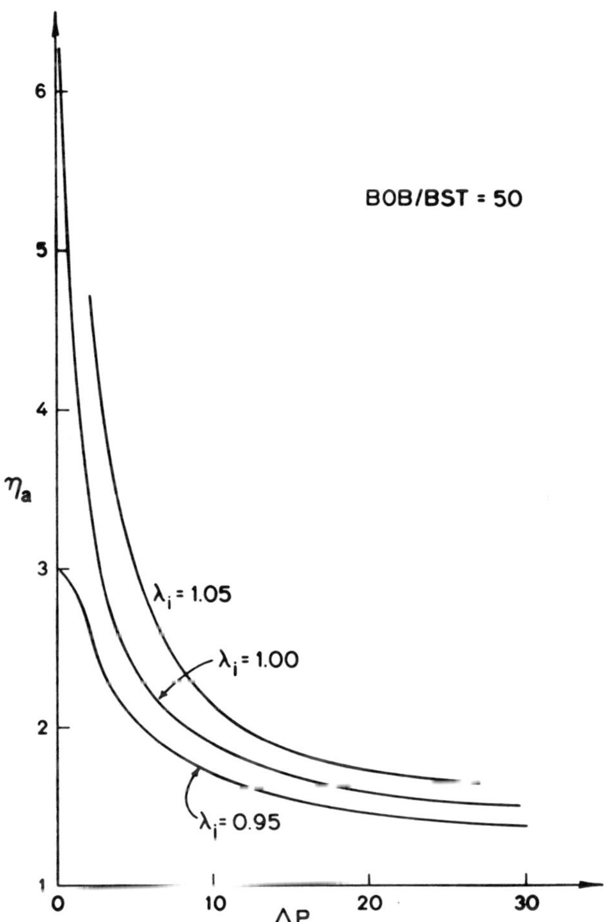

Fig. 11. Relative apparent viscosities computed for elastic red blood cell model as a function of dimensionless pressure drop and various values of initial diameter ratio λ_i.

the differences of η for the various initial diameter ratios are less at higher pressure drops than for the lowest pressure drops, where the velocity is small and the cells are not very much deformed.

The results in Figure 11 are plotted against pressure drop at constant λ_i, which is a different basis than the results for elastic spheres in Figure 5, where η_i is plotted vs λ_i for constant parameter A. However, if the results of Figure 5 for elastic spheres are replotted on the same basis as Figure 11, a qualitatively similar picture emerges. At low velocity (low values of A) there is little elastic deformation, and hence a higher apparent viscosity results. The apparent viscosity decreases with increasing velocity. This may also be qualitatively compared to the variation of the apparent viscosity of blood at normal hematocrit, (in a large-diameter tube or Couette viscometer), which also decreases with increasing shear rates. This is due in part to rouleaux formation and disaggregation [Chien, 1975] but is also due in part to an an elastic deformation mechanism similar to that demonstrated here for capillary flow.

REFERENCES

Batchelor GK, Green JT (1972): The determination of the bulk stress in a suspension of spherical particles to order c^2. J Fluid Mech 56:401–427.

Chien S (1975): Biophysical behavior of red cells in suspensions. In Surgenor DM (ed): "The Red Blood Cell." New York: Academic Press,vol II, pp 1031–1133.

Evans EA (1978): Measurement of RBC deformability: Theoretical analysis (this volume).

Fitz-Gerald JM (1069): Mechanics of red-cell motion through very narrow capillaries. Proc Soc B 174:193–227.

Goldsmith HL (1978): Physical and mathematical models of blood flow: Experimental studies. (this volume).

Hochmuth RM (1978): Measurements of RBC deformability: Experimental studies. (this volume).

Lighthill MJ (1968): Pressure-forcing of tightly-fitting pellets along fluid-filled elastic tubes. J Fluid Mech 34:113–143.

Skalak R, Chen PH, Chien S (1972): Effect of hematocrit and rouleaux on apparent viscosity in capillaries. Biorheology 9:67–82.

Skalak R, Tözeren A, Zarda PR, Chien S (1973): Strain energy function of red blood cell membranes. Biophys J 13:245–264.

Tözeren H, Skalak R (1978): The steady flow of closely fitting incompressible elastic spheres in a tube. J Fluid Mech 87:1–16.

Zarda PR, Chien S, Skalak R (1977): Interaction of viscous incompressible fluid with an elastic body. In Belytschko T, Geers TL (eds): "Computational Methods for Fluid-Structure Interaction Problems." New York: American Society of Mechanical Engineers, pp 65–82.

9

Physical and Mathematical Models of Blood Flow: Experimental Studies

Harry L. Goldsmith and Takeshi Karino

In the human body, blood consisting essentially of a 40–45% suspension of 8.3 μm diameter red corpuscles is subjected to pulsatile flow and circulates through a network of vessels whose diameters range from 25 mm to 5 μm. Obviously, the mechanics of the motion is quite different in the large arteries and veins, where the particulate nature of the suspension and the associated non-Newtonian flow behavior are generally unimportant, from that in the arterioles, capillaries, and venules, where the behavior of the individual corpuscles to a large extent determines the observed flow characteristics of the blood. Experimental studies of models of blood flow have thus ranged all the way from pressure-flow measurements of Newtonian fluids in motion through large straight, rigid or flexible tubes, or through bifurcations, bends, or constrictions, to observations of the flow and deformation of models of red blood cells in tubes of the same, or smaller diameter. Even in the large vessels, however, there are local fluid mechanical events, such as flow disturbances at bifurcations, bends, and obstructions, that appear to be involved in atherosclerotic disease and thrombus formation, and that can only be understood in terms of the motion of the individual corpuscles.

The present paper describes in vitro experimental studies concerned with the movements and interactions of individual blood cells and model particles in suspensions undergoing tube flow at volume concentrations > 20% when particle–particle and particle–wall interactions to a large extent determine the flow characteristics. The aim of the investigations was to relate the microscopic flow behavior of the suspended particles to the macroscopic flow properties of the suspensions and of blood. Two areas of interest to such microrheological studies are discussed below. The first deals with steady, slow flow in tubes in which the

Erythrocyte Mechanics and Blood Flow, pages 165—194
© 1980 Alan R. Liss, Inc., 150 Fifth Avenue, New York, NY 10011

ratio of particle to vessel diameter is of the order of $1:10$ — ie, corresponding to
that existing in the smallest arteries and veins. The tube Reynolds numbers

$$\mathrm{Re_t} = \frac{2\overline{U}R_o\rho}{\eta}$$

$\overline{U}$ being the mean tube linear fluid velocity, R_o the tube radius, ρ and η the
respective suspension density and viscosity, are very small (< 0.1). By contrast,
the second part deals with blood cell flow behavior in vessels many times larger
than the red cells at much higher flow rates corresponding to $\mathrm{Re_t}$ from 10 to 200.
Here, we are particularly concerned with the problem of platelet thrombus formation and wall adhesion at sites such as bifurcations and stenoses, where the flow
is disturbed and separation of streamlines from the wall, with attendant vortex
formation, can occur.

PHYSICAL MODELS OF BLOOD FLOW IN SMALL VESSELS

Suspensions and Techniques

The flow of blood through small arteries has been modeled using suspensions
of rigid spheres [40, 66, 67], rigid discs [40], emulsions of liquid drops [18, 75],
and reconstituted biconcave ghost cells [20–22, 26]. In order to observe the
detailed particle flow behavior in the interior of the suspensions of model particles,
it is necessary to render them transparent to transmitted light by matching the
refractive indices of suspended and suspending phases. A small volume of visible
tracer particles of the same size and shape is then added. In the case of the
already transparent ghost cells, normal red cells, platelets, or white cells from
the same donor are added as tracer particles.

Measurements of the velocity distribution in whole blood have also been made
in small glass tubes using photo-optical methods [5, 6], a laser-Doppler technique
[43], as well as the double-slit photometric device [1, 17]. In addition, it should
be noted that visual observations of red cell flow behavior have been made in a
cone and plate viscometer [62, 63] and the ghost cell technique used in flow
through rectangular channels [2].

In the work with transparent suspensions of model particles and ghost cells
the usefulness of the observations is considerably increased when one is able to
track the visible tracer particles in flow up or down the tubes. In suspensions or
emulsions of macroscopic model particles in tubes from 2 to 10 mm diameter
this is achieved by means of a traveling microscope [21, 28]. In the case of
colloidal size emulsions and ghost cell suspensions, the tubes, having diameters
$< 200\ \mu$m, are themselves moved with a velocity matched to that of the particle
being tracked, with a traveling microtube apparatus [21, 25, 31, 74]. With these

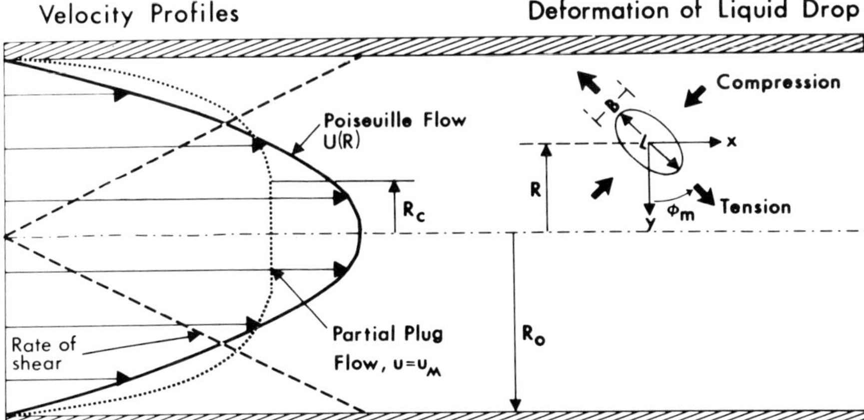

Fig. 1. Left: Flow in the median plane of a circular tube showing the parabolic, Poiseuille velocity profile of a Newtonian liquid, U(R) – solid line – and the blunting of the profile typical of a concentrated suspension at the same volume flow rate with partial plug flow in a core of radius R_C. In Poiseuille flow, the shear rate (dashed line) increases from 0 at the axis to a maximum at the wall. Right: The deformation of a Newtonian fluid drop by fluid stresses (compressive in one quadrant and tensile in the other) into an ellipsoid of length L and breadth B oriented at an angle ϕ_m to the Y-axis of the flow field (from Gauthier, Goldsmith and Mason [18]).

techniques, it is then possible to measure not only the velocity distributions within the tubes but also to follow a particle over distances from 100 to 500 tube diameters while recording its translational and rotational motion, and in the case of emulsions and red cells, the changes in shape due to deformation.

Effects due to High Particle Concentration

On increasing the particle volume concentration in a suspension subjected to steady tube flow, two important effects come to light:

Blunting of the velocity distribution. In very dilute suspensions, provided the particle radius $h \ll R_o$, the tube radius, the particle velocity u(R) in the axial direction at a radial distance R from the tube center (Fig. 1) is equal to that of the fluid, U(R), given by the familiar parabolic distribution:

$$U(R) = \frac{2Q}{\pi R_o^4} (R_o^2 - R^2)$$

$$- U(0) \left[1 - \frac{R^2}{R_o^2} \right]$$

where Q is the volume flow rate and U(O) is the centerline fluid velocity (Fig. 1).
As the concentration of particles is progressively increased and the average
interparticle distance is reduced to the same order of magnitude as the particle
diameter, the velocity distribution becomes blunted from the parabolic.
The velocities u(R) and U(R) continue to be equal, but a core of radius
R_C develops in the tube center (Fig. 1) in which the particle velocities u_M are
maximum and measurably constant, being less than the centerline velocity U(0)
in Poiseuille flow at the same volume flow rate. The onset of this region designated
as partial plug flow [40] occurs at a volume concentration, c, which decreases as
the ratio of particle to tube radius, b/R_o, is increased. Due to blunting of the
velocity profile, the wall shear rate, G_M, is greater than that in Poiseuille flow,
$G(R_o)$, at the same volume flow rate:

$$G(R_o) = \frac{4Q}{\pi R_o{}^3}$$
$$= \frac{2U(0)}{R_o}$$

Displacements of particle paths. Due to collisions between particles traveling
on adjacent streamlines, the particle paths are subject to fluctuations in directions
normal to that of the flow. The amplitude and frequency of these displacements
increases with increasing c as the encounters between particles change from
occasional two-body collisions in very dilute suspensions [29], to continuously
occurring multi-body collisions at $c > 15\%$.

In emulsions and in ghost cell suspensions, a third effect is observed at concentrations above 30%, when due to the combined effects of shear and particle
crowding there is a continual deformation of droplets or red cells in the suspension
outside the core of partial plug flow [18, 21, 26].

Rigid Spheres and Emulsion Droplets: Effects of Deformation

The most striking difference between concentrated suspensions of rigid spheres
or discs and emulsions of deformable liquid drops at low Reynolds numbers is
that the velocity distribution for the rigid particles at a given c and R_o is independent of flow rate, whereas for the emulsions the degree of blunting decreases with
increasing Q until the parabolic profile is obtained. This is illustrated in Figure
2A and 2B, in which the velocity distributions are plotted in dimensionless form:
$u(R)/u_M$ vs R/R_o. The concentrated suspensions of rigid spheres and discs thus
behave as quasi-Newtonian fluids with the measured pressure drop per unit
length of tube directly proportional to Q [40] and the apparent viscosity in a
given tube a constant. The emulsions, like mammalian blood, behave as pseudoplastic fluids, with the apparent viscosity decreasing with increasing Q [49].

Moreover, as illustrated in Figure 3, the degree of blunting at a given ratio of

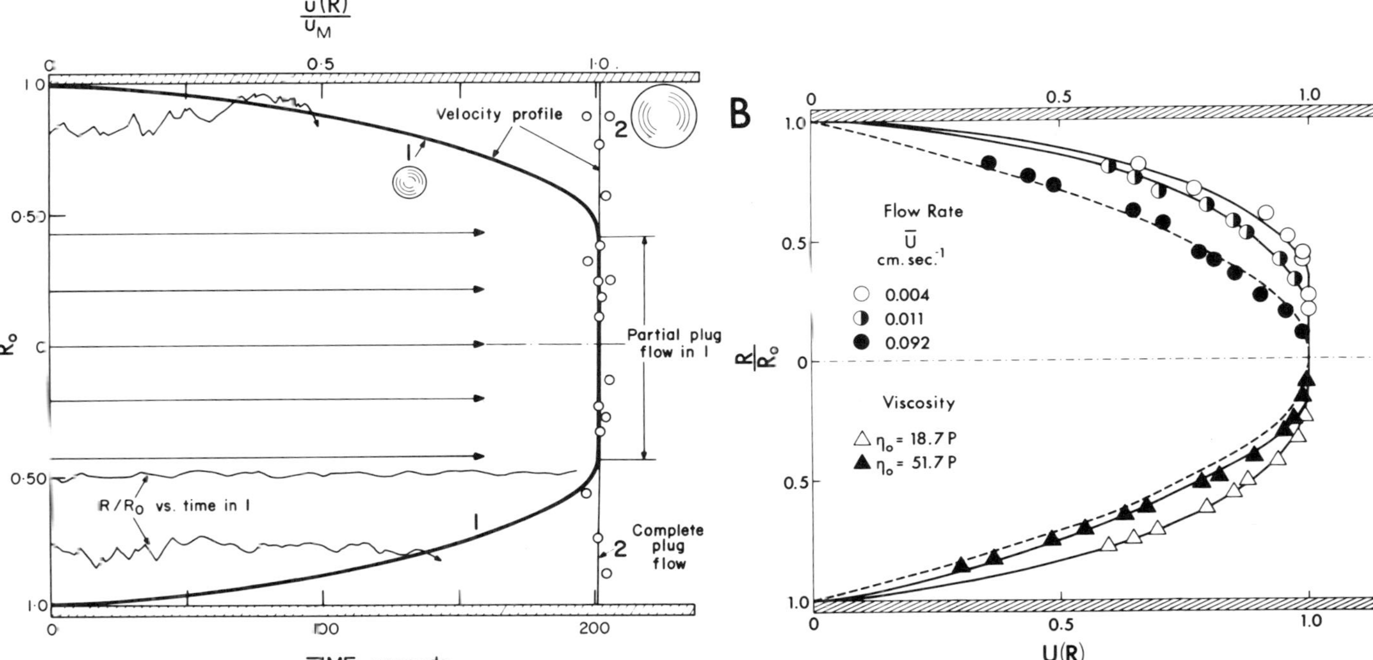

Fig. 2. Dimensionless plot of the velocity distributions in a concentrated suspension of rigid spheres, c = 0.33 (A) and in an emulsion of liquid drops, c = 0 37 (B). A. As the mean ratio of sphere to tube radius is increased from 0.056 in curve 1 to 0.112 in curve 2, the degree of blunting increases from partial to complete plug flow. The lines are the best fit of experimental point shown only for curve 2. Part A also shows the variations in radial position with time for a tracer sphere in the suspension giving the profile 1. The radial displacements decrease in magnitude with decreasing R and disappear in the region of plug flow (from Goldsmith and Mason [29]).
B. Effect of flow rate (upper) and suspending phase viscosity (lower) on the fluid velocity distribution U(R) in the emulsion (the particle, u(R), and fluid, U(R), velocity distributions are known to be equal). As Q and/or η_0 are increased, the profile approaches the parabolic, shown by the dashed line (from Gauthier, Goldsmith and Mason [18]).

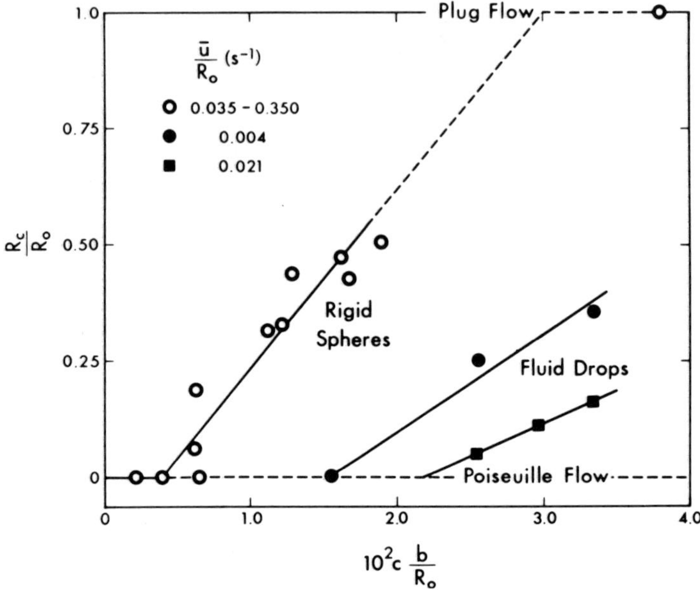

Fig. 3. Plot of the degree of blunting of the velocity distributions, as given by R_c/R_o, against the function b/R_o for rigid spheres [40] and fluid drops in the polyglycol-silicone oil emulsion [18] (from Gauthier, Goldsmith and Mason [18]).

particle to tube radius, b/R_o, and concentration is much greater in a suspension of rigid spheres than in an emulsion of uncharged, macroscopic deformable polyglycol droplets in silicone oil. At a given flow rate, this difference increases as the shear deformation of the droplet increases — eg, by increasing the suspending fluid viscosity (Fig. 2B). It is also of interest to note that a suspension of rigid spheres at $b/R_o = 0.112$ and $c = 38\%$ exhibits a complete plug flow ($R_c/R_o = 1.0$), whereas an emulsion at this concentration and comparable b/R_o shows only partial plug flow.

The non-Newtonian behavior of emulsions may be qualitatively explained by the ability of the drops to squeeze past each other in the particle-crowded suspension. Thus, it is observed that the drops, which in dilute suspensions are deformed into prolate ellipsoids in the flow (Fig. 1), were distorted into irregular shapes at high concentrations [18]. This effect will increase with increasing shear stress $\tilde{\tau} = G(R)\eta$ and concentration. In rigid sphere suspensions the particles must pass around each other and, provided they are uniformly distributed in the tube, at a given b/R_o, c, and Q the energy dissipation is greater. A comparison of particle paths at the same concentration does indeed show that the radial fluctuations of rigid spheres are greater than those of deformable droplets [18].

For the emulsions the mechanics of flow is further complicated by the radial migration of the droplets away from the tube wall. In very dilute suspensions, the particles migrate to the tube axis at a rate that increases with degree of deformation of the drops into ellipsoids [28]. The geometric deformation, in turn, increases with the rate of shear and the suspending medium viscosity η_o as well as with increasing drop size [29]. In concentrated suspensions the inward migration can only proceed a certain distance from the wall, since it is balanced by an outwardly directed dispersive force resulting from the crowding of particles in the core of the tube. This results in the two-phase flow of a peripheral particle-free or depleted layer surrounding a particle-enriched core. The relative diameters of the two phases will depend on particle size and concentration as well as η_o and Q. Such a flow regime itself contributes to the degree of blunting of the velocity profile [18].

In suspensions of rigid spheres and discs inward radial migration from the tube wall is only observed at much higher Reynolds numbers [4, 41, 65] and is then due to inertia of the fluid [4, 9, 10]. Here, the particle Reynolds number

$$Re_p = b \left(\frac{b}{R_o}\right)^2 \bar{U}\rho/\eta_o$$

is used as a criterion for the onset of inertial effects rather than the tube Reynolds number. At $Re_p > 10^{-3}$ migration of rigid spheres becomes appreciable. At these Re_p, experiments in concentrated suspensions of rigid spheres have shown that the mean thickness δ of the particle-free layer is much lower than that found at $Re_p < 10^{-5}$ in the polyglycol oil emulsions at comparable b/R_o and initial concentration in the tubes (Fig. 4).

Work has also been reported on the flow of concentrated emulsions of microscopic, 2 μm diameter, charged amyl acetate droplets in aqueous glycerol through 100 μm diameter glass tubes [75]. Here, as shown in Table I, there is still a flow rate-dependent velocity distribution, but the reduction in R_c with increasing Q is much smaller than in the macroscopic emulsions. This reflects, in part, the very much smaller shear deformation of the tiny droplets in the 0.045 P viscosity glycerol, as compared to that of 2 mm drops of polyglycol oil in the 19 or 52 P viscosity silicone oil. In addition, interactions between the charged amyl acetate spheres resulted in the presence of aggregates, and their progressive break-up with increasing shear rate would also contribute to the smaller dependence of R_c on Q.

The Ghost Cell Suspensions

These suspensions closely resemble those of the parent red cells, as is evident in Figure 5 from the plot of the apparent relative viscosity against c and shear rate, measured in a cone and plate viscometer. The points are scattered about the best-fit curve drawn through the values for the red cells. Usami and Chien [73]

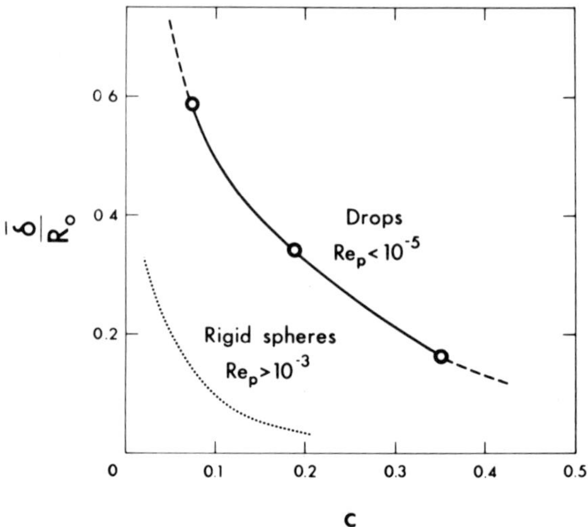

Fig. 4. The width δ of the particle free layer at the tube periphery plotted against the initial suspension concentration for rigid spheres ($b/R_O = 0.075$) and polyglycol drops in silicone oil ($b/R_O = 0.110$). In the case of the rigid spheres, migration of particles away from the tube wall occurred as a result of inertial effects at $Re_p > 10^{-3}$. In the emulsion, migration was due to interactions of the deformed drops with the tube wall [29] (from Gauthier, Goldsmith and Mason [18]).

have also shown a close correspondence in the viscometric properties of red cells and the derived ghost cells prepared by slow osmotic lysis.

The behavior of the tracer red cells in the ghosts parallels many of the observations made in the model suspensions described above. Thus, blunting of the particle velocity distribution from the parabolic is found in 60 μm to 150 μm diameter tubes at $c > 20\%$ (Fig. 6). (It should be noted that in Figure 6, unlike Figure 2B, the dimensionless plots of velocity vs radial distance show the Poiseuille flow distribution (dashed curves) plotted as $U(R)/u_M$, so that the center-line parabolic velocity = $U(0)/u_M > 1$.) As in the emulsions, the radius of the core of partial plug flow decreases with increasing Q and increases with increasing c and b/R_o. The degree of blunting is somewhat lower than that found in the amyl acetate emulsions at comparable c, despite the higher b/R_o (Table I). As with whole blood, it is possible to flow suspensions through 100 μm diameter tubes at cell volume concentrations as high as 90%.

These results appear to reflect the remarkable and continually changing deformation of the tracer red cells (and presumably the ghost cells themselves) that is observed at concentrations $> 30\%$. The distortions of red cells from the biconcave shape are shown at $c = 40\%$ in Figure 7 (upper part) in flow through a 100 μm tube at $u_M = 239$ μm s^{-1}. It is evident that the cells no longer undergo

TABLE I. Comparison of Velocity Profiles in Concentrated Suspensions of Model Particles and Red Cell Ghosts

System	c %	R_0 mm	b/R_0	Q $\text{mm}^3\,\text{s}^{-1}$	$2\bar{U}/R_0$* s^{-1}	R_c/R_0	$u_M/U(0)$
Rigid	34	4.0	0.056	7.1	0.035	0.43	0.73
spheres	34	4.0	0.056	71.0	0.350	0.43	0.73
[40]	38	4.0	0.030	35.6	0.177	0.31	0.78
	41	4.0	0.030	35.6	0.177	0.32	0.77
Large	37	10.0	0.091	12.9	0.004	0.35	0.72
drop	37	10.0	0.091	65.8	0.021	0.16	0.83
emulsion [18]	37	10.0	0.091	290	0.092	0	1.00
Colloidal	50	0.055	0.02^a	2.56×10^{-3}	4.98	0.31	0.81
size	50	0.055	0.02	16.6×10^{-3}	31.0	0.26	0.81
emulsion [75]	70	0.055	0.02	0.55×10^{-3}	1.12	0.45	0.70
	70	0.055	0.02	16.3×10^{-3}	27.8	0.36	0.73
Ghost cell	50	0.075	0.05^b	0.74×10^{-3}	1.10	0.22	0.74
suspensions	50	0.075	0.05	43.1×10^{-3}	64.0	0.15	0.83
[26]	75	0.075	0.05	0.78×10^{-3}	1.16	0.27	0.72
	75	0.075	0.05	24.5×10^{-3}	36.4	0.17	0.82

*Mean tube shear rate in Poiseuille flow at the same volume flow rate $Q = \pi R_0^2 \bar{U}$.
^{a}Approximate mean value.
^{b}Calculated using the mean ghost cell major diameter = 7.5 μm [26].

regular rotations with periodically varying angular velocities as in dilute suspensions [25], but instead remain roughly aligned in the direction of flow for considerable periods of time while continually changing shape. Periodically the cells assume a more rounded, irregular shape and lie across the flow as they attempt to rotate. As c is further increased to 80% (Fig. 7, lower part), the proportion of time cells are aligned with the flow increases, and the major axis of the corpuscles increases in length. It should be noted that such deformation occurs at local shear rates $< 7 \text{ s}^{-1}$ corresponding to shear stresses < 0.3 dyne cm^{-2}, at which the isolated red cell in plasma rotates without deforming [25]. No cell deformation is seen in the region of partial plug flow. The distortion of the red cells from the biconcave shape must therefore be attributed to the combined effects of particle crowding and shear.

Ghost Cell Suspensions: Radial Dispersion

As with the model particle systems undergoing tube flow, the continual interactions of the ghost cells outside the central core of uniform velocity results in radial fluctuations of the paths of tracer particles that leads to their radial dispersion as well as that of the suspending liquid in the tube. Measurements of

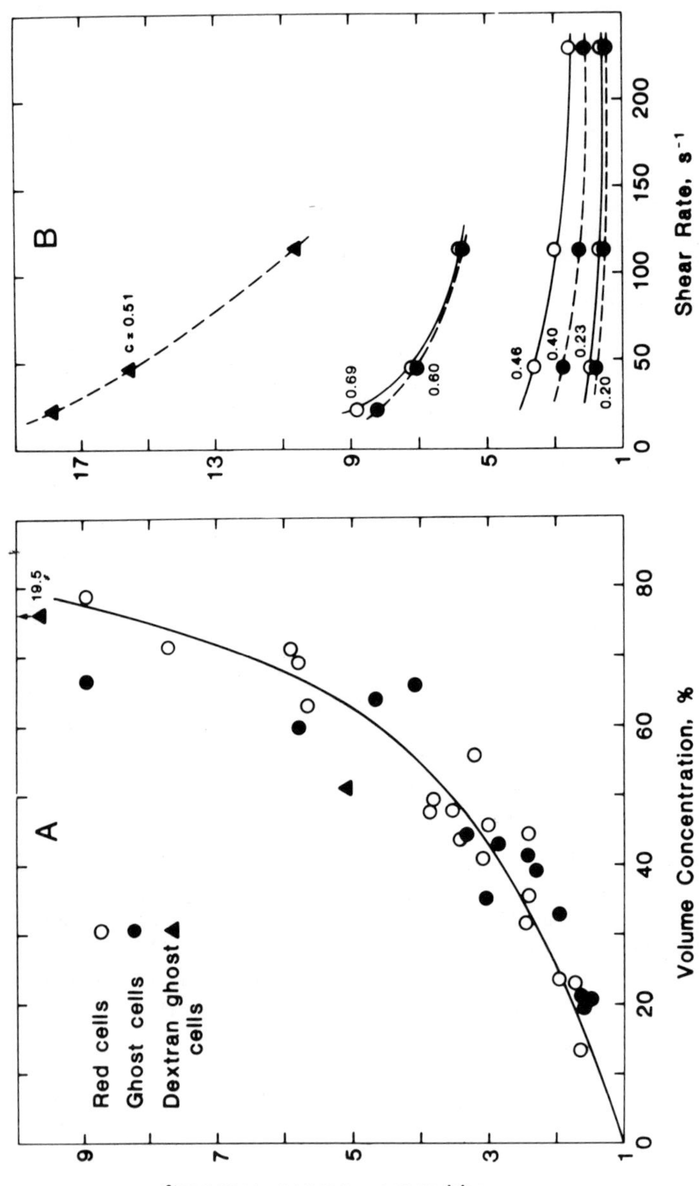

Fig. 5. Comparison of the apparent relative viscosity of red cell ghosts with that of normal red cells as a function of volume concentration at $G = 115$ s^{-1} (left) and as a function of shear rate at various volume concentrations (right). The measurements were carried out in a Brookfield cone and plate microviscometer.

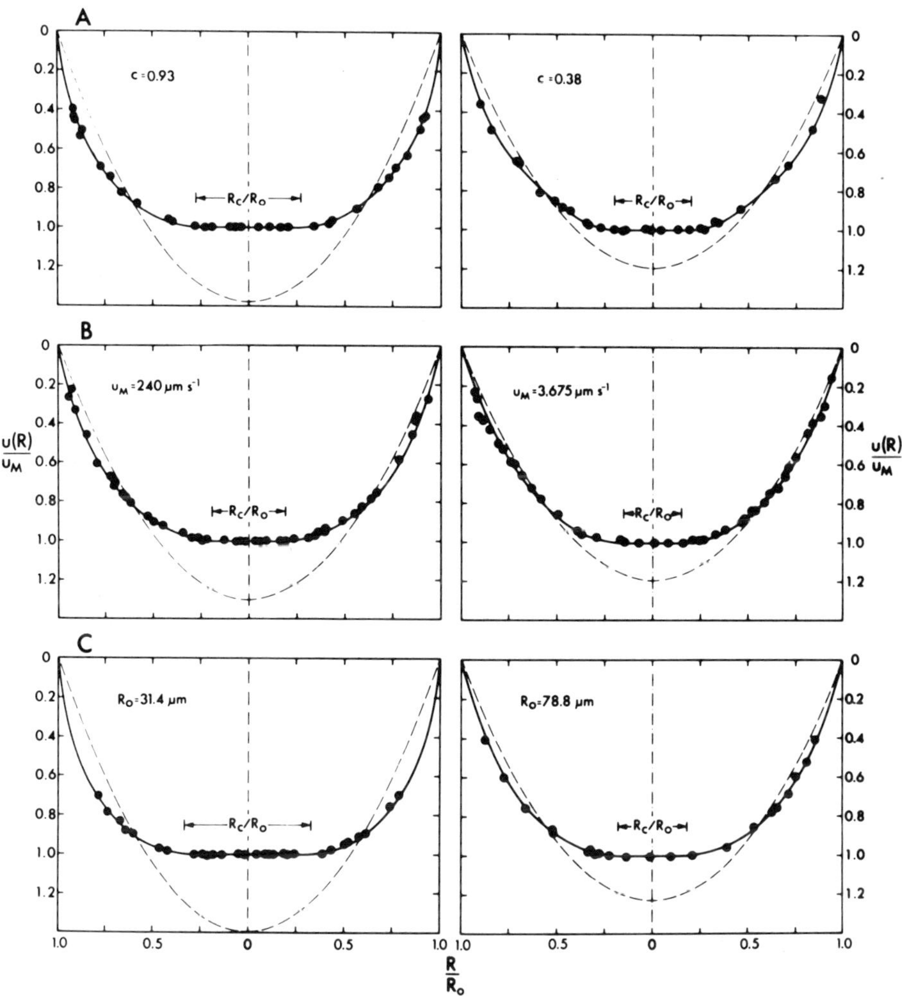

Fig. 6. Dimensionless particle velocity distributions in concentrated ghost cell suspensions showing the effects of volume concentration (A), flow rate (B), and tube radius (C).
A. $R_O = 40\ \mu m$, $Q = 7.5 \times 10^{-3}\ mm^3\ s^{-1}$ (left) and $8.8 \times 10^{-3}\ mm^3\ s^{-1}$ (right).
B: $R_O = 43\ \mu m$, $c = 0.43$, $Q = 0.88 \times 10^3\ mm^3\ s^{-1}$ (left) and $12.8 \times 10^{-3}\ mm^3\ s^{-1}$ (right).
C: $c = 0.60$, $2\bar{U}/R_O$ = mean shear rate in the tube = $11.7\ s^{-1}$ (left) and $9.0\ s^{-1}$ (right).

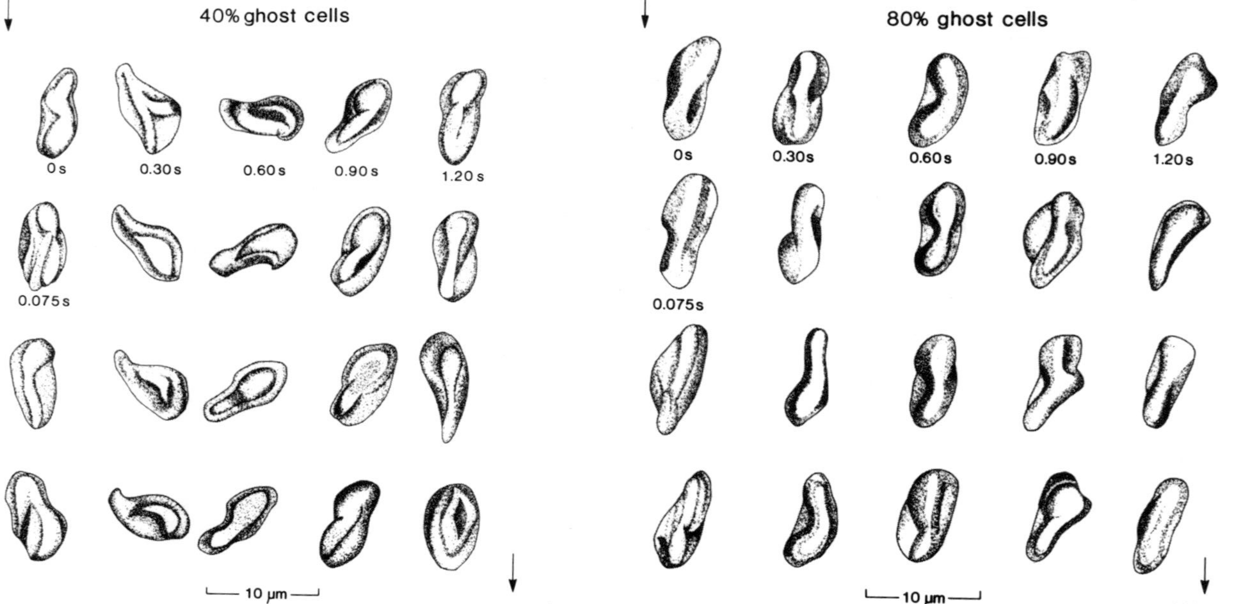

Fig. 7. Deformation of a human tracer red cell in 40% and 80% ghost cell suspensions under going flow in a 100 μm diameter tube. The tracings of the particles were taken from cine films at 0.075-second intervals. The red cells were located at $R \cong 0.7R_0$ and the local shear rate was $\sim 7 \text{ s}^{-1}$.

the paths of tracer red cells, 2 μm latex spheres [21, 22, 26], polymorphonuclear granulocytes (PMNs) [23], and platelets in the ghost cell suspensions, have been carried out over a range of c from 10% to 70%. The magnitude of the particle radial displacements at first increases with c and decreases again at c > 50%, due to particle crowding. At a given c, the displacements are greatest at radial distances between 0.5 R_o and 0.8 R_o. The amplitude of the displacements appears to be related both to particle deformability and size, being somewhat greater for red cells than for the PMN (Fig. 8) and greater for the 2 μm spheres and platelets than for the red cells.

By analogy with Brownian translational diffusion, a radial dispersion coefficient, D_r, obtained from the mean square radial distance $\overline{\Delta R^2}$ traveled by the particle in a time interval Δt

$$D_r = \frac{\overline{\Delta R^2}}{2\Delta t}$$

can be calculated [21, 75]. This does not imply that the motions are considered to be random, as in Brownian diffusion, even though the paths in Figure 8 have the appearance of erratic random traces. The fluctuations in R are due to multi-body collisions determined by the local velocity gradient, cell concentration, and deformation. Similar plots have been obtained in rigid sphere suspensions [40] (Fig. 2A), and there the paths of the tracer spheres were shown to be reversible in time and space; both the translational and rotational coordinates of a single sphere, and hence all the complex configurations of the interacting particles, were conserved [29]. In the amyl acetate emulsions and in the ghost cell suspensions this cannot be the case, since there is Brownian diffusion, sedimentation, and deformation of the suspended particles.

At local shear rates varying from 5 to 20 s^{-1} and c from 20% to 70%, the values of the calculated radial dispersion coefficient ranged between 10^{-8} to 2×10^{-7} cm^2 s^{-1} for red cells and 2 μm microspheres. These values are one to two orders of magnitude greater than the Brownian diffusion coefficients for the blood cells at the experimental temperature of 25°C..

As a consequence of the radial dispersion of the red cells, white cells, and platelets, the frequency of wall encounters of the corpuscles is greatly enhanced. In very dilute suspensions it is likely that inwardly directed forces resulting in radial migration of cells away from the vessel wall, due either to particle deformation and/or inertia of the fluid (depending on the Reynolds number [29, 30]), would prevent the cells from making contact with the wall. In the presence of the ghost cells, as is evident in Figure 8, the particles are propelled both to and from the boundary. Even if there were, on the average, a small concentration of cells at the vessel periphery, because of multibody interactions, some particles would still collide with the wall.

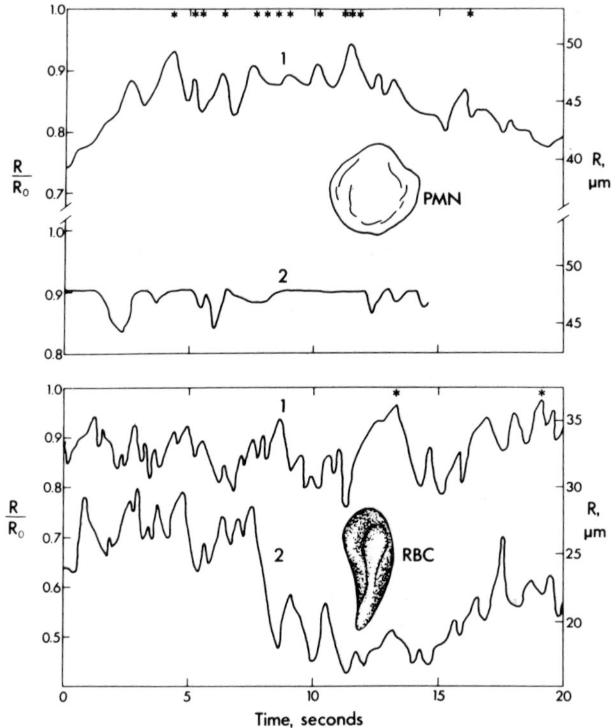

Fig. 8. The radial position (relative to the tube radius R_O) of a tracer human polymorphonuclear leukocyte (PMN) and a red cell (RBC) in a 40% ghost cell suspension, $R_O = 53$ μm (upper) and 38 μm (lower). Particles being tracked along the paths (1) in the periphery of the tube collided with the wall at times indicated with an asterisk. The slightly deformed PMN, and the considerably deformed RBC were drawn from actual cinemicrographs and are typical of the shapes seen during flow (from Goldsmith and Karino [23]).

In the case of platelets, there is much evidence to support the concept of an enhanced diffusion coefficient in blood and an increased cell—wall interaction. An increased platelet diffusivity has been shown by Turitto et al [72], who measured the axial dispersion of dog platelets in 50% blood undergoing laminar tube flow using the classical Taylor technique. Adhesion of platelets to various surfaces and subendothelium has been shown to be markedly enhanced by the introduction of red cells into the suspending medium [5, 12, 33, 70]. This physical role of the red cells has also been held responsible for the adhesion of platelets in the glass bead column test [77].

Ghost Cells as Models of Erythrocytes

There remains the question of whether the ghost cell suspension is in fact a
good physical model for blood. To judge from the viscometric data, it would
appear that the ghost cells are as deformable as the parent red cells and that
some aggregation, akin to rouleaux formation, must be present at low shear rates
(and is indeed often visible under the phase-contrast microscope). Yet the internal
viscosity of the ghost cells (presumed equal to that of the hemolysate fluid) is
approximately five times lower than that of the parent red cells. Some recent
results from this laboratory indicate that ghosts prepared by hemolysis in hypotonic buffered 250,000 mol wt dextran solutions (having a presumed internal
viscosity of 5 centipoise) show no significant differences in the velocity distributions at a given R_o, c, and Q from suspensions of ghosts obtained by hemolysis
in ordinary hypotonic buffer. On the other hand, as shown in Figure 5, the
apparent relative viscosity of the dextran ghosts is appreciably greater at a given
c and shear rate than that of the normal ghosts and parent red cells.

MODELS OF BLOOD FLOW AT STENOSES AND BIFURCATIONS

The discussion that follows is concerned with the effects of flow and fluid
mechanical stress on the genesis and development of thrombosis. It has long been
suspected that mechanical factors play a role in thrombus formation. Thus, regions
of disturbed blood flow at sites where there are rapid changes in the rate or direction of fluid motion, such as bifurcations, sharp bends and stenoses, have been
held responsible for bringing about injury to endothelium and blood cells, resulting in the aggregation and adhesion of the corpuscles to the wall. Platelet thrombi
in vivo have in fact been observed at sites downstream of bifurcations [19, 44,
58], stenoses [52], implanted rings in the vena cava [32], and venous valves
[11, 59]. The deposition of platelet aggregates at such sites has also been detected
in vitro using extracorporeal shunts of collodion [61], plastic [57], and glass [16].
Turbulence in blood flow has also been thought to be a factor in initiating
thrombosis. In vitro experiments using extracorporeal or implanted expansion
flow devices [68, 69] have been cited as evidence for a relation between turbulence
and thrombus formation. Often, however, what is described as turbulent flow is
in fact no more than flow separation — ie, a region where fluid circulates in a vortex
(eddy) situated between the forward-flowing mainstream and the vessel wall. The
role of flow separation and the existence of stagnation points in such regions
have been emphasized by several investigators [13, 14, 22, 24, 34, 35, 60].
Through the development of computational techniques, giving solutions of the
finite difference forms of the partial differential equations describing the fluid
motion, it has now become possible to simulate the steady or pulsatile flow of
blood through channels of almost any geometry. It is not surprising, therefore,
that there has been a considerable effort by researchers in the various engineering

disciplines to compute the flow patterns and distributions of fluid shear rates in liquids flowing through models of various bifurcations and stenoses. However, to render the problem amenable to computer simulation, one has had to ignore the particulate character of the blood, and to regard it as a homogeneous liquid whose flow behavior can be predicted from the equations of motion (Navier-Stokes equation) and continuity. Solutions have been given for flow through a tube having a bell-shaped, two-dimensional axisymmetric constriction or dilatation [46], through two-dimensional bifurcations [15, 36, 64], and through sudden expansions of lumen [50, 51].

The above treatments yield useful information on the distribution of velocity, shear stress, and pressure (and therefore data on the existence, size, and location of vortices), all variables that can affect the function of the blood cells. They cannot, however, give information on the complex series of events that occur at the microscopic level — ie, the involvement of the cells in thrombus formation and wall interactions. In order to deal with these phenomena we have studied the behavior of blood cells in flow through models of bifurcations [48] and T-joints [45, 48], past obstructions [76], and downstream of symmetrical stenoses (sudden expansions of lumen [37]). Some results obtained in the annular vortex formed at the corners of a sudden tubular expansion are given below.

Red Cells and Rigid Spheres in an Annular Vortex

A series of investigations has been carried out on particle behavior in the annular vortex formed at concentric, abrupt tubular expansions of small dimensions in which the motions of human blood cells and latex spheres $< 50 \mu m$ diameter could be viewed through a microscope [23, 38, 39]. Using an expansion of a 150 μm into a 500 μm tube, the paths and orientations of normal and gluteraldehyde-hardened human red cells (HBC) in dilute suspensions within the vortex were determined at Reynolds numbers, Re_t (based on the upstream tube radius), from 0 to 75.

At the lower Re_t, and in the almost circular orbits close to the vortex center, the cells traveled with virtually constant linear velocities, whereas in the outer, more elliptical orbits, the velocities varied considerably, being high in the forward-flowing portion of the vortex, low in the reverse-flow region, and remaining low until forward motion was resumed. The orbits traced out by the HBC over a single revolution shown in Figure 9 at Re = 37.8 were in good agreement with theory [50], which predicts that fluid in the vortex circulates in closed streamlines.

When the HBC were followed over several revolutions, they were seen to migrate spirally outward, finally rejoining the mainstream fluid. Eventually, the vortex became depleted of cells, and in order to study this phenomenon it was necessary to give small pulsed disturbances to the flow, thereby refilling the orbits with cells. This outward migration of particles was observed with normal red cells as well as with HBC, with platelets, and with latex spheres of diameters

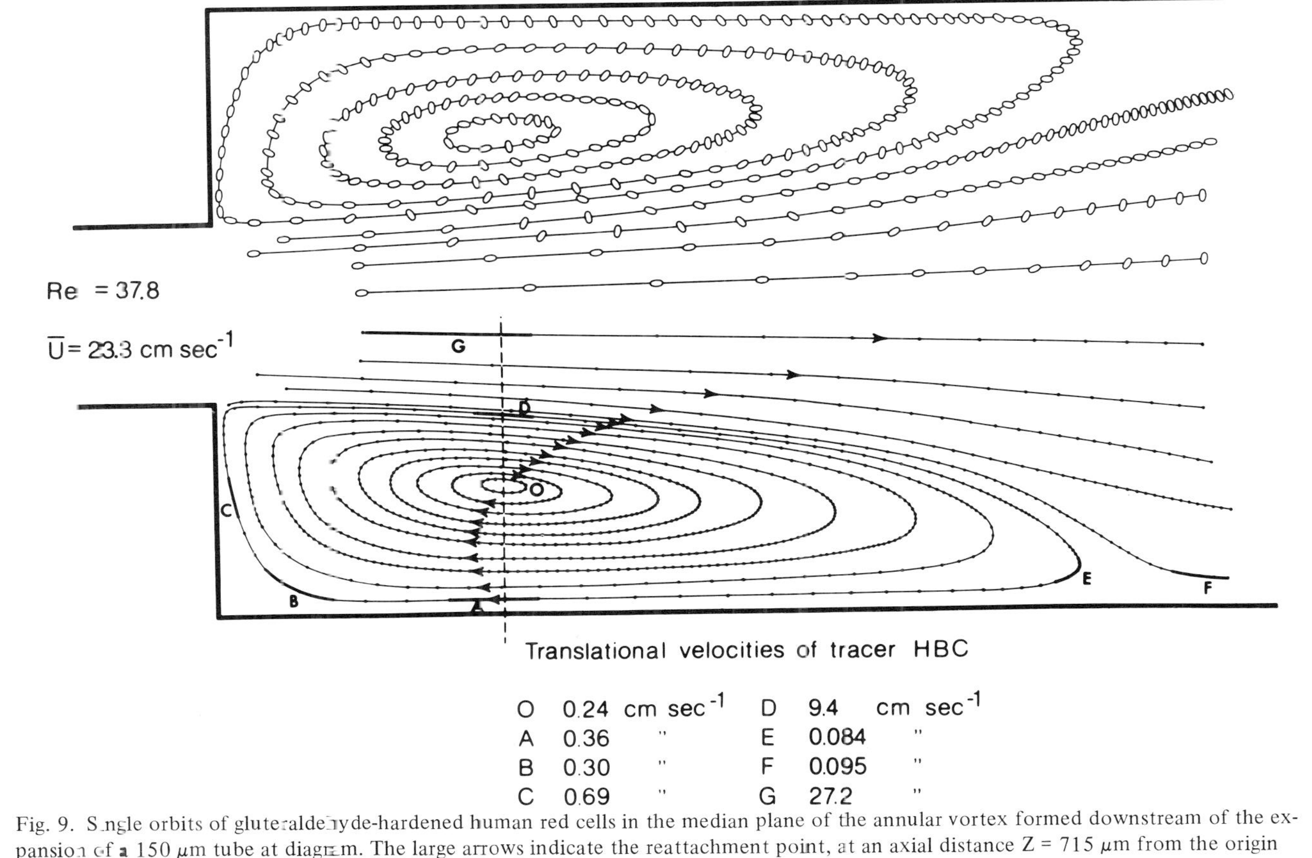

Fig. 9. Single orbits of gluteraldehyde-hardened human red cells in the median plane of the annular vortex formed downstream of the expansion of a 150 μm tube at diagram. The large arrows indicate the reattachment point, at an axial distance Z = 715 μm from the origin of the expansion. The letters A to G and O indicate the positions at which the particle velocities listed were measured. The arrows on the streamlines indicate schematically the velocity distribution along the dashed line PQ. In the reverse flow region it is close to parabolic with the maximum situated at a distance y = 65 μm from the wall (from Goldsmith and Karino [37]).

$< 20 \ \mu$m. There was no difference between the behavior of neutrally buoyant particles and spheres or blood cells lighter or heavier than the suspending fluid. In contrast, the larger latex spheres and aggregates of HBC $> 30 \ \mu$m diameter remained in the vortex at all Re_t and tended to migrate into orbits close to, or at the center of, the vortex. Measurements of the residence times of cells and spheres in the vortex (defined as the interval between the onset of flow with the vortex filled with particles and the last sphere rejoining the mainstream) showed that there existed a critical Re_t (~ 11 for an HBC, but only ~ 5 for a 20 μm latex sphere), above which one or two particles remained indefinitely in the vortex circulating in an equilibrium orbit. As Re_t was further increased, more and more spheres remained within the annular vortex and there were frequent collisions, some of which resulted in the formation of non-separating doublets of spheres, which because of the increase in particle size migrated into the center; others resulted in the elimination of one of the spheres from the vortex.

Thus, at sufficiently low Reynolds numbers, the vortex in steady flow acted as a region in which larger particles and aggregates could accumulate, while the smaller particles eventually rejoined the mainstream of the suspension. The same particle-size dependent migration (aggregates remaining in the vortex) was observed in pulsatile flow at frequencies of ½ to 3 cycles/second. The cells here described spiral orbits of continually changing diameter, and the vortex moved in phase with the upstream fluid velocity [37].

Red cells in plasma at hematocrits from 15% to 45% were also observed to migrate out of the vortex in steady flow at low Reynolds numbers, sometimes resulting in a completely cell-free annular region (Fig. 10), at other times leaving behind a few large rouleaux circulating in equilibrium orbits near the center. At higher Re_t an equilibrium red cell concentration was established in the vortex, as cells both entered and left the annular region.

To explain the particle size and Reynolds number dependent migration of particles in the vortex it was necessary to postulate the existence of both inwardly and outwardly directed forces. Since the observed effects were certainly not due to centrifugal forces, an explanation in terms of particle interactions with the tube wall and the effects of fluid inertia was found [37].

Lateral particle migration due to inertia has been described in dilute suspensions of rigid spheres and cylinders [4, 29, 65] and of human red cells [21, 30] undergoing Poiseuille flow. In neutrally buoyant systems the particles are observed to migrate inward from the wall and outward from the axis to an eccentric equilibrium position at $R \sim 0.6 \ R_o$. A theory by Cox and Hsu [10], which gives analytic solutions for the migration velocity of a rigid sphere in a flow field near a single vertical plane wall, was qualitatively applied to particle behavior in that section of the vortex in which flow is virtually parallel to the tube wall. In the reverse flow region the theory predicts that, depending on their initial distance, y, from the wall, spheres of radius, b, migrate either inward or outward toward

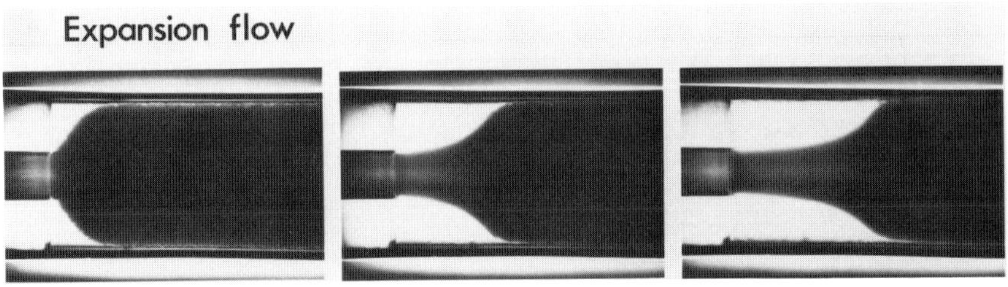

Expansion flow

Fig. 10. Photomicrographs of a 20% suspension of red cells in heparinized plasma subjected to steady flow through the 150 μm tube into 500 μm tubular expansion. The Reynolds number increases from left to right, the values being 1.5, 9.0, and 13.5. The pictures show the existence of a virtually cell-free zone corresponding to the region of the vortex (from Karino and Goldsmith [37]).

an equilibrium position about ¼ of the distance from the wall to the vortex center. The rates of migration are proportional to the particle Reynolds number and $(b/y)^2$. In the forward-flowing portion of the vortex, and theory predicts only the outward migration. It was shown that the existence and observed location of equilibrium orbits is qualitatively in accord with these predictions. However, the exact balance between inward and outward migration rates could not be forecast.

There is another mechanism that may be expected to result in the inward displacement of the circulating particles within the vortex and that would operate at the ends of the vortex where flow is in the radial direction. Here, because of the close proximity of the vessel walls, a sphere traveling in an outer orbit is deflected from the fluid streamline to follow a path nearer the vortex center.

Platelets in an Annular Vortex

As observed with the latex spheres, while circulating in the orbits of the vortex, platelets formed aggregates that migrated to the center and remained there. These observations led to a study of platelet aggregation in the vortex [38] using platelet-rich plasma (PRP) or washed platelets suspended in Tyrodes-albumin solution containing apyrase [57].

The results showed that in a given suspension the formation and growth of platelet thrombi occurred only in a narrow range of Re_t lying between 1.5 and 17, depending on the suspending medium. Furthermore, there were large differences in the degree and rate of aggregation in the various suspensions used. The most rapid and extensive aggregation was found in heparinized PRP, where microscopic examination of the suspension, even before it was subjected to flow, revealed the presence of a small number of loosely bound thrombi consisting of two to six cells. Some stages in the formation of aggregates are shown in Figure 11.

After filling the vortex with suspension, collisions between platelets circulating on adjacent orbits resulted in the formation of tiny aggregates of two to five cells.

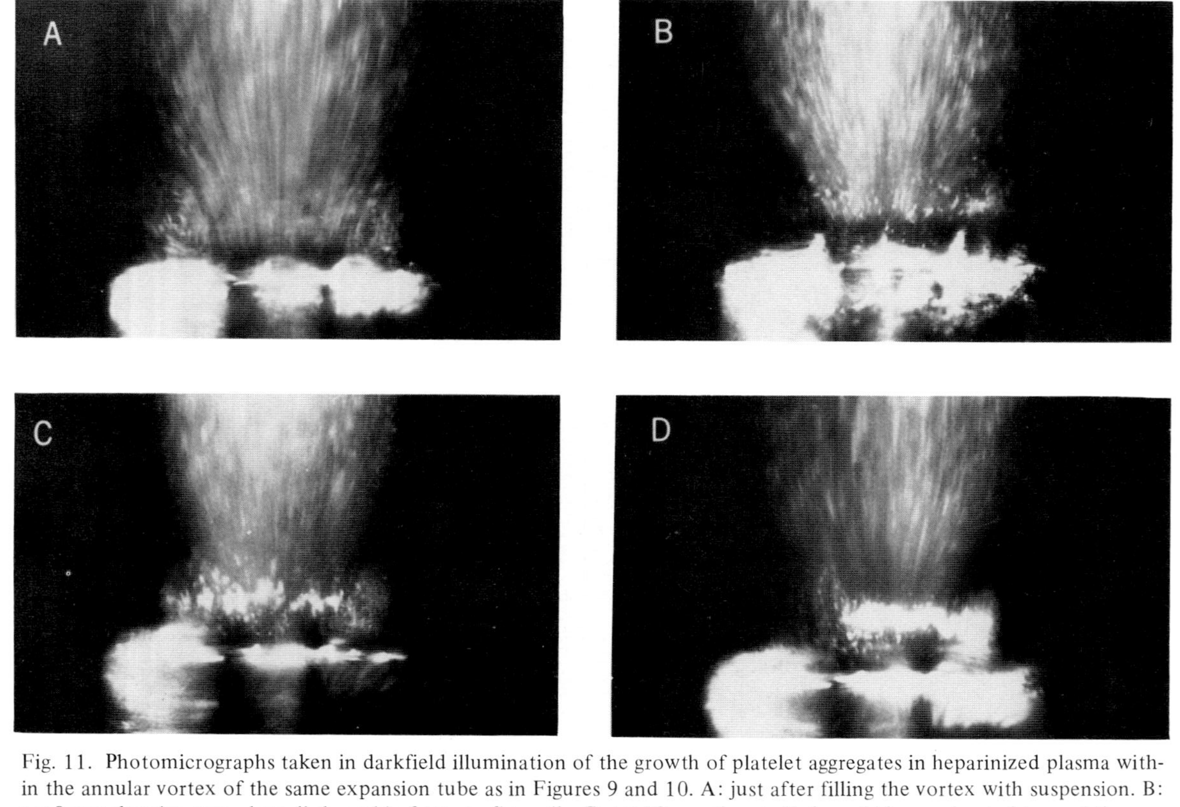

Fig. 11. Photomicrographs taken in darkfield illumination of the growth of platelet aggregates in heparinized plasma within the annular vortex of the same expansion tube as in Figures 9 and 10. A: just after filling the vortex with suspension. B: t = 8 sec; showing several small thrombi of two to five cells. C: t = 18 sec; the small ghrombi have migrated toward the vortex center and have grown to produce several large aggregates. D: t = 35 sec; finally one large aggregate has formed, which stretches more than half-way around the annululs along the vortex center line (from Karino and Goldsmith [38]).

Unlike the single platelets not involved in the formation of thrombi, which migrated outward and eventually rejoined the mainstream, the aggregates moved inward toward the vortex center. There they grew in size as they collided with other aggregates arriving from the periphery. Eventually, one aggregate, or sometimes several large aggregates from 100 to 600 μm long and 30 to 50 μm wide, were formed within 10 to 30 s, and stretched around the annulus along the vortex center line. Such aggregates consisted of a large number of smaller platelet thrombi loosely held together in the flow, and disintegrated into their constituent parts if the flow was disturbed, or if the flow rate increased or decreased beyond the narrow range of the Reynolds number within which aggregation occurred.

In citrated PRP the degree of aggregation was much reduced, except if the suspension was activated with ADP at a concentration below the threshold required to produce a reaction in a Payton aggregometer, when the large particles seen in the heparinized PRP were again formed.

The degree of aggregation observed in the washed platelet suspensions was even less than that in the PRP, although the addition of fibrinogen to the Tyrodes-albumin increased the number of 10–20 μm diameter aggregates formed in the vortex. Here, the addition of subthreshold concentrations of thrombin [27] again resulted in the formation of one very large aggregate.

Recent work by Mortin and associates [53, 54], using dog's blood in a stagnation-point flow chamber surrounded by a forward-facing step, has also demonstrated the formation and growth of platelet aggregates in the vortex, here situated at the corner of the step. At the same time, it was observed that many red cells migrated out.

Wall Adhesion of Platelets

When the effect of red cells on platelet aggregation in heparinized and citrated whole blood was studied, it was found that platelet thrombi containing red cells were observed to form and grow at low Re and were much larger in heparinized blood than in citrated blood. In both suspensions, some of the thrombi adhered to the tube wall — a phenomenon never seen in PRP and whose occurrence led to an investigation of platelet adhesion in the vortex. The initial adhesion of platelets to the tube walls in and downstream of the vortex was studied under well-defined conditions [39]. The experiments were carried out in an expansion of an 0.92 mm tube opening into a 3.00 mm diameter tube, in which there was no appreciable migration of red cells out of the vortex in the time of a run (one or three minutes' duration). The tube walls were coated with collagen fibers [8], and the adhering platelets were fixed and counted under a high-power microscope. Suspensions of washed human platelets in Tyrodes-albumin containing apyrase (to prevent aggregation), to which washed red cells were added, were used.

The experiments demonstrated convincingly that, both in steady and pulsatile flow, a localization of platelet adhesion occurred within the vortex and to a lesser extent just downstream of the reattachment point. Close to the reattachment point, the adhesion was at a minimum. This is illustrated in Figure 12 for a suspension of 5×10^5 platelets/μl in blood of 20% hematocrit. There was also a pronounced increase in the number density of adhering platelets (number of platelets per unit area of wall) as the hematocrit was increased, as shown in Figure 13. Except at the highest Re, the mean number density in the vortex was greater than that downstream, where the velocity distribution was fully developed.

It was surprising, however, that as Re_t increased the adhesion peak in the vortex decreased and flattened out considerably, the mean number density decreasing also (Fig. 12). At the same time the downstream adhesion number density remained almost constant as the wall shear rate quadrupled. The effect of superimposing an oscillatory flow on the steady flow was to level off the adhesion minimum and the secondary peak.

The above results are inconsistent with a diffusion controlled platelet adhesion, which should show an increase in adhesion number density with increasing shear rate [47, 71]. In that case, the rate of adhesion in the vortex is governed by the velocity of the fluid carrying the platelets along the wall — ie, a wall shear rate-dependent, diffusion-controlled adhesion. The fluid velocity and the shear rate do, in fact, exhibit a peak in the reverse-flow region of the vortex close to the wall, but this peak, which flattens out with increasing Re, grows larger at the same time [39]. Rather, the explanation for the observed localization of the adhesion appears to lie in a consideration of the flow patterns and the fact that only those platelets that are carried by the curved streamlines to within one particle radius of the surface will interact (collide) with the collagen fibers, and this applies on both sides of the reattachment point. The flux of platelets entering the adhesion region increases in going from the reattachment point toward the origin of the expansion. It then decreases again as the radial component of velocity of the streamlines changes sign and the cells are carried away from the wall. The exact position of the adhesion maximum will depend on the collision efficiency (the proportion of collisions resulting in adhesion). The effect will be most marked at the lower Re, where the streamlines are more curved throughout the vortex. Downstream of the reattachment point a similar argument applies, except that here the streamlines eventually become parallel to the wall, so that the adhesion number density is expected to decrease to an asymptotic value, as experimentally found (Figs. 12 and 13).

On the basis of the above considerations, one would predict relatively higher platelet adhesion, not only at sites where there is flow separation (adhesion peaks on either side of the reattachment point) such as downstream of mural thrombi and stenoses, but also in regions of disturbed flow, at bifurcations, T-joints, and curved segments. In the case of bifurcations, a secondary crossflow brings blood

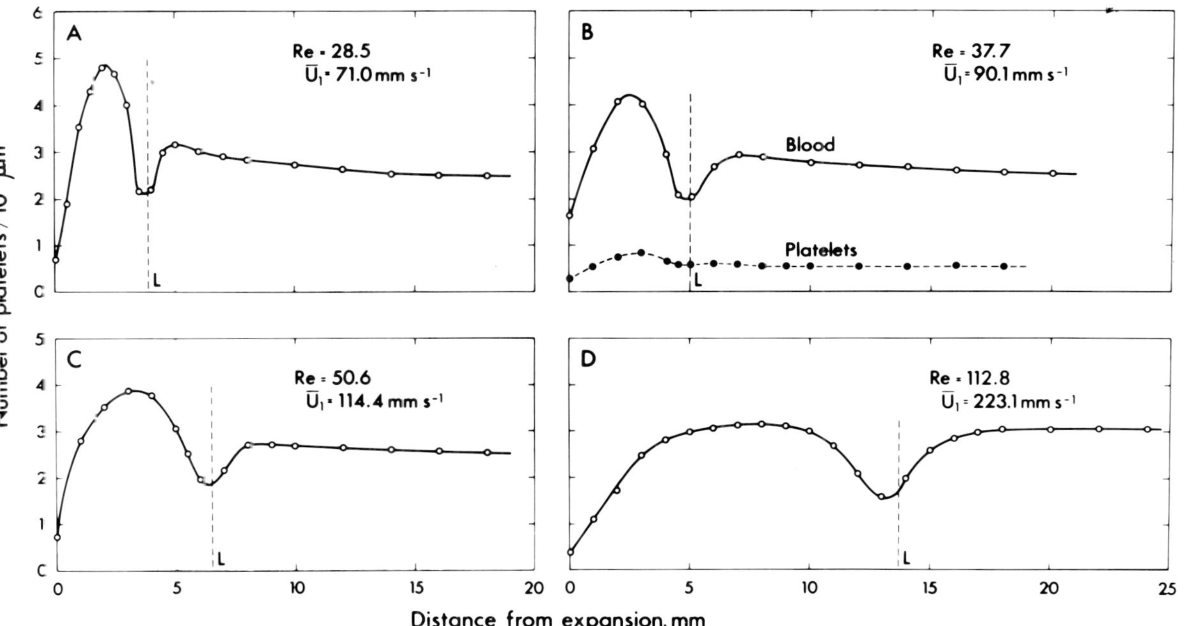

Fig. 2. Measured number density of adhering platelets as a function of axial distance Z from the origin of an expansion of an 0.92 mm into a 3.00 mm tube. The results were obtained with washed platelets in Tyrodes-albumin solution (5×10^5 cells/μℓ) in blood of 20% hematocrit. The dashed line shows the position of the reattachment point Z = L. Platelet adhesion is clearly greater in the annular vortex than downstream, except at Re = 112.8 The curve labeled "Platelets" was obtained at 0% hematocrit (from Karino and Goldsmith [39]).

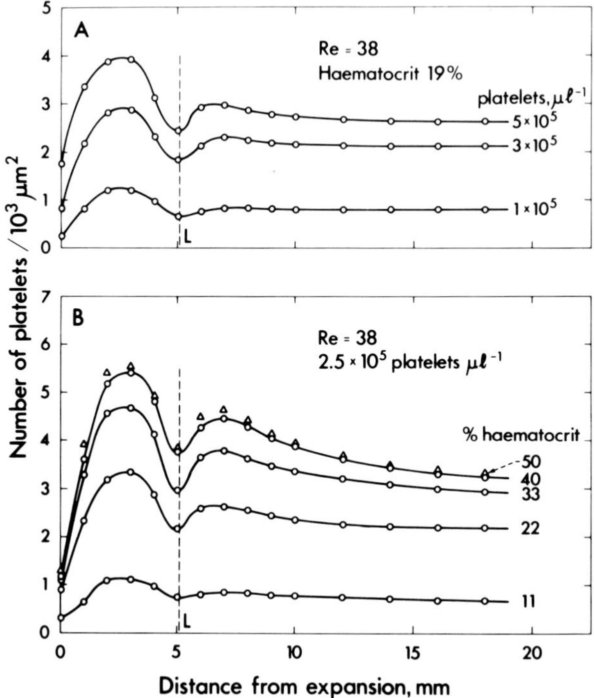

Fig. 13. Plot as in Figure 12 in the same tubular expansion showing the effect of platelet concentration on adhesion in reconstituted blood at 19% hematocrit (A), and of hematocrit on adhesion at a platelet concentration of 2.5×10^5 $\mu\ell^{-1}$ (B) (from Karino and Goldsmith [39]).

cells close to the walls of the daughter vessels, along streamlines having pronounced radial velocity components [3, 48]. There is also radial flow toward the wall on either side of the apex of the bifurcation, where there is a stagnation point.

Work is underway on mapping out the streamlines in flow through T- and Y-bifurcations in which one of the daughter tubes is partially occluded [45]. At a given parent tube Reynolds number Re_o, a vortex develops at the entry of the occluded branch (2) when the ratio of flow rates $Q_1 : Q_2$ in the daughter tubes exceeds a certain value [42, 48]. In the example shown in Figure 14, a large vortex forms at the wall near the entry of partially occluded daughter branch 2, and there is also a small vortex at the wall close to the entry of the side daughter branch 1, which is fed by particles coming from the larger vortex. It is proposed to measure platelet adhesion, both as a function of axial and circumferential position on the tube wall, in order to test the above hypothesis relating adhesion

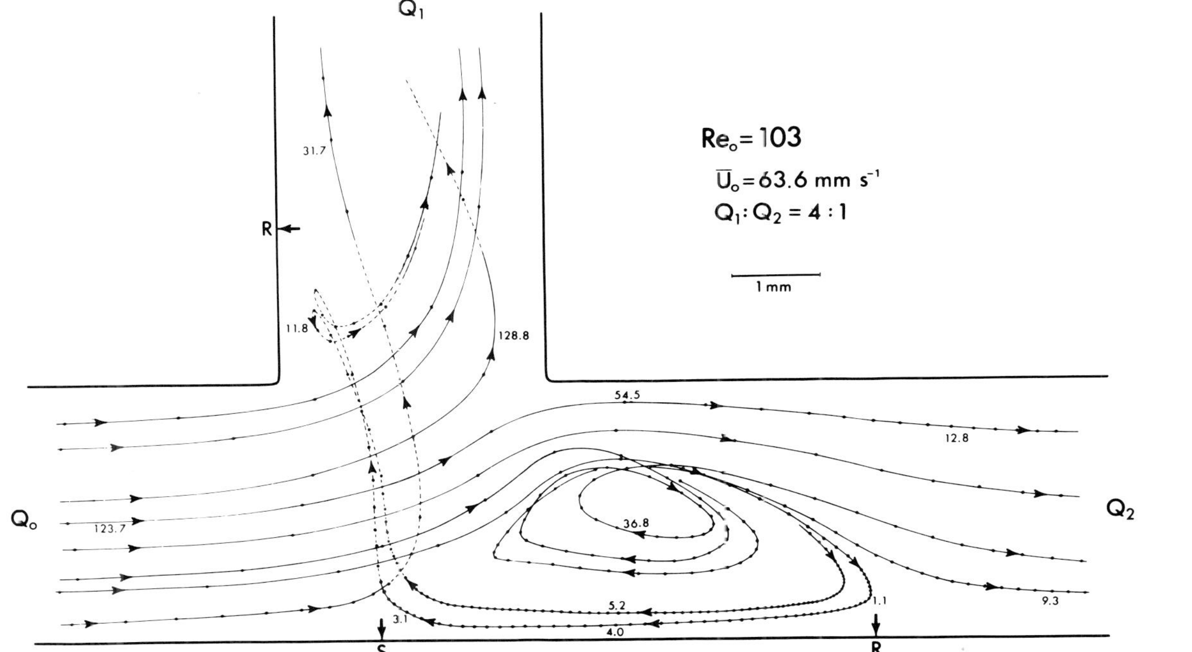

Fig. 14. Flow patterns in the common median plant of a T-bifurcation as indicated by the paths of neutrally buyoyant at 50 μm polystyrene spheres in aqueous glycerol. The suspension enters at the left at a mean velocity U_0, and the daughter branch, 2, is partially occluded so that 80% of the flow leaves through the side branch, 1. The bifurcation was constructed by fitting and glueing together two pieces of 3 mm diameter glass tubing, thereby ensuring a very low radius of curvature at the junction. The arrows at S and R indicate the respective separation and reattachment points of the flow. The points are experimental, showing particle positions at intervals of 22 milliseconds, except in the outermost orbit of the main vortex where the intervals are 44 milliseconds; the numbers indicate the linear velocities in mm s^{-1}. The solid lines represent particles traveling close to the median plane, the dashed lines those that are closer to the tube walls. It should be noted that all particles entering the smaller vortex at the entry of the side branch originate from the larger vortex. There is a pronounced secondary flow in the side branch (from Kwong, Karino and Goldsmith [45]).

peaks to curvature of streamlines near the tube wall. As found in the annular
vortex [39], Müller-Mohnssen [55], using a T-branch, has observed that in
whole blood red cells adhere preferentially near the reattachment point of the
flow where there is a stagnation region, in contrast to the platelets that are found
to adhere upstream of this region.

CONCLUSION

We have dealt with two quite different areas of modeling blood flow, and in
both cases emphasis was laid on the flow behavior of the corpuscles. In the small
tubes this was because, for the red cells, their interactions with each other and
the wall to a large extent determine the mechanics of the motion. In the large
tubes and at high flow rates, it was because the blood cells are involved in throm-
boembolic events that appear to occur preferentially at sites of disturbed flow and
flow separation. The techniques and approach of microrheology were therefore
emphasized, and the knowledge obtained from theory and experiment with suspensions of model particles [29] was applied to blood.

These are in vitro studies, and the physical models of blood flow have, of
necessity, been considerably simplified from that of the in vivo circulation.
Meaningful results, however, can only be expected when the biological system is
modeled step by step, in order of increasing complexity. This is the approach of
microrheology, which has proved to be so useful in helping to understand the
overall flow properties of a variety of materials from a detailed study of their
constituent elements.

ACKNOWLEDGMENTS

This work was supported by grant MT 1835 from the Medical Research Council
of Canada and a grant from the Quebec Heart Foundation. Harry Goldsmith is
Research Associate of the Medical Research Council of Canada and Takeshi Karino
a Research Fellow of the Canadian Heart Foundation. The authors gratefully
acknowledge the technical assistance of Jean Marlow and Diane Chajczyk and
thank Pamela Lilley for typing the manuscript.

REFERENCES

1. Baker M, Wayland H: On-line volume flow rate and velocity profile measurement for
   blood in microvessels. Microvasc Res 7:131–143, 1974.
2. Blackshear PL, Forstrom RJ, Dorman FD, Voss GO: Effect of flow on cells near walls.
   Fed Proc 30:1600–1609, 1971.
3. Brech R. Bellhouse BJ: Flow in branching vessels. Cardiovasc Res 7:593–600, 1973.
4. Brenner H: Hydrodynamic resistance of particles at small Reynolds numbers. In Drew TB,
   Hoopes JW, Vermeulen T (eds): "Advances in Chemical Engineering," vol 6. New York:
   Academic Press, 1966, pp 287–438.

5. Brash JL, Brophy JM, Feuerstein IA: Adhesion of platelets to artificial surfaces: Effect of red cells. J Biomed Mater Res 10:429–443, 1976.

6. Bugliarello G, Kapur C, Hsiao GCC: The profile viscosity and other characteristics of blood flow in a non-uniform shear field. In Copley AL (ed): "Proceedings of the Fourth International Congress in Rheology," Part 4. New York: Wiley, 1964, pp 351–370.

7. Bugliarello G, Sevilla J: Velocity distribution and other characteristics of steady and pulsatile flow in fine glass tubes. Biorheology 7:85–107, 1971.

8. Cazenave J-P, Packham MA, Mustard JF: Adherence of platelets to a collagen-coated surface: Development of a quantitative method. J Lab Clin Med 82:978–990, 1973.

9. Cox RG, Brenner H: The lateral migration of solid particles in Poiseuille flow. I. Theory Chem Eng Sci 23:147–173, 1968.

10. Cox RG, Hsu SK: The lateral migration of solid particles in a laminar flow near a plane wall. Int J Multiphase Flow 3:201–222, 1977.

11. Diener L, Ericsson LE, Lund F: The role of venous valve pockets in thrombogenesis. A postmortem study in a geriatric unit. In Shimamoto J, Numano N (eds): "Atherogenesis." Amsterdam: Excerpta Medica Foundation, 1969, pp 125–131.

12. Feuerstein IA, Brophy JM, Brash JL: Platelet transport and adhesion to reconstituted collagen and artificial surfaces. Trans Am Soc Artif Intern Organs 21:427–434, 1975.

13. Fox JA, Hugh AE: Localization of atheroma. A theory based on boundary layer separation. Br Heart J 28:388–399, 1966.

14. Fox JA, Hugh AE: Static zones in the internal carotid artery: Correlation with boundary layer separation and stasis in model flows. Br J Radiol 43:370–376, 1970.

15. Friedman MH, O'Brien VO, Ehrlich LW: Calculations of pulsatile flow through a bifurcation. Circ Res 36:277–285, 1975.

16. Fry FJ, Eggleton RC, Kelly E, Fry WJ: Properties of the blood interface essential to successful artificial heart function. Trans Am Soc Artif Intern Organs 11:307–317, 1965.

17. Gaehtgens PA, Meiselman HJ, Wayland H: Velocity profiles of human blood at normal and reduced hematocrit in glass tubes up to 130 μ diameter. Microvasc Res 2:13–23, 1970.

18. Gauthier FJ, Goldsmith HL, Mason SG: Flow of suspensions through tubes–X. Liquid drops as models of erythrocytes. Biorheology 9:205–224, 1972.

19. Geissinger HD, Mustard JF, Rowsell HC: The occurrence of microthrombi on the aortic endothelium of swine. Can Med Assoc J 87:405–408, 1962.

20. Goldsmith HL: Microscopic flow properties of red cells. Fed Proc 26:1813–1820, 1967.

21. Goldsmith HL: Red cell motions and wall interactions in tube flow. Fed Proc 30:1578–1588, 1971.

22. Goldsmith HL: The flow of model particles and blood cells and its relation to thrombogenesis. In Spaet TH (ed): "Progress in Hemostasis and Thrombosis," vol. 1. New York: Grune & Stratton, 1972, pp 97–139.

23. Goldsmith HL, Karino T: Microscopic considerations: The motions of individual particles. Ann NY Acad Sci 283:241–255, 1977

24. Goldsmith HL, Karino T: Mechanically induced thrombo-emboli. In Hwang NHC, Gross DR, Patel DJ (eds): "Quantitative Cardiovascular Studies–Clinical and Research Applications of Engineering Principles." Baltimore: University Park Press, 1978, pp 289–351.

25. Goldsmith HL, Marlow J: Flow behaviour of erythrocytes. I. Rotation and deformation in dilute suspensions. Proc R Soc Lond B 182:351–384, 1972.

26. Goldsmith HL, Marlow JC: Flow behavior of erythrocytes. II. Particle motions in concentrated suspensions of ghost cells J Colloid Interface Sci (In press).

27. Goldsmith HL, Marlow JC, Yu SK: The effect of oscillatory flow on the release reaction and aggregation of human platelets. Microvasc Res 11:335–359, 1976.

28. Goldsmith HL, Mason SG: The flow of suspensions through tubes. I. Single spheres, rods and discs. J Colloid Sci 17:448–476, 1962.
29. Goldsmith HL, Mason SG: The microrheology of dispersions. In Eirich FR (ed): "Rheology, Theory and Applications," Vol IV. New York: Academic Press, 1967. pp 85–250.
30. Goldsmith HL, Mason SG: Some model experiments in hemodynamics. IV. In Hartert HH, Copley AL (eds): "Theoretical and Clinical Hemorheology." New York: Springer, 1971, pp 47–59.
31. Goldsmith HL, Mason SG: Some model experiments in hemodynamics. V. Microrheological techniques. Biorheology 12:181–192, 1975.
32. Gott VL, Ramos MD, Najjar FB, Allen JL, Becker KE: The in vivo screening of potential throboresistant materials. In Hegyeli RJ (ed): "Artificial Heart Program Conference." Washington, DC: US Government Printing Office, 1969, p 181.
33. Grabowski EF, Friedman LI, Leonard EF: Effect of shear rate on platelet diffusion and adhesion to a foreign surface. Ind Eng Chem Fund 11:224–232, 1972.
34. Gutstein WH, Schneck DJ: In vitro boundary layer studies of blood flow in branched tubes. J Atheroscler Res 7:295–299, 1967.
35. Gutstein WH, Schneck DJ, Marks JO: In vitro studies of local blood flow disturbance in a region of separation. J Atheroscler Res 8:381–388, 1968.
36. Hunt WA: Calculation of pulsatile flow across bifurcations in distensible tubes. Biophys J 9:993–1005, 1969.
37. Karino T, Goldsmith HL: Flow behaviour of blood cells and rigid spheres in an annular annular vortex. Phil Trans R Soc Lond B 279:413–445, 1977.
38. Karino T, Goldsmith HL: Aggregation of human platelets in an annular vortex distal to a tubular expansion. Microvasc Res 17:217–237, 1979.
39. Karino T, Goldsmith HL: Adhesion of human platelets to collagen on the walls distal to a tubular expansion. Microvasc Res 17:238–262, 1979
40. Karnis A, Goldsmith HL, Mason SG: The kinetics of flowing dispersions. I. Concentrated suspensions of rigid particles. J Colloid Interface Sci 22:531–553, 1966.
41. Karnis A, Goldsmith HL, Mason SG: The flow of suspensions through tubes. V. Inertial effects. Can J Chem Eng 44:181–193.
42. Kreid DK, Chung C-H, Crowe CT: Measurements of the flow of water in a "T" junction by the LVD technique. Paper 74-WA-BIO 1, Bioengineering Division, Winter Annual Meeting, ASME, New York.
43. Kreid DG, Goldstein RJ: In "Proceedings of the Symposium on Flow – Its Measurement and Control in Science and Industry." Pittsburgh, PA: Instrument Society of America.
44. Kwaan HC, Harding F, Astrup T: Platelet behaviour in small blood vessels in vivo. In Brinkhous KM et al (eds): "Platelets: Their Role in Hemostasis and Thrombosis." Thromb Diath Haemorrh Suppl 26, Stuttgart: Schattauer, 1967, pp 208–220.
45. Karino T, Kwong HMM, Goldsmith HL: Particle flow behaviour in models of branching vessels: I. Vortices in 90° T-junctions. Biorheology 16:231–247, 1979.
46. Lee J-S, Fung Y-C: Flow in nonuniform small blood vessels. Microvasc Res 3:272–287, 1971.
47. Levich VG: "Physicochemical Hydrodynamics, Englewoods Cliffs, New Jersey: Prentice-Hall, 1962, pp 112–116.
48. Levine R, Goldsmith HL: Particle behaviour in flow through small bifurcations. Microvasc Res 14:319–344, 1977.
49. Lissant KJ: In "Emulsions and Emulsion Technology," Vol 6. New York: Marcel Dekker, 1974, chapter 1.
50. Macagno EO, Hung T-K: Computational and experimental study of a captive annular eddy. J Fluid Mech 28:43–64, 1967.

51. Macagno EO, Hung T-K: Computational study of accelerated flow in a two-dimensional expansion. J Hydraulic Res 8:41–64, 1970.
52. Mitchell JRA, Schwartz CJ: The relationship between myocardial lesions and coronary artery disease. II. A select group of patients with massive cardiac necrosis or scarring. Br Heart J 25:1–24, 1963.
53. Morton WA, Parmentier EM, Petschek HE: "A Study of Aggregate Formation in a Region of Separated Blood Flow." Research Report 401. Boston: Avco Everett Research Laboratory, 1974.
54. Morton WA, Parmentier EM, Petschek HE: Study of aggregate formation in region of separated blood flow. Thromb Diath Haemorrh 34:840–854, 1975.
55. Müller-Mohnssen H: Experimental results to the deposition hypothesis of atherosclerosis. Thromb Res 8:553–566, 1976.
56. Mustard JF, Murphy EA, Rowsell HC, Downie HG: Factors influencing thrombus formation in vivo. Am J Med 33:621–647, 1962.
57. Mustard JF, Perry DW, Ardlie NG, Packham MA: Preparation of suspensions of washed platelets from humans. Br J Haematol 22:193–204, 1972.
58. Packham MA, Rowsell HC, Jørgensen L, Mustard JF: Localized protein accumulation in the wall of the aorta. Exp Mol Pathol 7:214–232, 1967.
59. Paterson JC, McLachlin J: Precipitating factors in venous thrombosis. Surg Gynecol Obstet 98:96–102, 1954.
60. Rodkiewicz CM: Localization of early atherosclerotic lesions in the aortic arch in the light of fluid flow. J Biomech 8:149–156, 1975.
61. Rowntree LG, Shionoya T: Studies in experimental extracorporeal thrombus formation. Method for direct observation of extracorporeal thrombus formation. J Exp Med 46:7–18, 1927.
62. Schmid-Schönbein H, Wells RE Jr: Rheological properties of human erythrocytes and their influence upon the "anomalous" viscosity of blood. Ergeb Physiol 63:146–219, 1971.
63. Schmid-Schönbein H, Gosen JV, Heinich L, Klose HJ, Volger E: A counter-rotating "rheoscope chamber" for the study of the microrheology of blood cell aggregation by microscopic observation and microphotometry. Microvasc Res 6:366–376, 1973.
64. Schreck RM: Laminar flow through bifurcation with applications to the human lung. PhD Thesis, Northwestern University, Evanston, Illinois, 1972.
65. Segré G, Silberberg A: Behaviour of macroscopic rigid spheres in Poiseuille flow. II. Experimental results and interpretation. J Fluid Mech 14:136–157, 1962.
66. Seshadri V, Sutera SP: Concentration changes of suspensions of rigid spheres flowing through tubes. J Colloid Interface Sci 27:101–110, 1968.
67. Seshadri V, Sutera SP: Apparent viscosity of coarse, concentrated suspensions in tube flow. Trans Soc Rheol 14:351–373, 1970.
68. Smith RL, Blick EF, Coalson J, Stein PD: Thrombus production by turbulence. J Appl Physiol 32:261–264, 1972.
69. Stein PD, Sabbah HN: Measured turbulence and its effect on thrombut formation. Circ Res 35:608–614, 1974.
70. Turitto VT, Baumgartner HR: Platelet interaction with subendothelium in a perfusion system: Physical role of red blood cells. Microvasc Res 9:335–344, 1975.
71. Turitto VT, Baumgartner HR: Platelet deposition on subendothelium exposed to flowing blood: Mathematical analysis of physical parameters. Trans Am Soc Artif Intern Organs 21:593–600, 1975.
72. Turitto VT, Benis AM, Leonard EF: Platelet diffusion in flowing blood. Ind Eng Chem Fund 11:216–223, 1972.
73. Usami S, Chien S: Shear deformation of red cell ghosts. Biorheology 10:425–430, 1973.

74. Vadas EB, Goldsmith HL, Mason SG: The microrheology of colloidal dispersions. I. The microtube technique. J Colloid Interface Sci 43:630–648, 1973.
75. Vadas EB, Goldsmith HL, Mason SG: The microrheology of colloidal dispersions. III. Concentrated emulsions. Trans Soc Rheol 20:373–407, 1976.
76. Yu SK, Goldsmith HL: Behavior of model particles and blood cells at spherical obstructions in tube flow. Microvasc Res 6:5–31, 1973.
77. Zucker MB, Rifkin PL, Friedberg NM, Coller BS: Mechanism of platelet function as revealed by the retention of platelets in glass bead columns. Ann NY Acad Sci 201:138–144, 1972.

10

Behavior of Abnormal Erythrocytes in Capillaries

P. L. La Celle

The erythrocytic properties important for the passive adaptation of the volume configuration to the flow requirements of the microcirculation include intrinsic features — ie, membrane and cytoplasmic parameters — and extrinsic factors such as cell shape [1, 2]. The cellular deformation as result of shear forces is net behavior representing the aggregate of contributions or limitations of both intrinsic and extensive factors and cannot be defined explicitly. However, from the pragmatic standpoint it is presumed that cellular deformability determines the transit of capillary circulation and other restricted, specialized portions of the circulation such as splenic cord—sinus region and that the erythrocyte's surface features, as well as those of endothelial cells of the flow channels, are significant to this cell's net rheologic behavior.

The capacity of the erythrocytic membrane to deform is considered important to shear elongation of the cell in bulk flow and in transmission of shear force to the cell cytoplasmic contents; deformation of the cell at narrow apertures — eg, during marrow egress and transit of splenic cords to sinus, in entry phenomena at bifurcations, and in limiting diameter capillaries. The membrane, extremely thin in comparison to cellular dimensions, undergoes shear deformation at constant surface area in response to external force; analyses of this membrane resistance to shear deformation and rate of deformation have led to estimates of elastic shear modulus and viscosity in elastic deformation. These are surface properties of an essentially two-dimensional material. The constant area constraint is important to note, and the normal erythrocyte membrane tolerates only approximately 3% increment in area before failure occurs [3], at forces orders of magnitude greater than required for membrane or cellular deformation.

The geometry of the cell contributes to cellular deformability, largely as result of the excess membrane area for the enclosed volume which allows changes in volume configuration without development of large tensions in the membrane; modification of the biconcave shape toward a spherical shape — ie, reduction of surface area to enclosed volume — adversely affects deformability, since relatively

Erythrocyte Mechanics and Blood Flow, pages 195—209

small shear forces will induce tension in the membrane. Factors that produce relative sphericity or increase cytoplasmic viscosity above its normal low value [4] reduce cellular deformability, and in the extreme could produce cells of sufficient rigidity that, intuitively, abnormal flow behavior in the microcirculation would be expected.

Most of the observations of the membrane and cellular deformability characteristics of individual cells have been made in steady-state experiments, and little information exists concerning the dynamics of cyclic force application, behavior at points of rapid transition (eg, entrance into small apertures), and the relative contribution of membrane elastic response in these circumstances. If the time scale of cellular deformability and achievement of new equilibrium is an appreciable fraction of capillary transit time or time for entrance in a narrow channel, the extrinsic factor of shape may dominate considerations of membrane elastic properties and cytoplasmic viscosity in the case of normal cells. Erythrocytes from disorders in which shortened erythrocytic life-span may be demonstrated are presumed to have abnormal characteristics of cellular deformability and membrane surface properties or abnormal metabolism leading to decreased deformability or altered surface if no adverse changes of the flow system are evident. Alternatively, relative differences of vascular endothelial surface, enhanced immune activity, increased phagocytic activity in the monocyte–phagocytic system (reticuloendothelial system), or toxic effects on the cell membrane may be important contributors to reduced survival. At present no information exists to define the minimal changes in membrane viscoelastic properties, failure characteristics, reduction in cellular deformability as result of relative sphericity or increased cytoplasmic viscosity, or time dependence of these characteristics that would be sufficient to cause disease (defined as decreased erythrocyte life-span). Furthermore, no detailed information concerning the propensity of such erythrocytes to adhere to endothelium is available. Thus, putative mechanisms lack direct experimental confirmation. The purpose of this paper is to review some experimental observations of the relation between cellular deformability and flow characteristics of erythrocytes from sickle cell disease, hereditary spherocytosis, thalassemia, and diabetes mellitus in light of these questions about the significance of membrane and cellular factors for cell rheology.

EXPERIMENTAL METHODS

Micropipette Technique

Aspiration of portions of the erythrocyte membrane from the flat or "dimple" region of cells into glasss micropipettes of approximately 1 μm internal diameter permits characterization of the membrane elastic properties. This method, employed by Evans [1], Evans and La Celle [2], and more recently by Waugh [5], yields a value of the order of 10^{-2} dynes/cm for the shear modulus of surface elasticity μ in normal erythrocytes, employing Evan's analysis of force-deformation behavior of the membrane. Recently Hochmuth [6] has used micropipettes to extend the whole cell, and by observation of the time course of the cell's return from deformed to equilibrium biconcave disc shape, was able to obtain a value of the shear modulus, the time for recovery of the biconcave shape, and hence the coefficient for surface viscosity, η. In the present experiments we have used this micropipette aspiration technique to obtain the ΔP required to induce standard deformation of the membrane and to calculate values for μ. The Hochmuth method has been used to determine the time constant for recovery of deformed whole cells. The aspiration was done at slow rates — ie, deformation essentially at equilibrium conditions.

To determine the cellular deformability, erythrocytes were aspirated into larger micropipettes, 2.7–3.0 μm in internal diameter, to characterize the ΔP required to draw the entire cell, regardless of surface area to volume relation or shape into the pipette [7]. The pipette in this experiment is a standard physical channel for both control and abnormal cells, and thus the ΔP required to aspirate an abnormal cell is a measure of *relative* cellular deformability.

Capillary Flow Experiments

The simplest experiment consists of observation of flow rate of a single erythrocyte in a narrow glass capillary of limiting diameter (approximately 2.8 μm) as a function of applied pressure over a 200 μm length [8]. The narrow capillary provides the maximum resistance to flow and allows comparison of abnormal cells with control erythrocytes. In the same capillary the time course of acceleration or deceleration of a cell may be determined from analysis of video-tape records of the experiment.

In vivo flow experiments were performed in capillaries of the transilluminated cremaster muscle of mouse by the method described by Baez [9]. Adult white mice (wt 35–45 gm) were anesthetized with urethane-pentobarbital, and exposed muscles were supported by a plastic ring in a microscope stage chamber such that the preparation could be visualized at 1,000–1,500 X with a long working distance water objective. The muscle was bathed in phosphate buffer containing KCl, 4.5 mM; $MgCl_2$, 1 mM. Dimensions were derived by use of calibrated micrometers, and cell movements were calculated from video-tape records.

Measurement of Tissue Oxygen

Oxygen tensions of muscle cells were obtained by calibrated oxygen microelectrodes, an Ag–AgCl reference electrode and picoammeter. The electrodes were calibrated by measurement of PO_2 of blood samples vs values obtained by a blood gas analyzer [10].

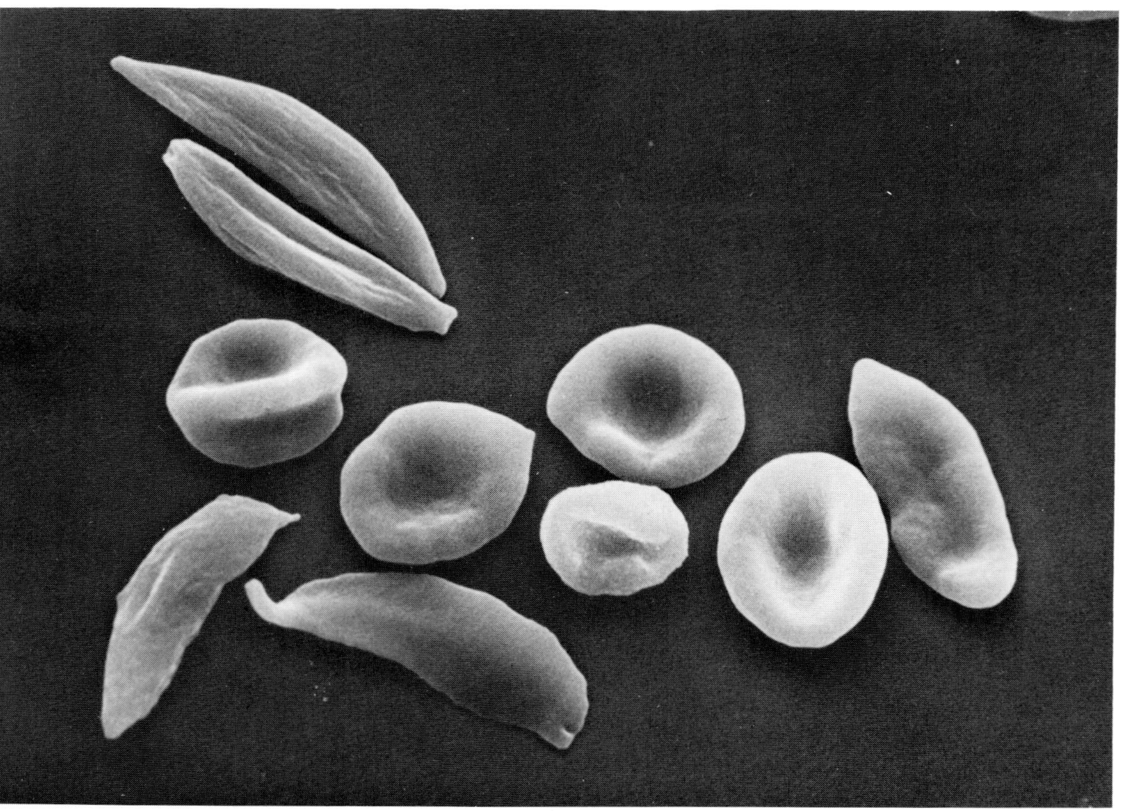

Fig. 1. Sickle cells at normal oxygen tension. Irreversibly sickled (ISC) forms contrast with reversibly sicklable cells (RSC) in the center of the field.

RESULTS

Sickle Cells

Figure 1 illustrates the appearance of oxygenated sickle cells, revealing the heterogeneity that extends from the irreversibly sickled cell (ISC) to the relatively normal biconcave, disc-shaped cells. Figure 2 indicates the proportion of reversibly sickled cells (RSC) over the physiologic range of oxygen tension: it is noteworthy that at $PO_2 = 40$ torr (venous blood) some sickling is observed in a significant number of sickle cells (HbSS) and that in the heterozygous condition (HbSA), some sickling is observable at 20 mm Hg. In Table I data for membrane deformation of sickle cells shows increase in rigidity of membranes in intact sickle erythrocytes, in contrast to minimal differences previously observed in sickle cell ghosts (essentially hemoglobin-free membranes) [8]. The measure of cellular deformability showed clear loss of deformability in the reversibly sickled cell (RSC) at low oxygen and indicated no change in ISC as oxygen tension was lowered. (Thin ISC, narrower than the micropipette, were aspirated at low pressures, $\Delta P = 2$ mm H_2O.) It should be stressed that there is a broad range of values for RSC with no obvious correlation between RSC shape and deformation pressures until the relatively rigid, final "holly leaf" shape occurs.

No significant effect on flow behavior of sickle cells in a standard muscle capillary was noted as a function of pH (Table II). Increasing pH to 8.0 in the

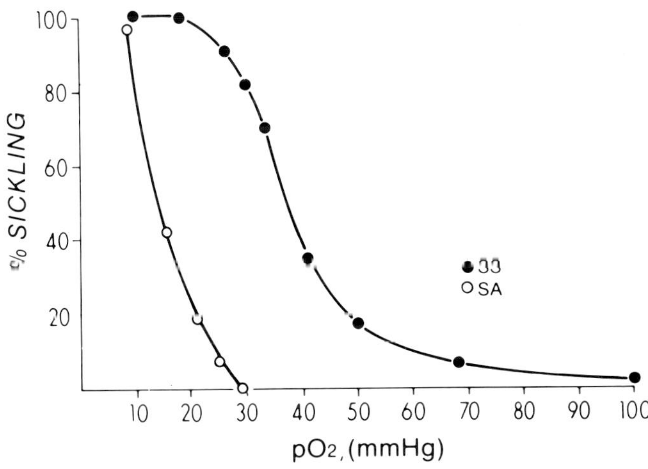

Fig. 2. Proportion of sickled forms vs PO_2. Sickled forms were enumerated in two blood samples equilibrated at each oxygen tension. Open circles represent homozygous state (HbSS) and closed circles the heterozygous state. Sickling was defined as minimal detectable change in morphology from the normal biconcave shape.

TABLE I. Deformability of Sickle Cells

	Pressure, mm H_2O			
	Membrane deformation[a]		Cellular deformation[b]	
	$PO_2 = 100$	$PO_2 < 20$	$PO_2 = 100$	$PO_2 < 20$
Control cell	1.4 ± 0.1	2.1 ± 0.3	5.9 ± 0.1	8.1 ± 1.1
Reversibly sickled cell (RSC)	1.6 ± 0.3	25 ± 5.5	6.9 ± 0.3	79 ± 22
Irreversibly sickled cell (ISC)			2 to ∞	no change

[a]Pipette diameter: 1.0 μm
[b]Pipette diameter: 2.7 μm

TABLE II. Effect of Capillary pH on Sickle Cell Flow

	Velocity (μ/sec)		$\Delta P = 1,000$ dynes/cm^2
pH	Control	ISC[a]	RSC[b]
7.4	41 ± 10 SE	45 ± 11 SE	33 ± 12 SE
7.0	38 ± 9	42 ± 15	35 ± 13
6.5	35 ± 11	41 ± 12	30 ± 12
6.0	37 ± 11	41 ± 15	33 ± 11

[a]Irreversibly sickled cells.
[b]Reversibly sickled cells.

perfusing solution and bathing solution also had no effect on this system. These pH data are similar to those in rigid glass capillaries, where no changes in diameter occur.

Muscle cell PO_2 did not change appreciably around the straight capillaries of this muscle system as PO_2 of the animals' inhaled air, that of the bathing solution, and that of perfusion solution were lowered (Table III); only when the bath O_2 was lowered appreciably so that the bath acted as a sink, did flow decrease and muscle cell PO_2 fall off toward zero. Thus, under the condition resembling the in vivo situation of the human — ie, venous PO_2 of 40 torr — the cells RSC and ISC flow normally and appear to deliver oxygen effectively. It should be noted that this is not an exhaustively critical experiment, since some O_2 diffuses to the tissue from the bath. However, in the most severe case, bath $PO_2 = 10$ torr, perfusate $PO_2 = 25$ torr, flow velocity was normal, and oxygen delivery was approximately normal.

At the low flow rates in these studies, some adherence of RSC to capillary

TABLE III. Muscle Cell PO_2 During Capillary Perfusion With Reversibly Sickled Cells

Cell type	Air	Bath	Perfusion	Velocity (μm/sec)	Muscle PO_2 (mm Hg)
Control (HbAA)	> 100	100	100	52 ± 17	17 ± 5
RSC[a]	100	100	100	45 ± 18	20 ± 8
RSC	40 ± 5	50 ± 6	50 ± 5	36 ± 12	18 ± 5
RSC	40 ± 5	$0-10$	25 ± 5	$30 \rightarrow 0$	$20 \rightarrow 4$

[a]Reversibly sickled cells.

endothelium occurred. As PO_2 was lowered it was not evident whether RSC or ISC tended to initiate capillary obstruction.

Poikilocytes

The distorted shape of the poikilocyte, occurring in a number of disorders, was interpreted by Rous [11] to indicate increased red cell destruction. The poikilocyte is typical of thalassemia (Fig. 3), in which an imbalance of α and β globin chain synthesis leads to accumulation of masses of denatured excess globin in the cell and enhanced destruction of cells. Rigid intracellular material appears to be "pitted" from the cell during transit in the spleen. Surviving portions of such fragmented cells often have a teardrop shape. However, since poikilocytes are seen in a number of disorders, the occurrence of intracellular rigid masses, as in thalassemia, is not the sole basis for poikilocyte formation; rather it may be that any process in which sustained low force is placed on the cell may lead to permanent shape change to the poikilocyte form. Table IV records the cellular deformability of patients' cells and the poikilocytes within those populations and shows clearly that only minor reduction in cellular deformability is associated with the poikilocytic shape. In Table V, thalassemia major erythrocytes and poikilocytes from each patient's blood appear to be somewhat less deformable than normal erythrocytes, and the requirement for transfusion appears to be increased (shorter interval between transfusions) as deformability is reduced.

The flow velocity of poikilocytes from thalassemic blood is normal and the membrane deformability — ie, pressure differential to produce a hemispherical bulge into a 1.0 μm micropipette — is similar to control. Measurement of membrane deformability around the cell periphery, including the narrow "teardrop" region, indicates no regions of altered elasticity.

Hereditary Spherocytes

These cells are stomatocytic (cup-shaped or uniconcave) and have significant numbers of spherocytic forms, as shown in Figure 4. The relative sphericity results

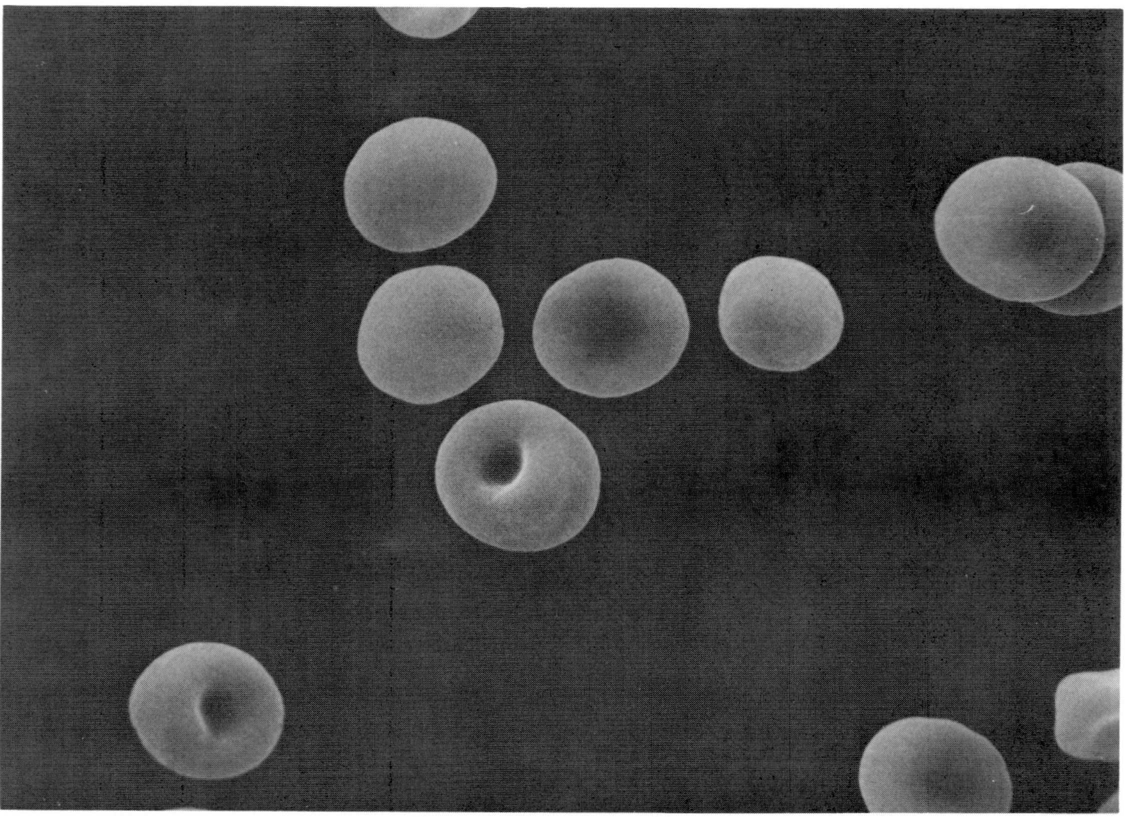

Fig. 3. Hereditary spherocytes. The predominant shape is stomatocytic, with reduced surface area to volume ratio, compared to normal.

TABLE IV. Cellular Deformability of Poikilocytes*

| | ΔP (mm H_2O $\pm$ SE) | | |
Source of cells	Control cells	All patient cells	Poikilocytes
Myeloid metaplasia #1	3.9 ± 0.2	–	6.4 ± 1.4
#2	4.1 ± 0.1	4.1 ± 0.2	7.0 ± 1.1
#3	3.8 ± 0.2	4.4 ± 0.1	10.2 ± 1.9
#4	3.9 ± 0.3	5.1 ± 0.1	–
Acute myelocytic leukemia	3.8 ± 0.2	9.1 ± 2.0	9.7 ± 2.2
Thalassemia minor (no splenomegaly)	3.8 ± 0.2	–	4.2 ± 0.2
Hemoglobin Koln	4.0 ± 0.2	4.9 ± 0.6	9.9 ± 2.6
Thrombocytic thrombocytopenic purpura	5.8 ± 0.3	21.6 ± 5.4	26.8 ± 16.3

*Pipette diameter: 3.0 μm

TABLE V. Cellular Deformability of Thalassemic Erythrocytes*

| Thalassemia major patient | Transfusion requirement | ΔP (mm H_2O)[a] | |
		Random cell sample	Poikilocytes
1	5-week interval	15.2 ± 1.4 N = 52	23.0 ± 11 N = 16
2	3-week interval	30.7 ± 9.8	68.8 ± 19
3	–	9.9 ± 0.9	18.0 ± 6
4	8-week interval	51.8 ± 19.8	85.2 ± 41
5	–	9.5 ± 0.4	13.1 ± 6

*Pipette diameter: 2.7 μm.
[a]ΔP, control cells: 8.7 ± 0.4.

in an increased force requirement for aspiration into a standard glass micropipette (Table VI). Preliminary experiments show that the shear modulus of surface elasticity of the hereditary spherocyte membrane is similar to normal membranes ($\sim 10^{-2}$ dynes/cm); however, the tension at which membrane failure occurs is less than in normal cells. This decrease in "tensile strength" corresponds to the well-known observation that hereditary spherocytes are more fragile than normal. It should be noted that the decreased cellular deformability develops in this cell

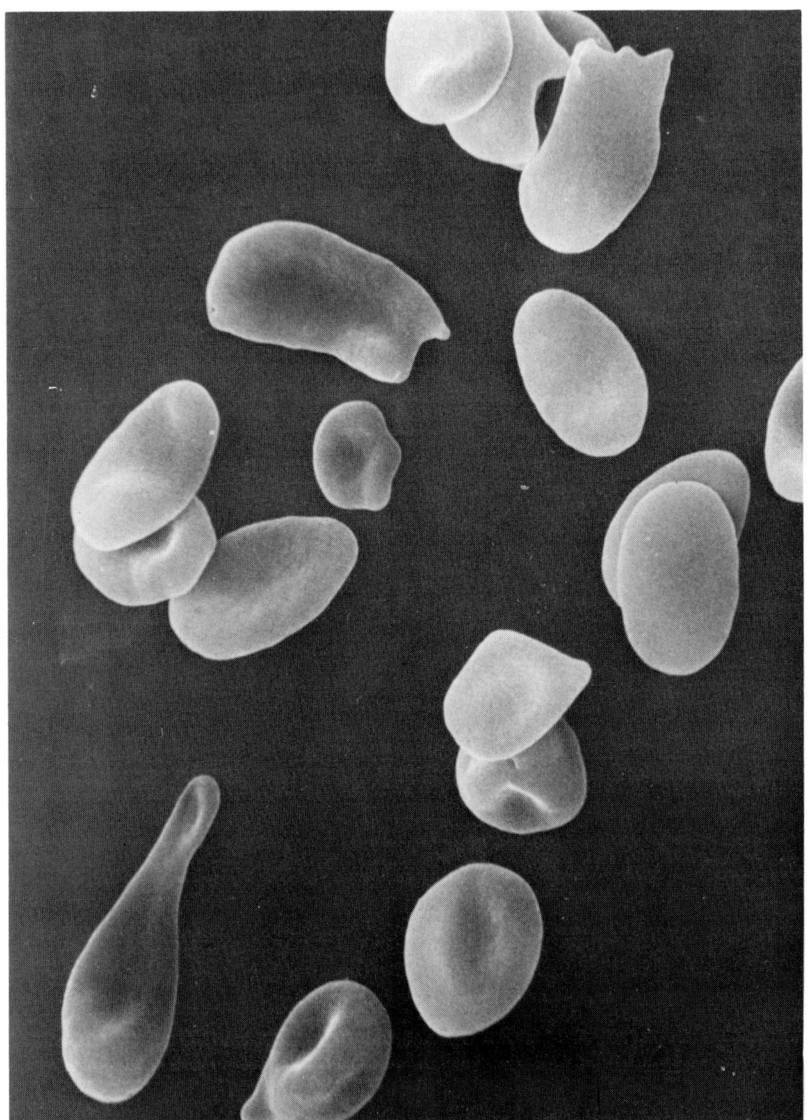

Fig. 4. Thalassemic erythrocytes. The pokilocyte (teardrop-shaped cell) and great heterogeneity of cell shape typify this disorder.

TABLE VI. Cellular Deformability of Hereditary Spherocytes

	ΔP (mm H_2O)	
Patient	Intact cell	Ghost
1	40.4 ± 13.6	37.0 ± 14.8
2	48.2 ± 20.7	30.7 ± 9.4
3	28.6 ± 6.7	22.1 ± 4.8
4	32.6 ± 19.4	–

TABLE VII. Erythrocyte Deformability in Diabetes Mellitus*

		Deformation pressure (mm H_2O)			
	n	1st deformation	15	50	100
Control cell	25	2.9 ± 0.9	4.1 ± 0.9	4.0 ± 1.1	4.4 ± 0.9
Diabetic cell (pH = 7.4)	10	4.2 ± 1.0	4.6 ± 1.4	4.5 ± 1.4	4.5 ± 1.7
Diabetic cell (pH = 6.5)	10	4.0 ± 1.3	–	–	4.2 ± 2.1

*Pipette diameter: 2.8 μm; blood glucose: 215 mg% mean.

after its maturation in the bone marrow; earlier forms have normal cellular deformability characteristics [12].

Diabetes

The appearance of the erythrocyte in the disorder diabetes mellitus is normal, and conclusive evidence for the contribution of a cellular abnormality to the perceptibly increased bulk viscosity of blood is not available. Some preliminary data do suggest decreased cellular deformability, however [13]. Our data do not support the conclusion that the erythrocyte deformability is abnormal or that membrane elasticity is decreased, as displayed in Table VII. These data reflect the ΔP to achieve repeated cellular deformation in a 2.8 μm micropipette; the ΔP for the one hundredth deformation is similar to control. Relaxation of deformed diabetic cells, using the Hochmuth method, indicates viscoelastic properties similar to control: $T_c = 0.108 \pm 0.018$ sec vs 0.123 ± 0.025 in the control cells.

Flow behavior of the diabetic cells in glass capillaries of 2.8–3.5 μm paralleled that of normal cells in experiments in which cells were observed to move 100 μm and return at measured ΔP and time determined to one hundredth of a second. In these experiments the pressure was adjusted, and corresponding time was observed for the first cycle and the seventy-sixth cycle (Table VIII). When the pressures are normalized, the time for excursion in diabetic cells for the seventy-sixth excursion is $\sim$ 1.87, slightly less than for control. These data represent eleven cells from three diabetic patients.

In animals made diabetic by streptozotocin treatment no alteration in tissue oxygen levels was observed (Table IX). Since the rat erythrocyte is smaller in diameter, a 2.3 μm micropipette was used to determine cellular deformability. PO_2 was measured in abdominal (rectus) muscle. The objective of the experiment was to determine whether an abnormality in a physiologic parameter, tissue oxygen at rest, would be observed in the disordered metabolic state and/or as result of putative changes in blood rheology. The data suggest that no difference in oxygen transport accompanies the induced diabetic state, but the variability of PO_2 in individual cells in the control suggests that small differences of mean PO_2 would not be detected by this method. (The variability is with the cells, since the electrode produces reproducible results in calibration.)

Spectrin-Deficient Mouse Cells

A mouse mutant that has marked hemolysis has been found to have erythrocytes with marked reduction of the principal extrinsic membrane protein, spectrin [14]. It was postulated that the relative lack of spectrin reduces the solid material properties of the membrane and thus reduces the capacity to withstand shear forces. The relative lack of spectrin to afford a "cytoskeleton" would be

TABLE VIII. Erythrocyte Flow Behavior in Diabetes Mellitus

	Measured pressure	Time for 100 μm $\times$ 2
Control erythrocyte	P_1: 277 $\pm$ 6.8	T_1: 1.95 $\pm$ 0.42
	P_{76}: 225 $\pm$ 7.1	T_{76}: 2.01 $\pm$ 0.37
Diabetic	P_1: 176 $\pm$ 6.3	T_1: 2.35 $\pm$ 0.15
	P_{76}: 178 $\pm$ 7.3	T_{76}: 2.36 $\pm$ 0.13

TABLE IX. Erythrocyte Deformability and O_2 Delivery in Streptozotocin Diabetic Rats*

		Deformation pressure (mm H_2O)	$\sim$ Skeletal muscle cell PO_2 (mm Hg)
Control cell	54	12.6 $\pm$ 2.1	17 $\pm$ 5
Diabetic cell	51	11.8 $\pm$ 3.4	21 $\pm$ 7

*Pipette diameter: 2.3 μm.

expected to result in a membrane dominated by the properties of the lipid bilayer, and indeed the cells are spherical, not disc-shaped. Attempts to aspirate membrane from these cells resulted in production of long filaments fragmenting from the cell surface and disruption of the membrane at pressures as low as 200 dynes/cm^2, two orders of magnitude less than control. When cells were equilibrated in hypertonic medium (600 mOsm) to increase relative surface area by reduction of intracellular volume, the spherical shape persisted, suggesting the dominant role of the lipid. In comparison with normal reticulocytes ($>$ 90% normal reticulocytes, corresponding to the high proportion of reticulocytes in the mutant) the data were similar.

COMMENTS

These experiments indicate that although significant differences in membrane elastic properties and cellular deformability may be demonstrated in vitro, flow behavior of abnormal cells from disease states is similar to that of normal human erythrocytes in relatively narrow glass or in vivo capillary systems. These experiments are limited in the very fact that the experimental capillary systems employed are simple systems that do not reflect the complexity of geometry and dynamics of flow typical of most of the microcirculation, and thus the interpretation must be limited to the conclusion that abnormal erythrocytes, once in a narrow vessel, may have the same flow pattern as normal cells. It appears likely that in the in vivo situation, relatively rigid erythrocytes do not approach and enter very small vessels but are shunted to higher velocity larger vessels; or if they enter, they may be moved along in the manner observed for leukocytes [15]. At bifurcations relatively spherical cells, poikilocytes, reversibly sickleable cells, and irreversibly sickled forms do tend to cause occlusion, but these aggregations appear to be temporary even at low perfusion pressures. This is of interest from the clinical perspective, since abnormal cells do not generally cause abnormal flow patterns and clinical symptoms in relatively straight capillary systems such as in muscle.

Oxygen tension in muscle is not altered importantly even by presence of relatively rigid, partially deoxygenated sickle cells, illustrating that oxygen delivery occurs quite normally when flow is maintained. It might be predicted that, in light of capillary spacing and flow rate, this should be the case and that only marked reduction of capillary hematocrit or velocity would result in tissue hypoxia.

Sickle cells have normal membrane elasticity, judged from studies of hemoglobin-free ghosts, and the change of membrane properties during deoxygenation appears to be related to the stiffness of the adjacent membrane-related hemoglobin. In the rigid state of the ISC and deoxygenated RSC, the rigid shape dominates any contribution of membrane alteration in terms of cellular deformability. However, possible membrane changes, expressed functionally as increased adhesivity, could be an important contributor to the phenomenon of capillary occlusion. The postulated increase in adhesivity deserves further study and may be important in consolidation of a functional occlusion caused by other factors.

The poikilocytic shape may be interpreted as a result of protracted stress with resulting plastic flow of the erythrocyte membrane; the shape is not specific to individual clinical disorders. The poikilocyte indicates increased destruction of cells by a fragmentation of membrane, not necessarily implying a disorder of the membrane itself, but rather suggesting extracellular circumstances such as altered splenic function or obstructions in small vessels, which contribute to increased destruction. The lack of strong correlation of cellular deformability to in vivo life-span suggests that the reduced deformability is not critical. In thalassemia better correlation of these factors may reflect the degree of abnormal architecture of flow channels in the reticuloendothelial system, particularly bone marrow and spleen, with consequent cell destruction.

The membrane constituents in the disorder hereditary spherocytosis are considered generally to be qualitatively normal; however, the membrane permeability to ions is increased, as is mechanical fragility of the membrane. The normal elastic shear modulus suggests grossly normal structural arrangement of the protein elements and resultant elastic behavior; thus, the increased resistance to deformation appears to relate to the extrinsic factor of relative sphericity (reduced surface area to volume ratio). Since cellular deformability of bone marrow precursor cells is similar to normal erythrocyte precursors, the abnormality leading to the spherical shape of the hereditary spherocyte is expressed as the cell circulates. In light of the known mechanical fragility, it appears likely that the sphericity results from repeated local fragmentation and spontaneous resealing of the membrane during trauma in the circulation. The relatively rigid spherical cells have shortened life-spans, approximately in proportion to reduction in deformability.

The reported changes in cellular deformability of erythrocytes in diabetes are relatively small, judged from flow of cell suspensions through 5 μm apertures in polycarbonate filters, and it is likely that both aggregation and relative sphericity contributed to the observation. The washed cells deformed at equilibrium in a medium of pH = 7.4 in the present experiments do not exhibit abnormal cellular deformability, and estimates of the shear modulus in elastic deformation are normal. These observations suggest that there is no intrinsic abnormality of the erythrocyte in diabetes and that alteration of plasma proteins accounts for the abnormal rheology in this disease. The rheologic abnormality is relatively small, suggesting that the vascular component may be a more important determinant of pathophysiology of the disorder.

ACKNOWLEDGMENTS

This work was supported by USPHS grants HL16421 and HL18208 and is based in part on work performed under contract with the U.S. Department of

Energy at the University of Rochester Department of Radiation Biology and Biophysics and has been assigned report number UR-3490-682.

REFERENCES

1. Evans EA: New membrane concept applied to the analysis of fluid shear and micropipette-deformed red blood cells. Biophys J 13:941–954, 1973.
2. Evans EA, La Celle PL: Intrinsic material properties of the erythrocyte membrane indicated by mechanical analysis of deformation. Blood 45:29–43, 1975.
3. Evans EA, Waugh R, Melnik L: Elastic area compressibility modulus of red cell membrane. Biophys J 16:585–595, 1976.
4. Cokelet GR, Meiselman HJ: Rheological comparison of hemoglobin solutions and erythrocyte suspensions. Science 162:275–277, 1968.
5. Waugh RE: Temperature dependence of the elastic properties of red blood cell membrane. Ph.D. Thesis, Duke University, 1977.
6. Hochmuth RM, Worthy PR, Evans EA: Surface viscosity of red cell membrane. AICNE Symposium, (in press).
7. La Celle PL: Alteration of deformability of the erythrocyte membrane in stored blood. Transfusion 9:238, 1969.
8. La Celle PL: Pathologic erythrocytes in the capillary microcirculation. Blood Cells 1:269–284, 1975.
9. Baez S: An open red cremaster muscle preparation for the study of blood vessels by in vivo microscopy. Microvasc Res 5:384–394, 1973.
10. La Celle PL: Oxygen delivery to muscle cells during capillary occlusion by sickled erythrocytes. Blood Cells 3:273–282, 1977.
11. Rous P: Destruction of the red blood corpuscles in health and disease. Physiol Rev 3:75–105, 1923.
12. Leblond PF, La Celle PL, Weed RI: Rheologie des erythroblasts et des erythrocytes dans la spherocytose congenitale. Nouv Rev Hematol 11:537–551, 1971.
13. Schmid-Schoenbein H, Volger E: Red cell aggregates and red cell deformability in diabetes. Diabetes 25:897–902, 1976.
14. Lux SE: Personal communication.
15. Schmid-Schoenbein GW, Zweifach BW: RBC velocity profiles in arterioles and venules of the rabbit omentum. Microvasc Res 10:153–164, 1975.
16. Bagge U, Branemark PI: White blood cell rheology. An intravital study in man. Adv Microcirc 7:1–17, 1977.

11

Reduced Erythrocyte Deformability and Vascular Pathology

Donald E. McMillan

Altered blood flow properties may affect the functional and anatomic properties of the vessels through which blood passes. This possibility is considered in a single aspect by examining the alteration in red blood cell flow properties in two disorders, diabetes mellitus and spherocytosis, and the anatomic changes known to occur in diabetes.

DIABETIC VASCULAR CHANGES

Diabetes mellitus is a chronic disorder characterized by hyperglycemia and associated with a number of additional metabolic changes. Over a period of years many diabetics develop complications not seen in nondiabetics. It has come to be generally held that these changes are sequelae of diabetes [1]. The changes have in common an involvement of the circulation. A sclerotic process, identified through pathological study as an increase in periodic acid-Schiff (PAS)-positive material, is regularly seen in vessel walls. In the eye this vascular pathology affects the retina, and is associated with hemorrhage into the vitreous, detachment of the retina, or macular degeneration, all of which lead to blindness. In the kidney, vascular and glomerular changes are associated with the development of proteinuria, loss appears to be related to the vascular changes. Widespread vascular sclerosis characterizes diabetic angiopathy in its advanced form [2].

When the electron microscope was introduced into microvascular investigation nearly two decades ago, it was found that the basement membrane of the capillaries in many tissues was thickened in diabetes even before serious changes could be recognized in the eyes, kidneys, or nervous system. This thickening of the basement membrane appeared to be specific to diabetes, being only occasionally seen in other disorders [3]. It occurs most regularly in the renal glomeruli [4] and in muscle [5], where it is seen so often that it distinguishes diabetic and nondiabetic adults. Since capillary basement membrane thickening is a regularly recognizable pathologic change requiring high magnification to be visualized, the

Erythrocyte Mechanics and Blood Flow, pages 211–228

term "diabetic microangiopathy" has come to be associated closely with it, even though evidence that the capillary thickening is actually responsible for the serious sequelae of diabetes is all but lacking. But there is an obvious parallel between sclerosis of somewhat larger vessels and thickening of the basement membrane. For instance, both are associated with PAS-positive histochemical changes [6].

HEMORHEOLOGIC CHANGES IN DIABETES

Because individual diabetics develop either marked or minimal microvascular changes as the duration of their disease lengthens, we have looked for factors that might predict their individual likelihood for developing microangiopathy. Alterations in the level of certain plasma proteins are associated with earlier development of recognizable diabetic sequelae — microaneurysms in the retina, loss of vibration sense in the toes or fingers, and the development of proteinuria as a manifestation of diabetic kidney disease. Fibrinogen and haptoglobin elevation and depression of serum albumin are more striking in diabetics who manifest early microangiopathic changes [7]. These protein changes are linked both to increased plasma viscosity and enhanced erythrocyte aggregation. While the protein changes could simply be markers of early microangiopathy, it appears much more likely that they are causally linked to vascular changes, because of the limited number of plasma protein changes linked to microangiopathy. Almost all plasma protein levels are changed in diabetes, but only those plasma protein changes that affect the viscosity of plasma and increase the aggregatability of erythrocytes are also linked to microangiopathy. Arguing against their role in the production of microangiopathy is a lack of specificity to diabetes. Similar plasma protein changes are found in other chronic disorders not accompanied by such striking vascular sequelae, such as in chronic rheumatoid arthritis.

A number of studies have suggested rheologic abnormality of erythrocytes in diabetes. The abnormality has been described as increased rigidity [8] or reduced deformability [9, 10] of erythrocytes, depending on the technique used in its assessment. The term "rigidity" was linked to the flow properties of blood in viscometry [8] and "deformability" to the ability of red cells to pass through 5μ nucleopore filters [9, 10]. These measurement techniques do not distinguish clearly between erythrocyte aggregation and erythrocyte deformation. Since any changes in erythrocyte deformability should have a complex effect on the flow properties of blood, a direct investigation of red cell flow properties was carried out using micropipettes and an artificial suspending medium. Deformability studies on erythrocytes have often been carried out in micropipettes [11] but commonly the pipettes used were 3μ in diameter, very close to the limiting size for entry of erythrocytes. The 3μ system is quite sensitive to changes in red cell surface to volume ratio. Because most muscle capillaries are $4-5\mu$ in diameter, we selected 4μ as more physiologic. We have found evidence of an abnormality of red cell deformability in diabetes which has a potential role in the development of vascular disease.

MICROPIPETTE STUDIES

Patients from three disease categories were selected for analysis: 1) overt diabetics, about half of whom required insulin; 2) individuals with hereditary spherocytosis; and 3) chronic rheumatoid arthritics who had been found to have plasma protein changes similar to those seen in diabetes. Red cells from these individuals were compared in paired studies to those from control subjects who were selected to be within a decade in age and, except in the earliest studies, were of the same sex. Red cells were prepared from blood collected in EDTA anticoagulant by washing with saline and then washing and suspending in albuminated Ringer's solution. Seven of each individual's red cells were studied in the same pipette. Peak pressure difference in millimeters of water (mm H_2O) was recorded after a standard oscillatory motion over a 130μ path and back in three seconds was established. A second observer recorded the peak-to-peak water pressure change when told by the first observer that a satisfactory motion had been established. Motion pictures were taken with a Bolex 16 mm camera and a Leitz trinocular microscope. The technique is illustrated and described in detail elsewhere [12].

The pressure requirements for standard motion in paired studies in 4μ pipettes are shown in Figures 1 and 2. The studies are compared as ratios to controls in Figure 3. It can be seen that in diabetes a 50% greater mean pressure was required for erythrocyte motion in the nine paired studies. In the six paired studies of hereditary spherocytosis, a mean 14% increase was found. No difference was found in rheumatoid arthritis in six paired studies. Studies on diabetic and control rats were also carried out. They showed the same abnormality as in diabetic humans, a 50% increase in pressure [12].

An unmeasurably low pressure was required to cause India ink-containing Ringer's solution to flow at the same rate as the erythrocytes [12]. Calculation of Poiseuille flow of Ringer's solution indicates that this pressure should exceed 1 mm H_2O. We therefore studied the performance of the water manometer used. It consists of two vertical glass tubes 3 mm in diameter, connected by a short piece of rubber tubing. The air reservoir system has a volume of approximately 600 ml. Its pressure to the manometer is conveyed by rubber tubing connected to the left manometer tube. With a change in air pressure, the fluid level in the two graduated glass tubes moves in opposite directions. The manometer was originally filled with water containing 0.3% India ink. This solution has a surface tension of 61 dynes/cm, measured using a deNuoy tensiometer. The manometer's performance was evaluated by changing the reservoir air volume. For each milliliter volume change, the manometer fluid moves 7.8 mm in opposite directions

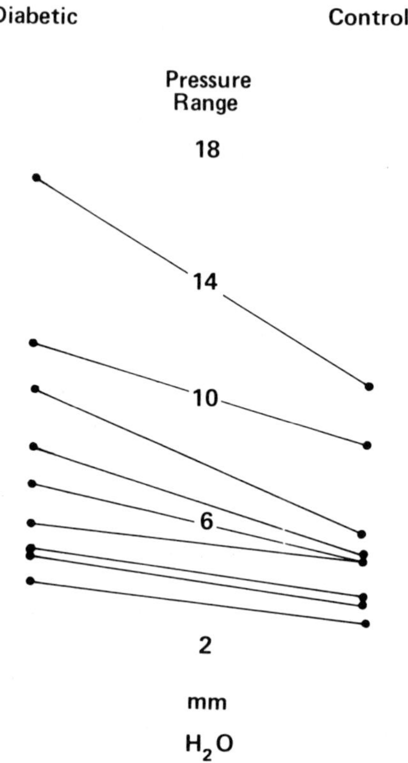

Fig. 1. The pressure fluctuation observed during cyclic motion of individual erythrocytes in a 4μ pipette is shown. In each study diabetic and control erythrocytes were compared. Values for seven individual cells were averaged and are given in millimeters of water. Each pair is connected by a line. In all nine studies the diabetic value exceeded the control.

in the two arms. Its response to three-second cycles of air volume change is smaller than its response to permanent stable changes in pressure of small magnitude. The system used in our original studies was found to require 1.1 ± 0.2 mm H_2O pressure to initiate measurable manometer motion (Fig. 4). If sodium dodecyl sulfate and salt are added in sufficient amounts to produce a 33 dynes/cm surface tension fluid with better wetting properties, the system has no such baseline pressure requirement (Fig. 5). Inspection shows that the contact angle is not maintained at zero degrees in the absence of detergent. The readings in Figures 1 and 2 are, therefore, too low by 1.1 ± 0.2 mm H_2O. This pressure is also addable to the unmeasurably low value obtained for motion of the India ink-

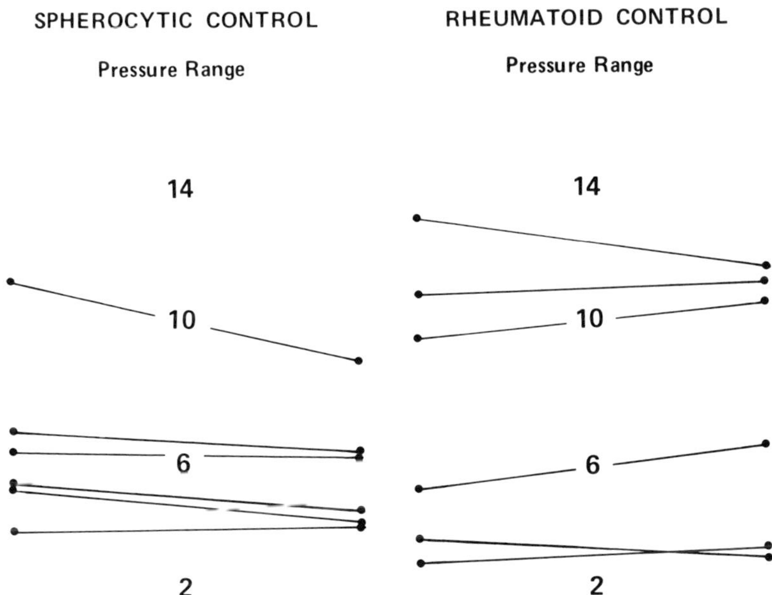

Fig. 2. The format of Figure 1 is followed in comparing studies of spherocytic vs healthy erythrocytes and rheumatoid arthritic vs healthy erythrocytes. The lines connect each paired study. There is a much less striking increase in flow resistance of spherocytes and no increase for rheumatoid red cells. The pressure is in millimeters of water.

containing Ringer's solution, which therefore may have been as much as 1.5 mm H_2O. The observed pressures in Figures 1 and 2 still represent actual increments produced by the erythrocytes.

Motion pictures of red cells moving back and forth in micropipettes document the striking changes in shape that develop with reversal of direction. Erythrocyte appearance is different at the two ends of the path traversed. The central fold is also narrower near the tip than at the other end, indicating that compression of the erythrocyte occurs during movement toward the tip and expansion occurs during movement away from the tip. Further evidence to support asymmetry during pipette movement was gained by placing two erythrocytes spaced about 10μ apart in the same pipette. Motion of the two erythrocytes required far less than a doubling of pressure. Indeed, different pipettes showed different increased pressure requirements. These observations indicate that not only is there bending of each erythrocyte from a discocytic into a folded pancake shape but that the

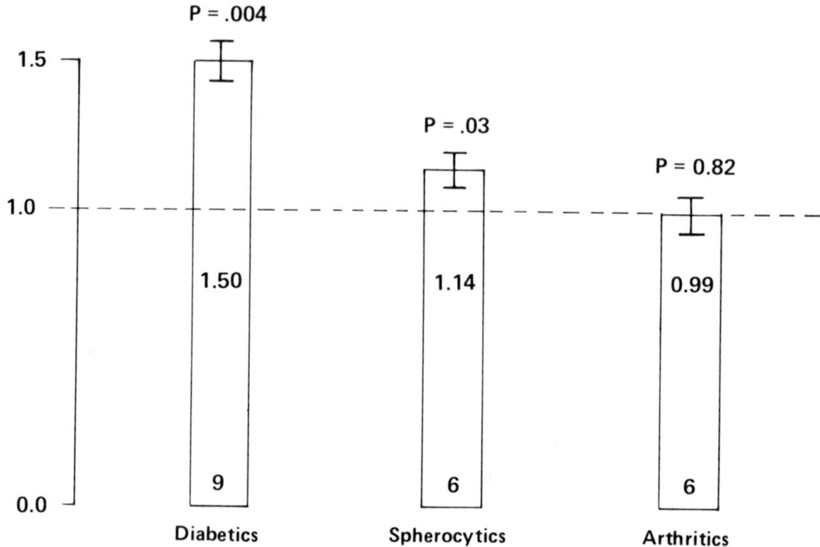

Fig. 3. The mean ratios of paired pressures are shown in histogram form with one standard error indicated by brackets. The data shown in Figures 1 and 2, treated as ratios and examined statistically, show a significant difference of 50% magnitude for diabetic erythrocytes. The increase is only 15% for spherocytes. A 1% decrease was observed for rheumatoid arthritic red cells. A vertical scale value of 1.0 represents pressure value of the control cell. The number of studies in each group is shown at the bottom of each histogram.

erythrocyte is required to streamline during each direction change, and to narrow and widen as it goes to and from the tip. Not only is the resistance to bending of the erythrocyte being tested, but also its resistance to flow. We set about to determine whether increased rigidity (decreased elasticity) or increased viscosity was responsible for the observed change. Following Hoeber and Hochmuth's technique [13], we ejected individual erythrocytes from 4μ pipettes while taking slow motion pictures (64 frames/second). Diabetic erythrocytes were compared to nondiabetic erythrocytes. In three studies it was shown that diabetic erythrocytes unfolded in 50% greater time than the nondiabetic erythrocytes. Rather than increased elastic energy storage, elevated flow resistance was present.

HEMOGLOBIN VISCOSITY STUDIES

Since hemoglobin Alc is regularly elevated in diabetic erythrocytes it became a prime candidate for increasing diabetic red cell flow resistance. Intrinsic viscosity studies were therefore carried out to compare hemoglobin from individual diabetic and control subjects.

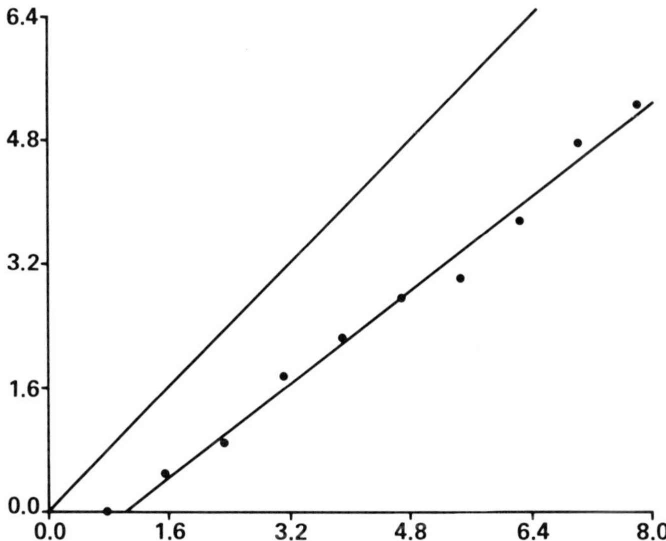

Fig. 4. The dynamic response of the water manometer used in the studies shown in Figures 1 through 3 is compared to its static response. Because the miniscus contact angle was imperfect, small changes in pressure were not measured adequately by the manometer. There is also a slight tendency by the manometer to underestimate larger pressure fluctuations in addition to its resistance to motion at very low pressure gradients.

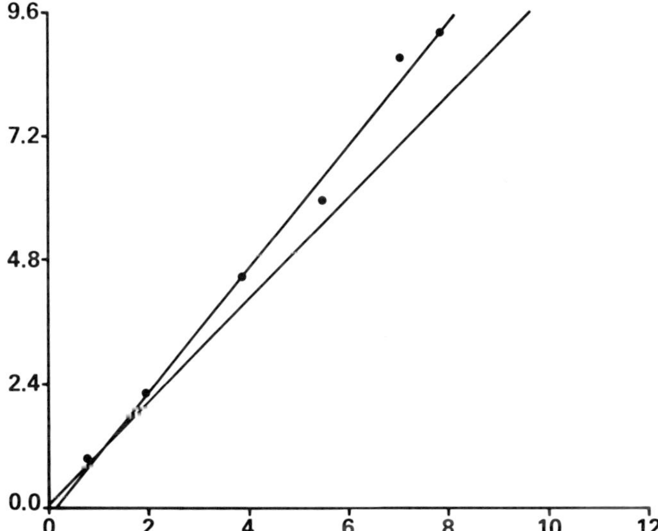

Fig. 5. The effect of the addition of a detergent which lowers manometer fluid surface tension and enhances wetting is shown. The resistance of the manometer to small changes of pressure disappears and there is actually an elevation of the observed readings above static values with large pressure changes. This action occurs presumably because of the high rate of motion of the fluid in the manometer. In the absence of detergent, the manometer underestimates the actual pressure; it slightly overestimates pressure in its presence.

Hemoglobin intrinsic viscosity analyses were performed with Cannon-Ubbelohde semi-micro viscometer at 37°C by a technique already described [14]. Intrinsic viscosity is determined as the slope of specific fluidity (1−1/relative viscosity) vs concentration. In contrast to plasma albumin [14], hemoglobin's regression line passed through the zero concentration value at a specific fluidity above 0.01. Studies of glass adsorption properties of hemoglobin solutions showed that this was the basis for the difference. For this reason, the regression was recalculated using increasing solvent time to obtain a line passing through the origin. The two intrinsic viscosities differ by only a small amount (Fig. 6). The fitted line slope should be a more accurate intrinsic viscosity estimation. In six diabetic and six control studies the intrinsic viscosity was found to be the same within the limit of resolution of the system (Fig. 7).

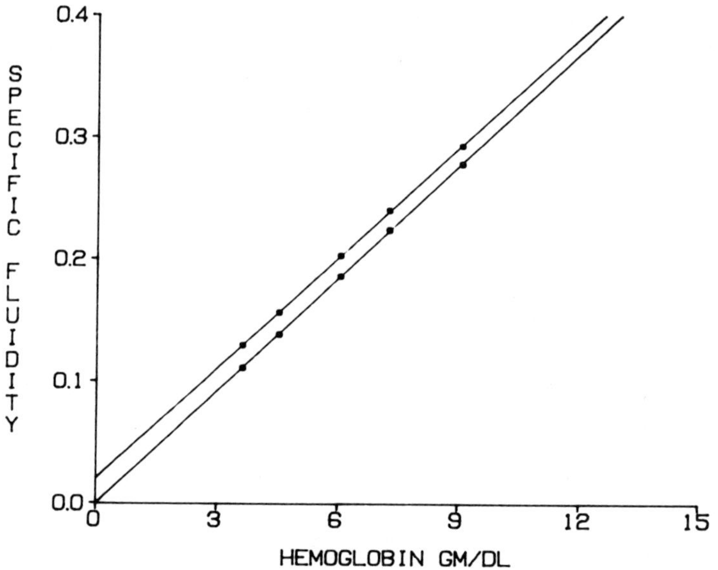

Fig. 6. The relation of specific fluidity to hemoglobin concentration is shown. With the original solvent time the regression line ordinate crosses the intercept well above zero. Recalculation using an appropriately higher solvent time allows the regression line to pass through zero. The curve remains linear and the slope increases very slightly. There is evidence that hemoglobin is adsorbed to the surface of the glass in the capillary during viscosity studies. Advancement of the solvent time appears to improve the estimation of intrinsic viscosity.

	Controls (6)	Diabetics (6)	t
Original	2.91 ± .05	2.96 ± .06	1.61
Fitted	2.97 ± .05	3.01 ± .06	1.59

ml / gm · mean ± S.D.

Fig. 7. Diabetic and control hemoglobin intrinsic viscosity mean values (six individuals in each group) are shown. The intrinsic viscosity measurement technique is highly precise, and mean values differ only slightly whether either the original data or the fitted data are examined. The diabetic increase, in addition to being minimal, is not statistically significant.

Because the intracellular concentration of hemoglobin is high, hemoglobin molecular interactions might be modified by the presence of glucose attached to hemoglobin to form hemoglobin AIc [15]. For that reason viscosity studies of hemoglobin from diabetic and nondiabetic cells were carried out at concentrations at and above physiologic range — 32 to 36 gm/dl. Hemoglobin was prepared using toluene, following the method of Antonini and Brunori [16]. It was concentrated in a Zeineh dialyzer-concentrator. The hemoglobin studied was shown to be undenatured by dilution followed by estimation of its intrinsic viscosity. The linear relation between specific fluidity and concentration found in the intrinsic viscosity studies cannot persist at high concentrations unless the viscosity of hemoglobin solutions becomes rapidly infinite (specific fluidity = 1.0) at about 34 gm/dl (Fig. 6). We observed development of curvilinearity at specific fluidity values about 0.53, corresponding to hemoglobin concentrations about 17.7 gm/dl (Fig. 8). The following empiric equation has been fitted by the least-squares method to the data.

$$\frac{1}{1 - \eta_{rel}} = [\eta]\, C_{\eta l} + (1 - [\eta]\, C_{\eta l}) \tanh \frac{[\eta]}{(1 - [\eta]\, C_{\eta l})}(C - C_{\eta l}) \tag{1}$$

Where $C_{\eta l}$ is the hemoglobin concentration at the beginning of nonlinearity, $[\eta]$ the intrinsic viscosity, and C the actual hemoglobin concentration. The equation applies only above $C_{\eta l}$, and only points above this value are fitted. Below $C_{\eta l}$ the equation is

$$1 - \frac{1}{\eta_{rel}} = [\eta]\, C \tag{2}$$

The equations can be used to describe relative viscosity directly.

Below $C_{\eta l}$

$$\eta_{rel} = \frac{1}{1 - [\eta]\, C} \tag{3}$$

Above $C_{\eta l}$

$$\eta_{rel} = \frac{1}{(1 - [\eta]\, C_{\eta l})\, [1 - \tanh\, [\eta]\, \dfrac{(C - C_{\eta l})}{1 - [\eta]\, C_{\eta l}}]} \tag{4}$$

Two typical fitted curves, the upper diabetic and the lower nondiabetic, are shown in Figures 8 and 9. Diabetic hemoglobin appears to flow slightly differently from normal hemoglobin at physiologic concentration, but this difference is clearly not sufficient to explain the 50% increase in the flow resistance offered by diabetic erythrocytes (Figs. 1 and 3).

FLOW EFFECT OF INCREASED RED CELL HEMOGLOBIN

The mean corpuscular hemoglobin concentration was determined (Coulter Model S) in the six hereditary spherocytics (Fig. 2). It is compared with a large healthy control group in Figure 10. In spherocytosis the mean corpuscular hemoglobin concentration was elevated above the normal range. Spherocyte resistance to flow in the 4μ pipette was 14% greater than the flow resistance of healthy erythrocytes (Fig. 2). Since hemoglobin concentration is the major determinant of intracellular viscosity, it is possible that elevated intracellular hemoglobin concentration in spherocytosis could affect the flow properties of spherocytes. The effect of increasing hemoglobin concentration on intracellular viscosity may be calculated using Equation 4. In six studies of pooled healthy subject hemoglobin the following values for constants were obtained: $[\eta] = 3.00$ ml/gm and $C_{\eta l} = 17.7$ gm/dl. Using these values, the mean control relative viscosity was 10.27, and the spherocytosis relative viscosity was 12.35. For a solvent viscosity of 0.7 centipoise at 37°C, these values correspond to intracellular viscosities of 7.19 and 8.65 centipoise, a difference of 20%.

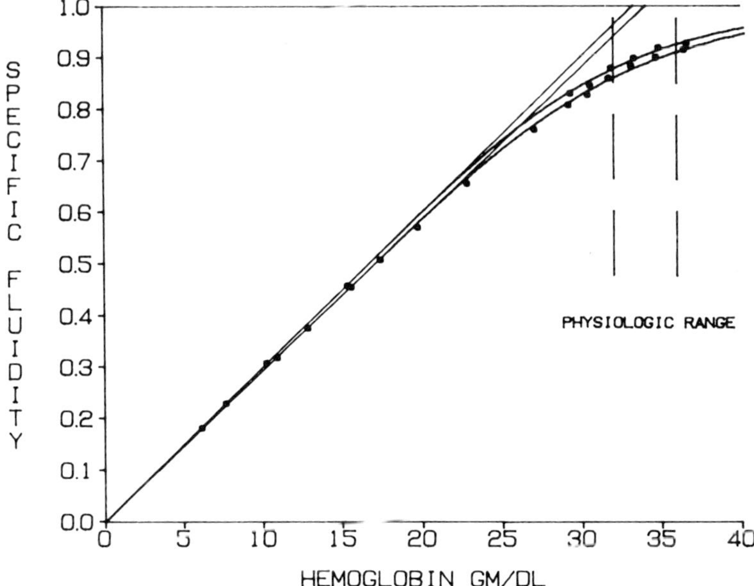

Fig. 8. The pattern shown in Figure 6 has now been extended to higher concentrations of hemoglobin. A hemoglobin pool derived from four diabetics is compared to a control pool. At high hemoglobin concentrations the fluidity relationship becomes curvilinear. Its curvature develops at a specific fluidity of 0.53. A hyperbolic tangent function has been used empirically to determine this nonlinearity point. When both the nonlinearity point and the intrinsic viscosity are known the viscosity concentration is completely described. The diabetic hemoglobin curve is just above the control curve.

Viscometry of erythrocyte suspensions was accomplished in a Gilinson-Dauwalter-Merrill (GDM) viscometer [17] constructed by this laboratory. In our GDM system an angle sensor (1 mV/μradian) is used to signal motion of the air-bearing table. Its signal is amplified in a solid-state system that supplies sufficient voltage to maintain minimum angular deflection. Amplifier gain can be varied. The required voltage is measured by a digital voltmeter (Tektronix DM501), which is queried by Tektronix 31 programmable calculator that averages 100 voltage readings taken five times each second. A Krieger Elrod [18] analysis leads ultimately to presentation of the data by an X-Y plotter. The system has been calibrated using standard viscosity oils. The viscometer differs from the original GDM viscometer in having an unscored gold-plated brass cylinder rather than a scored coin silver one. The grooves present on the outer cup are not as deep as those originally described. The response time of our system at low shear rate is more rapid, a maximum reading being obtained six to eight seconds after the motor is activated.

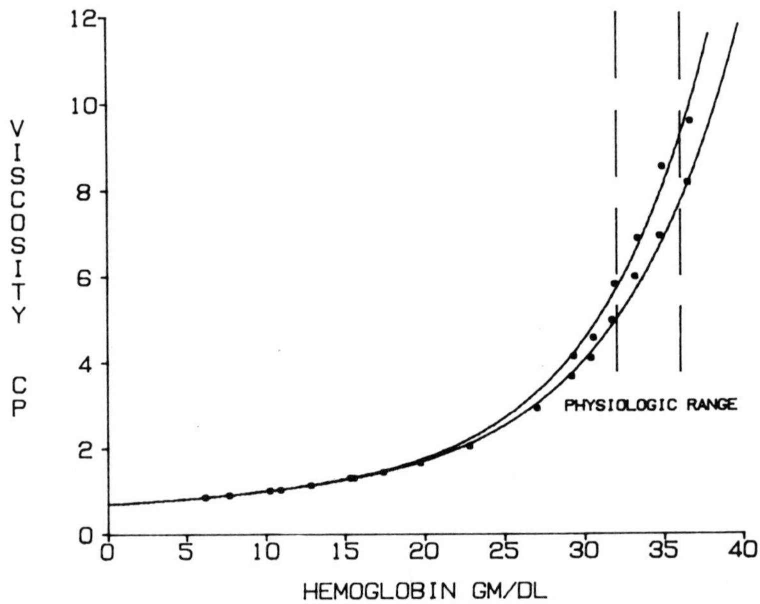

Fig. 9. The same values plotted in Figure 8 are now shown as viscosity vs concentration. Viscosity is shown as measured in centipoise at 37°C. The vertical broken lines define hemoglobin's physiologic concentration range as in Figure 8. It can be seen that there is a marked increment in viscosity from the lower to the upper end of the physiologic range of hemoglobin. A slight increase in diabetic hemoglobin's viscosity, compared to nondiabetic hemoglobin, is shown; a series of studies have shown that this increase is not statistically significant.

Control (63)	34.6 ± 0.9
Spherocytic (6)	36.2 ± 0.7

gm/dl   Mean ± S.D.

Fig. 10. The figure comparing the mean corpuscular hemoglobin concentration of spherocytes studied in Figure 2 and the hemoglobin concentration of red cells from 63 healthy subjects is shown. The spherocytes from the six subjects studied had the increase in corpuscular hemoglobin concentration well known to occur in spherocytosis.

The effect of raising corpuscular hemoglobin concentration was examined
by exposing erythrocytes to two different osmotic strength buffers in the GDM
viscometer. The results are shown in Figure 11. The buffers were similar in phosphate, potassium, glucose (100 mg/dl), and albumin content and differed only in
amounts of sodium chloride present. The 305 mOsm buffer had the same sodium,
potassium, and phosphate content as that used by Cokelet and Meiselman [19].
The 329 mOsm has more sodium chloride. There was a regular elevation of viscosity in the shear rate range between 125 and 1 sec^{-1}, in such fashion that the
power law coefficient was unchanged. The mean corpuscular hemoglobin concentration was found to be 35.9 gm/dl at 305 mOsm and 37.2 gm/dl at 329
mOsm. The predicted viscosities are 8.35 and 9.72 cp at 37°C. Intracellular viscosity is therefore enhanced by 16%. At the same time the directly measured
blood viscosity enhancement shown in Figure 11 is 15% at 125 sec^{-1} and 11%
at 2.5 sec^{-1}.

ERYTHROCYTE LIPIDS

Studies were carried out measuring the lipid elements in erythrocyte plasma
membranes. Erythrocyte membrane cholesterol and phospholipid were studied
following extraction with ethanol. The cholesterol was measured colorimetrically
using the Liebermann Burchard reaction. Phospholipid was hydrolyzed by reflux
in perchloric acid in a 275°C sand bath, followed by phosphate analysis. Cholesterol level was found to be very similar in diabetic and control erythrocytes (Fig.
12). There was a mild decline in phospholipid content. It proved not to be statistically significant.

POSSIBLE BASIS FOR ALTERED RED CELL RHEOLOGY

Impairment of erythrocyte flow properties in the 4μ pipette was demonstrated
for both diabetic and spherocytic erythrocytes but not for rheumatoid arthritic
and healthy erythrocytes. In the case of spherocytic erythrocytes an increased
intracellular hemoglobin concentration was found that should produce an increase
in the intracellular viscosity, which might be sufficient to account for the observed
increase in resistance to flow. The relatively mild flow impairment of spherocytes
in the 4μ pipette should be contrasted to studies in 3μ pipettes [11]. In the latter
a far more striking flow impairment can be attributed to the spherocyte's reduced
surface-to-volume ratio. In the 3μ pipette the cell surface could be stretched by
increased intracellular pressure, and increased intracellular pressure would make
motion more difficult by compressing the space between the cell and the pipette
wall.

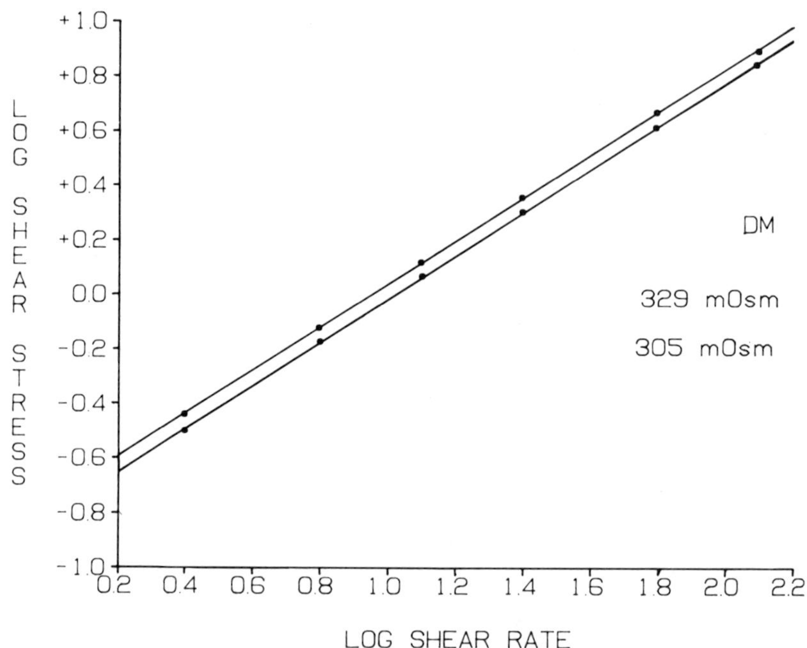

LOG SHEAR RATE

Fig. 11. A healthy individual's erythrocytes were suspended in buffers containing different concentrations of sodium chloride but similar amounts of phosphate and potassium. The increased osmolality produced is shown. It results in shrinkage of the erythrocytes and an increase in mean corpuscular hemoglobin concentration. The increase in intracellular viscosity (see Fig. 9) is detectable at high shear rate at constant (60%) hematocrit.

	Diabetic	Control
Cholesterol	1.24 ± .09 (11)	1.23 ± .05 (10)
Phospholipid	3.2 ± 0.3 (7)	3.4 ± 0.4 (6)

mg/ml cells - mean ± S.D. (N)

Fig. 12. Alcohol extraction was used to isolate the cholesterol and phospholipid from erythrocytes of diabetic and nondiabetic subjects. The cholesterol content was found to be quite similar in the two groups. Phospholipid showed a slight decline in diabetes compared to the control population but the decrease was not statistically significant. All values have been reported in mg/ml of red cells. Plasma cholesterol and phospholipid are usually reported as mg/dl. A comparison indicates that erythrocyte cholesterol content is lower than that of plasma (2.2 vs 1.2 mg/ml), while red cell phospholipid content is higher (2.5 vs 3.4).

In two other disorders there have been reports of impaired erythrocyte flow properties [20] . In β-thalassemia, hemoglobin concentration is lower than normal and the cells are smaller than normal, with a high surface-to-volume ratio, but an excess synthesis of alpha chains alters the cell's interior. The reduction in deformability of the sickle cell is attributable to both denaturation and membrane-binding of sickle hemoglobin [21] and permanent deformation of the plasma membrane induced by reversible hemoglobin crystallization [22] . Irreversible distortion of the plasma membrane makes some sickle cells far more resistant to pipette flow than others [20] .

What is the basis for reduced deformability in diabetes? Motion pictures demonstrated that the red cell membrane is quite deformed in the 4μ pipette. The deformation increases with change in direction of flow. The suddenness of change in direction of flow determines the degree of distortion of the membrane, suggesting that elastic, viscous, and inertial effects are capable of altering red cell membrane motion. Membrane motion could be seen to be slower in the motion pictures of diabetic erythrocytes.

In addition to changing shape with each alteration in direction, the red cell must decrease in its diameter as it moves toward the tip of the pipette to avoid impairing its flow. The expanding and collapsing of the erythrocyte in its passage back and forth in the pipette is resisted by both membrane and interior viscosity. The gap between the capillary wall and the plasma membrane would be narrowed by increased resistance to bending of the erythrocyte. This could be produced by increased viscosity of either the red cell interior or its membrane. Delayed return to discocyte shape following ejection from a micropipette indicates that no excess of elastic energy is stored in the diabetic erythrocyte. Indeed, the reduction suggests a greater viscosity. Although both membrane and interior changes could explain the diabetic findings, our hemoglobin studies make it unlikely that the interior viscosity is sufficiently elevated.

We therefore conclude that an increase in the membrane viscosity of the diabetic erythrocyte should be suspected. Techniques to measure erythrocyte membrane viscosity have been described by Evans and Hochmuth [23] , but we have not yet been successful in applying them to the diabetic erythrocyte.

POSSIBLE EFFECTS OF ALTERED ERYTHROCYTE RHEOLOGY

It is interesting to consider first what the effects of altered erythrocyte flow properties are on the bulk flow of blood. Erythrocyte deformation is no longer required as it is in the capillary. We have examined the effect of increasing intracellular hemoglobin concentration and found it to be detectable. Contributions during flow by streamlining of erythrocyte membranes are not as easy to assess. It appears quite possible that membrane viscosity changes principally affecting the rate of change of shape of erythrocytes could affect flow at very different shear rates than the intracellular viscosity changes.

The potential effect of alterations of the erythrocyte's interior or membrane viscosity on erythrocyte aggregation should also be considered. The first step in aggregation, erythrocyte doublet formation, is characterized by the need for each cell to accommodate its membrane to the other cell. It is possible that changes in membrane and interior viscosity might be detected by observing doublet formation. Any observed effect would apply to both aggregation and disaggregation of erythrocytes.

What would be the effect on blood vessels of a reduced deformability of erythrocytes in diabetes and spherocytosis? In diabetes there is a specific thickening of the capillary basement membrane, most striking in skeletal muscle [5] and in the renal glomerulus [4]. In skeletal muscle the erythrocyte must be deformed to pass through the capillary, a process that may enhance its ability to deliver oxygen. With deformation the erythrocyte compresses itself against the endothelial cell, delivering mechanical energy to it and to its underlying basement membrane [24]. The endothelial cell and its underlying basement membrane function together to maintain the endothelial junction as a barrier between the vascular and extravascular spaces. Enhanced mechanical intrusions of erythrocytes may stimulate thickening of basement membrane in proportion to the force delivered by the erythrocyte to the endothelial cell. If this occurs then reduced deformability of the diabetic erythrocytes could be a direct mechanical means to provoke basement membrane thickening. The existence of such a mechanism would mean that a milder but measurable thickening of basement membrane should occur in hereditary spherocytosis. One might also expect an enhancement of the leakage of high molecular weight substances from the vascular compartment into the extravascular space, as is seen in the eye in diabetes [25].

Since aggregation-enhancing plasma protein changes have been linked to clinically recognizable diabetic microangiopathy, which in turn is related to arteriolar disease, it is of interest to consider the potential effect of an aggregation-deformation interaction on arterioles. Intravascular aggregates occur only occasionally, but when one develops in the arterial system, because each arterial branch is smaller than its predecessor, the aggregate will ultimately block a small vessel until it is disrupted. The disruption is produced by the rising pressure gradient that develops when flow ceases. This pressure gradient must overcome the attractive force holding the erythrocytes together before the cells will separate. The rate at which they separate would be influenced by reduced deformability. Since the pressure continues to rise if disaggregation is delayed, the vessel wall will experience more strain. Since vessel strain induced by hypertension causes vessel wall sclerosis [26], it appears plausible that smaller vessels could become sclerotic in diabetes from the transient pressure elevations induced by erythrocyte aggregation combined with reduced deformability.

ACKNOWLEDGMENTS

The studies reported here were supported by the Doris Fay Palmer Trust, the Kroc Foundation, and the American Diabetes Association of Southern California Affiliate. The rotational viscometry was carried out by Mr. Jerzi Stocki and the capillary viscometry by Ms. Kathleen Gion. Dr. John LaPuma carried out the micropipette studies, which were supervised by Dr. Nyle G. Utterback, who was also responsible for supervision of the construction of the GDM discometer and for its modifications from the original design.

REFERENCES

1. Lundbaek K, Jensen VA: Long-term diabetes. In "The Clinical Picture in Diabetes Mellitus of 15–25 Years' Duration With a Follow-Up of a Regional Series of Cases." Copenhagen: Ejnar Munksgaard, 1953, pp 158–192.
2. McMillan DE: Deterioration of the microcirculation in diabetes. Diabetes 24:944–957, 1975.
3. Norton WL: Comparison of the microangiopathy of systemic lupus erythematosus, dermatomyositis, scleroderma, and diabetes mellitus. Lab Invest 22:301–308, 1970.
4. Osterby R: Early phases in the development of diabetic glomerulopathy. Acta Med Scand Suppl 574:13–80, 1975.
5. Siperstein MD, Norton W, Unger RH, et al: Muscle capillary basement membrane width in normal diabetic and prediabetic patients. Trans Assoc Am Physicians 79:330–347, 1966.
6. McGee WG, Ashworth CT: Fine structure of chronic hypertensive arteriopathy in the human kidney. Br J Ophthalmol 43:273–299, 1963.
7. McMillan DE: Plasma protein changes, blood viscosity, and diabetic microangiopathy. Diabetes 25:858–864, 1976.
8. Dintenfass L: Blood viscosity factors in severe nondiabetic and diabetic retinopathy. Biorheology 14:151–157, 1977.
9. Schmid-Schonbein H, Volger E: Red-cell aggregation and red-cell deformability in diabetes. Diabetes 25:897–902, 1976.
10. Barnes AJ, Locke P, Scudder PR, et al: Is hyperviscosity a treatable component of diabetic microcirculatory disease? Lancet 2:278–291, 1977.
11. La Celle PL: Alteration of deformability of the erythrocyte membrane in stored blood. Transfusion 9:238–245, 1969.
12. McMillan DE, Utterback NJ, La Puma J: Reduced erythrocyte deformability in diabetes. Diabetes 27:895–901, 1978.
13. Hochmuth TW, Hochmuth RM: Measurement of red cell modules of elasticity by in vitro and model cell experiments. J Basic Engin 92:604–609, 1970.
14. McMillan DE: A comparison of five methods for obtaining the intrinsic viscosity of bovine serum albumin. Biopolymers 13:1367–1376, 1974.
15. Bunn HF, Gabbay KH, Gallop PM: The glycosylation of hemoglobin: Relevance to diabetes mellitus. Science 200:21–27, 1978.
16. Antonini E, Brunori M: "Hemoglobin and Myoglobin in Their Reactions With Ligands." Amsterdam: North-Holland, 1971.

17. Gilinson PJ, Dauwalter CR, Merrill EW: A rotational viscometer using an A.C. torque to balance loop and air bearing. Trans Soc Rheol 7:319–331, 1963.
18. Krieger IM, Elrod H: Direct determination of the flow curves of non-Newtonian fluids. II. Shearing rate in the concentric cylinder viscometer. J Appl Physics 24:134–136, 1953.
19. Cokelet GR, Meiselman HJ: Rheological comparison of hemoglobin solutions and erythrocyte suspensions. Science 162:275–277, 1968.
20. La Celle PL, Kirkpatrick FH: Determinants of erythrocyte membrane elasticity. In Brewer G (ed): "Erythrocyte Structure and Function." New York: Alan R. Liss, Inc., 1975, pp 535–557.
21. Asakura T, Minakata K, Adachi K, et al: Denatured hemoglobin in sickle erythrocytes. J Clin Invest 59:633–640, 1977.
22. Lux SE, John KM, Karnovsky MM: Irreversible deformation of the spectrin-actin lattice in irreversibly sickled cells. J Clin Invest 58:955–963, 1976.
23. Evans EA, Hochmuth RM: Membrane viscoelasticity. Biophys J 16:1–11, 1976.
24. Simionescu N, Simionescu M, Palade GE: Structural basis of permeability in sequential segments of the microvasculature of the diaphragm. II. Microvasc Res 15:17–36, 1978 (Fig. 11).
25. Cunha-Vaz J, Faria De Abreu J, Campos AJ, et al: Early breakdown of the blood-retinal barrier in diabetes. Br J Ophthalmol 59:649–656, 1975.
26. Ooshima A, Fuller GC, Cardinale GJ, et al: Increased collagen synthesis in blood vessels of hypertensive rats and its reversal by antihypertensive agents. Proc Natl Acad Sci USA 71:3019–3023, 1974.

12

Microvascular Transit of Normal, Immature, and Altered Red Blood Cells in Spleen Versus Skeletal Muscle

Alan C. Groom

Careful observation of the motions of individual red cells in flowing blood shows that the cells are continually changing their shape. This is true when blood of normal hematocrit flows through tubes of comparatively large diameter, and arises because of cell—cell interactions [25]. The changes of shape are even more marked, however, when red cells pass through blood capillaries [6]; the diameters of these vessels are often smaller than the disc diameter of the red cell, and thus the alterations in shape occur in response to cell—vessel wall interactions. It is a very salutary experience to watch these changes occurring and to remember that this is the lot of the red cell throughout its entire life-span of four months. Clearly, the properties of deformability and nonadhesiveness (to other blood cells and vessel walls) are indispensable to the red cell. A loss of deformability may not be serious as regards the flow of blood in large vessels unless the hematocrit is high, in which case a "log-jam" may occur; ie, the viscosity becomes infinite [9, 30]. At the microcirculatory level, however, the constraints are more severe; unless the red cells show sufficient deformability and nonadhesiveness they are likely to become lodged in the capillaries, impairing the distribution of blood flow to the tissues.

Much has been written on the importance of the shape, deformability, and surface properties of red cells. Since the membrane area and the volume of a given cell cannot change significantly when it deforms, the ratio (surface area/volume) sets a limit to the amount of deformation that can occur [8, 61]. In consequence, even though the membrane be perfectly flexible and the cytoplasm be of low viscosity (6 cP approx [21]), there is a lower limit to the diameter of cylindrical tubes through which a given cell can pass without damage. For human red cells this limit is near 2.5 μm [39], roughly one-half the internal

Erythrocyte Mechanics and Blood Flow, pages 229–259

diameter of human capillaries. In point of fact, normal red cells passing through tubes of capillary diameter (less than the disc diameter of undeformed cells) offer very little resistance to flow [4, 41, 61]. The tendency of red cells to adhere to one another in the presence of fibrinogen [10, 11] does not seem to present a problem in healthy subjects, for although rouleaux form spontaneously in parts of the microcirculation when the flow is arrested, they break up instantly when it is restored [5, 20]. The question to which we must address ourselves is this: What changes in the physical properties of the red cell can be tolerated before impairment of microcirculatory flow occurs?

Many workers have studied the physical properties of red cells in vitro and have attempted to predict how alterations of these properties would affect red cell flow in the microcirculation. Our own premise has been that the spleen (Fig. 1) must contain the answer to the above question, for it appears to be a remarkably effective filter that clears abnormal cells or particles from the blood [38]. The mechanisms by which the spleen traps cells that are so slightly altered as to escape destruction elsewhere in the body are not fully understood. However, it seems clear that trapping occurs following changes in the physical properties of red cells. My colleagues and I have therefore made use of the investigations of other workers to determine the treatments necessary to alter red cell properties in various ways (eg, deformability, surface charge, etc). We have then studied the retention of such altered red cells in the spleen, as a means of determining what alterations in physical properties the body considers most undesirable. Following this we have investigated the delay or retention of altered red cells, deemed undesirable by the spleen, in the microvascular bed of skeletal muscle. This particular tissue was chosen since it represents 45% of lean body mass and is responsible for the

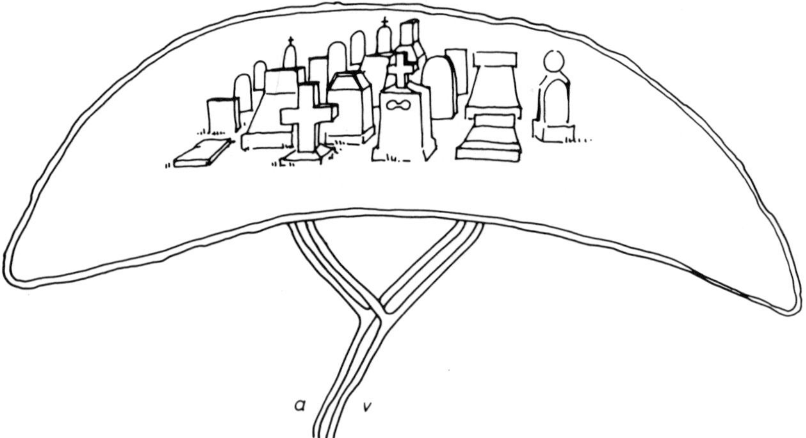

Fig. 1. Hematologists' view of the red cell function of the spleen: a graveyard.

external work of the body; it is clearly important that microvascular blood flow
to this tissue be not impaired.

The microvascular bed of the spleen is very unusual and, although many of
its features have been well described anatomically [63, 73, 75], little is known
about it from the hemodynamic standpoint. We considered it important to learn
something about the storage and transit of normal red cells in the spleen before
attempting to study the treatment meted out to abnormal and altered cells. In the
present report the same strategy will be followed, and we shall discuss the normal
intrasplenic distribution and flow of red blood cells before discussing the retention
of altered and abnormal red cells by the spleen and by skeletal muscle.

DISTRIBUTION AND FLOW OF RED BLOOD CELLS WITHIN
THE SPLEEN

Since we are interested in the storage and transit of red cells specifically, it is
obviously not possible to use for this investigation a tracer substance that is
confined to the plasma, nor a radioactive gas that is soluble in plasma, red cells,
and tissue. Moreover, since the splenic microcirculation is so very unusual and
complicated, we cannot assume that the distribution of radiolabeled microspheres
would be representative of that of red cells. The tracer substance would have to
be confined to red cells exclusively.

There are two ways in which we might investigate the passage of radiolabeled
red cells through the spleen:

1) Inject a bolus of labeled cells into the splenic artery, in vivo, and sample the
outflow from the splenic vein as a function of time. The problem would be that
recirculation of labeled cells would obscure the recognition of cells that took a
long time for a single transit.

2) Inject labeled red cells into the general circulation and, after they have
become uniformly distributed in the blood, isolate the spleen and perfuse it with
unlabeled blood from a donor animal, sampling the outflow as a function of time.
The advantage would be that such a "washout" method without recirculation
would allow proper investigation of the slow clearance of labeled red cells. The
problem would be that this method would require far more blood than several
donors together could provide.

Both of these methods involve the assumption that the preparative and
radiolabeling procedures have not altered the red cells in any way that would
affect, significantly, their transit through the vascular bed. Although this would
seem to be a reasonable assumption where most organs and tissues are concerned,
it may well not be true in the case of the spleen, an extremely sensitive detector
of red cell abnormalities. For this reason we considered the use of labeled cells
to be undesirable for this study.

We have used nonlabeled cells, instead, by the simple expedient of isolating the spleen and perfusing it with oxygenated, buffered, Ringer's solution (Fig. 2), sampling the outflow as a function of time [64]. This system has the advantage that we do not process the red cells first; the cells collected at the outflow are those that were contained in the spleen, in vivo, at the time of cannulation and are now simply being "washed out" by Ringer perfusion. Moreover, since this is a nonrecirculating system and the perfusate entering the organ is cell-free, we may quantitate without difficulty slow as well as fast clearance of red blood cells from the spleen. The cellular concentrations in successive samples of the outflow may be measured very precisely, over several orders of magnitude, by means of an electronic cell counter (eg, Celloscope: Particle Data, Inc.).

Kinetics of Red Cell Washout, and a Mathematical Model

Our experiments have been carried out using cats, anesthetized with sodium pentobarbital (40 mg/kg ip). The spleen was isolated surgically except for the main splenic vessels, and the animal was then heparinized (1,000 USP units/kg iv). After cannulation of the artery and vein, the organ was transferred to a lucite chamber, containing saline or mineral oil at 37°C, and was perfused with Ringer's solution of pH 7.4 equilibrated at 37°C with 5% CO_2 in O_2 (Fig. 2). This solution was prepared free of particulate matter by filtration through Millipore filters (GSWP 04700, 0.22 μm) three times before use. Perfusion was carried out under either constant-flow or constant-pressure conditions, the arterial pressure (80–100

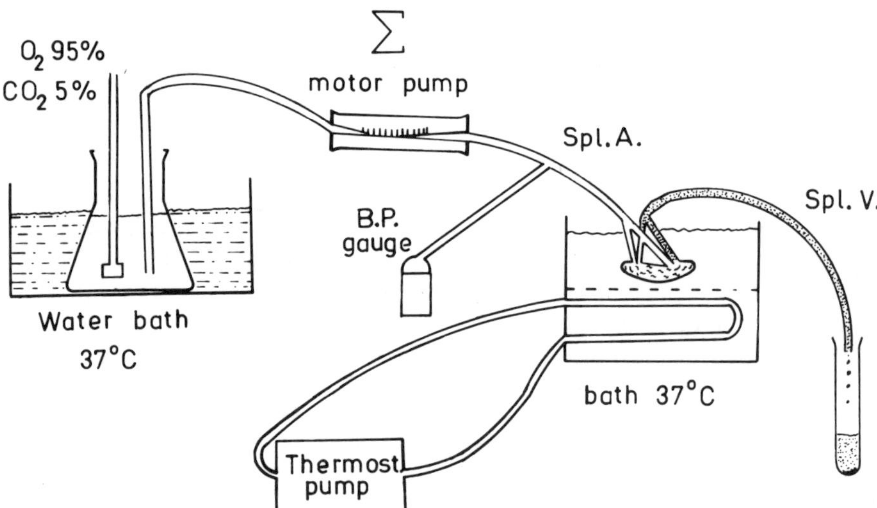

Fig. 2. Schematic diagram of perfusion system. From Levesque and Groom [47]: Washout kinetics of red cells and plasma from the spleen. Am J Physiol 231:1665–1671, 1976.

cm H_2O) being measured by means of a Statham transducer (P23DC), and the venous pressure being maintained from 0 to 8 cm H_2O. The outflow from the venous cannula was collected continuously in a measuring cyclinder except that, at a series of prearranged intervals, samples were collected in small glass tubes for measurement of red cell concentration.

These concentrations were plotted semilogarithmically versus the volume of fluid per gram splenic weight that had perfused the spleen (Fig. 3). A rapid decrease by an order of magnitude occurred, followed by a slower rate of decline over the next 1.5 orders of magnitude; this in turn was followed by an exceedingly slow but consistent fall of concentration throughout the interval from 1 to 3 liters of fluid perfused. Now, although a number of different continuous functions might be fitted successfully to a cell-washout curve, only in the case of a series of exponentials can the equations be derived as the outcome of a set of rational assumptions. A single exponential component represents washout of a single compartment, and the cell concentration, C_V, after perfusion by a volume of solution V will be related to that at V = 0 by the equation

$$\log C_V = \log C_0 - k \cdot V$$

where the slope k represents the reciprocal of the volume of the compartment.

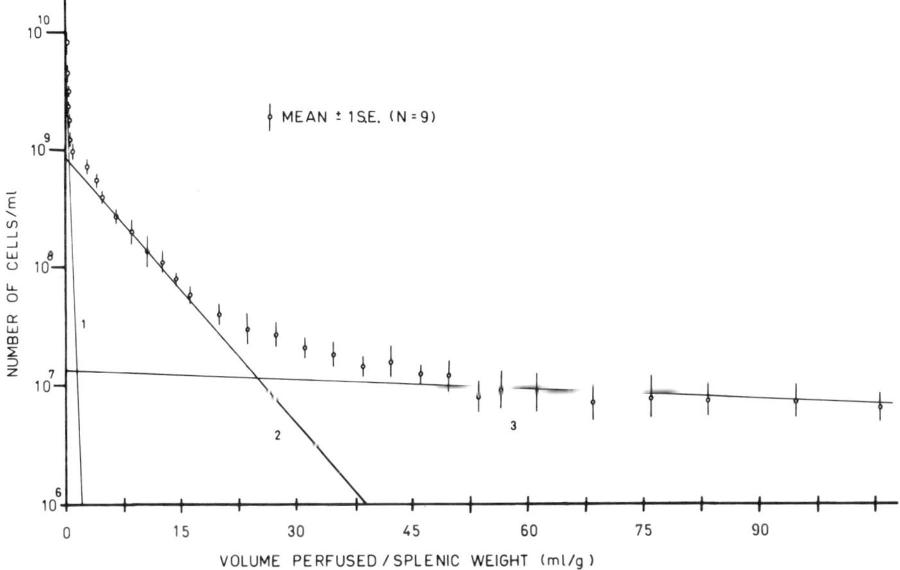

Fig. 3. Mean red cell washout curve from cat spleens during constant-flow perfusion with Ringer's solution. Cell concentration in outflow is plotted against cumulative volume of fluid perfused. Washout curve may be expressed as sum of three exponential components [1–3] (see text). From Levesque and Groom [47]: Washout kinetics of red cells and plasma from the spleen. Am J Physiol 231:1665–1671, 1976.

If a washout curve contains two exponential components, then the final part of the graph will be a straight line, whereas the earlier part will be curved continuously. To resolve a curve plotted semilogarithmically into its exponential components, a straight line is fitted to the terminal portion (Fig. 3); the absolute values of the differences between the ordinate values of the original curve and the straight line are then plotted semilogarithmically, and the procedure is repeated until only one straight line remains [62]. The straight lines obtained in this way are the exponential components of the original curve; the intercept and slope of each line represent the constant factor and rate constant of each exponential term. It is clearly important to establish that the washout curve does finally become a single exponential function of the volume perfused; ie, that the terminal portion on a semilogarithmic plot is linear. For the present case this was obviously so (Fig. 3). In each separate experiment, therefore, a straight line was fitted to this portion by the method of least squares, and the process of analyzing the curve into its exponential components was carried out. Overall mean values plus/minus standard errors were then computed for the various parameters concerned (Table I).

I have described these experiments and their analysis in some detail, because the whole idea of red cell washout and the interpretation of the washout curve present some difficulty at first acquaintance. What the curve tells us is that the washout of red cells can be resolved into three exponential components (Fig. 3, solid lines), each being characterized by a very different volume of perfusate per gram splenic weight needed to reduce the concentration of cells by one-half. These $V_{1/2}$ values turned out to be 0.067, 4.7, and 97 ml/gm, respectively, and on this basis we have designated these components, sequentially, as "fast," "intermediate,"

TABLE I. Analysis of Red Cell Washout From Cat Spleen During Perfusion With Ringer Solution (Fig. 3).

Parameter	(1) Fast	(2) Intermediate	(3) Slow
Intercept C_0:			
cells/ml	8.5×10^9	9.0×10^8	1.4×10^7
$\pm$ SE	0.7×10^9	0.4×10^8	0.7×10^7
Desaturation $V_{1/2}$:			
ml/gm	0.067	4.70	97
$\pm$ SE	0.007	0.33	3
Total flow (%)	90.3	9.6	0.15
RBC Store			
cells/gm	8.15×10^8	60.7×10^8	19.3×10^8
$\pm$ SE	1.1×10^8	2.2×10^8	5.9×10^8
ml/gm	0.033	0.242	0.077

Modified from Levesque and Groom [47]: Washout kinetics of red cells and plasma from the spleen. Am J Physiol 231:1665–1671, 1976.

and "slow." The fact that the sum of three exponential terms can describe completely the washout curve indicates that a simple model consisting of three compartments in parallel, is sufficient to approximate the washout processes from the spleen. The initial rates of cell washout from the compartments will be proportional to their perfusion rates. On this basis the fraction of the total initial concentration, in the venous outflow, of cells deriving from each compartment is equal to the fraction of the total flow of cells perfusing that compartment in the blood-perfused steady state immediately prior to washout. The cell store of such a compartment is equal to C_O/k, where C_O is the intercept (Fig. 3) and k is the rate constant. We may summarize by saying that our results suggest the red cells come from three distinct compartments (Table I) containing 9%, 69%, and 22% of the total cells and perfused by 90.3%, 9.6%, and 0.15% of the total inflow, respectively.

Vascular Compartments and Cellular Factors: Morphological Counterparts to the Mathematical Model

This is, of course, a simple equivalent model, sufficient to represent the kinetics but not necessary in the sense of being the only valid model. The question arises: Are there really morphological counterparts to the various "compartments" of the model? Evidence in the affirmative has come from three principal lines of investigation:

1) histological studies of the distribution of red cells within the spleen, made after different periods of washout [65];
2) examination of red cells collected from the venous outflow at different stages of the washout, from the point of view of cellular volume, specific gravity, and supravital staining [29, 67];
3) simultaneous studies of red cell and plasma washout [47].

Of particular interest to the theme of this paper is our demonstration of the nature of the slow compartment.

(1) **Histological studies.** Examination of histological sections of the spleen, stained with hematoxylin and eosin, showed that the red blood cells could be divided into three groups: free cells in vascular channels and in the spaces of the red pulp, cells adhering to reticulum fibres and macrophages, and red cells in the cytoplasm of macrophages. Perfusion of the spleen by only 1 to 2 ml/gm Ringer's solution is sufficient to reduce to zero (approx) the concentration of red cells comprising the fast compartment ($V\frac{1}{2} = 0.067$ ml/gm). When this was done the splenic sections showed a complete absence of red cells in vascular channels. Perfusion by a larger volume of Ringer's solution (up to 15 ml/gm) will reduce by a factor of ten the red cells comprising the intermediate compartment ($V\frac{1}{2} - 4.7$ ml/gm). Sections from such spleens showed marked absence of free red cells in the spaces of the red pulp, as well as complete absence of red cells in vascular channels. If the volume of Ringer's solution perfused is increased to 50–150 ml/gm, then the concentration of cells comprising the intermediate

compartment is reduced to zero, whereas that of cells comprising the slow compartment remains at roughly one-half the value in the normal blood-perfused spleen. Under these conditions no free red cells were found in either the vascular channels or the spaces of the red pulp; almost all the red cells seen were in close contact with either reticulum fibers or reticuloendothelial cells (we shall designate these as "bound" red cells), but a few were within the cytoplasm of macrophages. From the washout kinetics the total number of exchangeable red cells in the spleen after a given volume of Ringer perfusion could be calculated and, for the slow compartment phase, this number agreed, within experimental error, with the numbers of bound cells determined from histological sections. This confirmed quantitatively that the bound cells correspond to the slow compartment of the model derived from washout kinetics. Taken together, these histological studies of red cell distribution in the spleen indicate that the fast compartment consists of red cells in the splenic vessels; the intermediate compartment, of free red cells in the red pulp; and the slow compartment, of cells adhering to the fine structures of the red pulp [65]. By means of both optical and scanning electron microscopy the cells of the slow compartment were later identified as immature and abnormal cells [66, 68].

(2) **Cellular properties.** From the washout kinetics we may compute the percentages of the total red cells in samples of the outflow that come from each of the three compartments. These percentages are shown in Figure 4 as a function of the volume of Ringer's solution that had perfused the spleen. At certain stages of the washout fairly pure ($> 85\%$) samples of cells from each compartment may be obtained. The cellular densities of red cells collected from arterial blood and from the three compartments in turn were measured, in a series of experiments, using the phthalate ester method [16]. In addition, mean cellular volumes were determined by means of a Celloscope counter used in conjunction with a 128-channel pulse height analyzer (Nuclear Data N.D. 110). The mean densities and volumes of red cells are shown in Table II for cells from arterial blood and from each of the three compartments [29]. It is clear that the cells of the fast and intermediate compartments showed no significant difference from those of arterial blood, whereas those of the slow compartment were lighter and larger. This suggested that the cells comprising the fast and intermediate compartments were representative of those found in arterial blood, whereas those of the slow compartment might be younger cells predominantly [57]. In that case the red cells of the slow compartment might contain more than the usual proportion of reticulocytes.

We therefore studied the washout of reticulocytes from the spleen [67]. Samples of the outflow were collected at various stages of the washout and the percentages of reticulocytes in each sample was measured from smears treated with either Brilliant Cresyl Blue or New Methylene Blue stain. In Figure 5 these percentages are shown superimposed on the graph of Figure 4, and it is quite clear that the

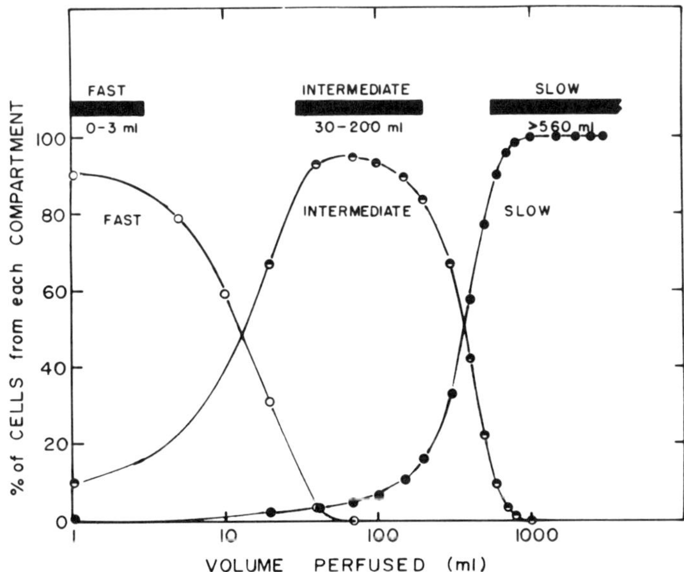

Fig. 4. Percentages of the total red cells in outflow samples, taken during washout of spleen with Ringer's solution, which come from each of the three compartments. Shaded regions indicate stages of the washout when fairly pure samples (> 85%) of cells from each individual compartment may be obtained. Reproduced by permission of National Research Council of Canada, from Groom et al [29] : Physical characteristics of red cells collected from the spleen. Can J Physiol Pharmacol 49:1092–1099, 1971.

TABLE II. Mean Densities and Volumes of Red Cells From Arterial[1] Blood Compared With Those of Cells From the Three Splenic Compartments (Fig. 4). Cells Suspended in Ringer Solution of 300 mOsmole/liter

Sample	Mean density ± SE	Mean volume ± SE (channel number)
Arterial	1.0932 ± 0.0006	42.18 ± 0.78
Fast	1.0918 ± 0.0007	41.35 ± 1.05
Intermediate	1.0905 ± 0.0010	42.92 ± 0.99
Slow	1.0868[a] ± 0.0010	44.32[a] ± 0.79

[a]Difference from arterial value significant (P < 0.01).
Modified from Groom et al [29]. Physical characteristics of red cells collected from the spleen. Can J Physiol Pharmacol 49:1092–1099, 1971.

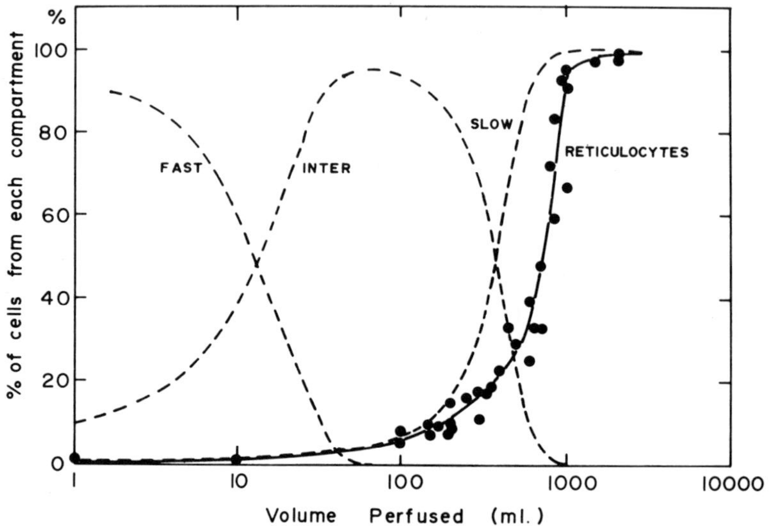

Fig. 5. Percentage reticulocytes in outflow samples, taken during washout of spleens with Ringer's solution, superimposed on Figure 4. The reticulocyte counts follow the same general shape as the slow compartment curve. Reproduced by permission of National Research Council of Canada, from Song and Groom [67] : Sequestration and possible maturation of reticulocytes in the normal spleen. Can J Physiol Pharmacol 50:400–406, 1972.

reticulocyte counts follow the same general shape as the slow compartment curve. The two curves overlap below 100 ml and above 2 liters perfused, but in between these limits the percentage of reticulocytes is always lower than that of slow compartment cells. A major part of this difference turned out to consist of basophilic red cells, such that the sum of reticulocytes plus basophilic cells in the outflow at any stage of washout is almost equal to the predicted number of cells from the slow compartment.

How did these reticulocytes come to be in the spleen? Were they produced there? Although we have found many rubricytes within the normal feline spleen, we have never found proerythroblasts [66] . The reticulocytes found in the spleen could therefore have originated from these rubricytes, but the rubricytes themselves must have come from somewhere else. If that be true, then since the maturing cells spend four days in the rubricyte stage and two days as reticulocytes [31] , we would expect the ratio of reticulocytes to rubricytes in the spleen to be approximately 1:1. However, the ratio we found experimentally was 75:1, almost two orders of magnitude larger than would occur if the reticulocytes were derived from rubricytes in the spleen. It therefore excludes the possibility that the high percentage of reticulocytes in the spleen might have originated from erythropoiesis within the organ itself.

It has been suggested that reticulocytes might be trapped in the spleen because of their reduced deformability [45]. We found no evidence of slow compartment cells retained in orifices too narrow for them to pass through; instead, we observed them adhering to fine structures of the pulp. It had been proposed some years ago that reticulocytes might be trapped in the spleen because of their increased stickiness [2, 37], and this harmonizes with what we have found. The structures of the red pulp form a network whose effective surface area is so very large that it would be well nigh impossible for cells with increased adhesiveness to pass through. Hence our finding that the slow compartment cells, which are reticulocytes and basophilic cells, appear as bound cells in the pulp.

What then would be the purpose of this sequestration of immature red cells in the spleen? The cells in question were not macrocytic reticulocytes produced in response to erythropoietic stress and retained for destruction [44, 69] or remodeling [22, 32]. These were normocytic reticulocytes whose total number was equivalent to 1.5 times the daily production. Their presence could be explained on the hypothesis that all the reticulocytes released from bone marrow are retained in the spleen for a one- to two-day period of maturation before being returned to the circulation. This suggests that the spleen functions as a nursery or, perhaps, a "finishing school" for immature red cells. The grounds for retention appear to be "abnormal" surface properties of these cells, raising the possibility that changes to the cell membrane could in general be a signal for removal from the circulation. Is this because such cells would adhere to the walls of capillaries and obstruct the flow?

It is clear that these "bound" cells may be released by continued perfusion of the organ with Ringer's solution. Is the rate of release dependent on the local fluid shear rate? That is to say, would the shape of the red cell washout curve, even though plotted as a function of the cumulative volume of perfusate, be different at high versus low flow rates (ml/min)? Our experiments indicate that there are, indeed, highly significant differences under these conditions in respect of the slow component only, the $V\frac{1}{2}$ for cells decreasing by a factor of two when the flow rate is increased threefold [50]. There is also a reduction in the number of bound cells in the pulp when the pH of the perfusate is lowered from 7.4 to 6.6, perhaps indicating that some cells of the slow compartment (which comprises a spectrum of immature and abnormal cells) are only weakly bound at normal pH and are released when ambient pH is lowered, joining the mass of free cells in the pulp [50]. The spleen in many species is highly contractile, and on contraction large numbers of immature red cells, which were previously adhering to reticulum fibers and other structures of the red pulp, are released into the outflow [24]. During a contraction the outflow rate from the red pulp must increase considerably, causing a large transient increase in local shear rate within the pulp, leading to the sudden detachment and expulsion of many bound cells. Conversely, if adhesion of red cells to splenic structures is an important mechanism of trapping, then the consequent retardation of flow through the pulp could well promote further sequestration of cells, as suggested by Holzbach et al [33].

(3) Vascular compartments. From the foregoing results we formulated the hypothesis that there are only two vascular compartments within the spleen: 1) the vessels constituting the fast pathway, and 2) the red pulp; the distinction between the intermediate and slow compartments is purely the result of a cellular factor. If this be true, then the washout kinetics of plasma should show only two components, corresponding to the fast and intermediate red cell components. Any result other than this would indicate the need for a different interpretation of the previous findings. We therefore studied the washout of ^{125}I-labeled albumin, injected into the circulation one hour before cannulation and perfusion of the spleen with Ringer's solution, alongside that of red cells [47]. The plasma washout curve (Fig. 6), which is the companion to Figure 3, covers four orders of magnitude, yet it consists of only two exponential components. The $V_{1/2}$ values

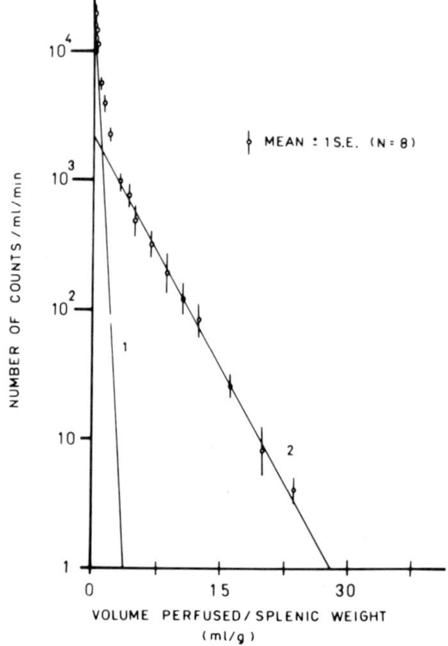

Fig. 6. Mean ^{125}I-labeled albumin washout curve from cat spleens during constant-flow perfusion with Ringer's solution. Radioactive concentration in outflow (counts/ml/minute) is plotted against cumulative volume of fluid perfused. Washout curve may be expressed as sum of two exponential components only [1, 2] (see text). From Levesque and Groom [47]: Washout kinetics of red cells and plasma from the spleen. Am J Physiol 231:1665–1671, 1976.

correspond closely to those of the fast and intermediate red cell components, and so also do the percentage flows of the compartmental models (Fig. 7). Clearly there exists no plasma counterpart to the slow component of red cells, indicating that the latter must be caused by some delay process peculiar to red cells.

Taking all these results together, the following picture emerges: The red cells comprising the slow compartment are immature cells that must have been accumulated from the blood flowing through the red pulp, by the process of adhesion to the fine structures of the pulp. When released from the bound state, these cells will join the free cells of the intermediate compartment in the pulp. For this reason, the slow compartment in Figure 7 has been shown as exchanging with the intermediate compartment, rather than being in parallel with it, as the washout kinetics, taken alone, would suggest. (In point of fact, however, the $V_{1/2}$ value for the intermediate compartment is 20 times smaller than that for the slow compartment, so that it would be difficult to distinguish between the two arrangements purely on the basis of cell washout kinetics.) It is obvious, then, that there are only two distinct vascular compartments within the spleen. From the histological studies at different stages of cell washout we believe these to be

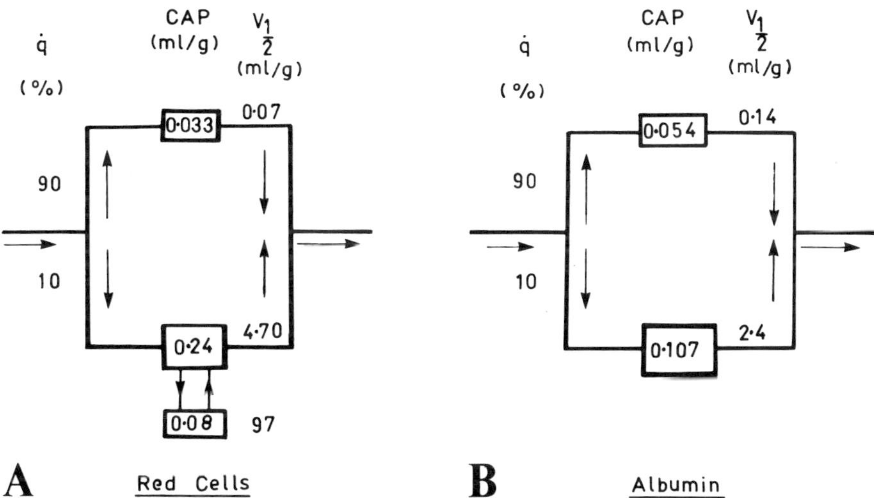

Fig. 7. A: Three-compartment model, derived from cell washout kinetics, for distribution of red cells in spleen. B: Two compartment model, derived from kinetics of ^{125}I-labeled albumin washout, for distribution of plasma in spleen. In both A and B, q̇ is flow to compartment (% of total inflow); CAP is capacity of compartment (milliliter cells or plasma per gram splenic weight); $V_{1/2}$ is desaturation half volume of compartment (milliliter perfusate per gram splenic weight). From Levesque and Groom [47]: Washout kinetics of red cells and plasma from the spleen. Am J Physiol 231:1665–1671, 1976.

1) the vessels, and 2) the red pulp. On this basis we may combine the two separate models for cells and plasma into one model for the morphological distribution of blood within the spleen (Fig. 8). For this purpose the cells of the intermediate and slow compartments, which are both in the red pulp, are lumped together.

Since the appropriate volumes of red cells and plasma are known, we may calculate the hematocrit of the blood in each vascular compartment. For the fast compartment this turns out to be 37%, almost identical to the mean value 35.8% ± 2.5 (SD) for femoral venous blood in this group of animals. The agreement between these two values supports the hypothesis that the fast compartment represents blood contained in the splenic blood vessels. For the larger compartment the hematocrit is 75%, suggesting very strongly that this compartment represents blood in the red pulp. It has been known for many years that in dogs the hemoglobin concentration in splenic blood is considerably higher than in peripheral blood and, more recently, that the hematocrit of blood drained from contracting excised dog spleens may reach 90% [56]. As there are no comparable data available for cat spleen, we used a splenic drainage procedure to provide to provide an independent check on the hematocrit values determined by the washout method.

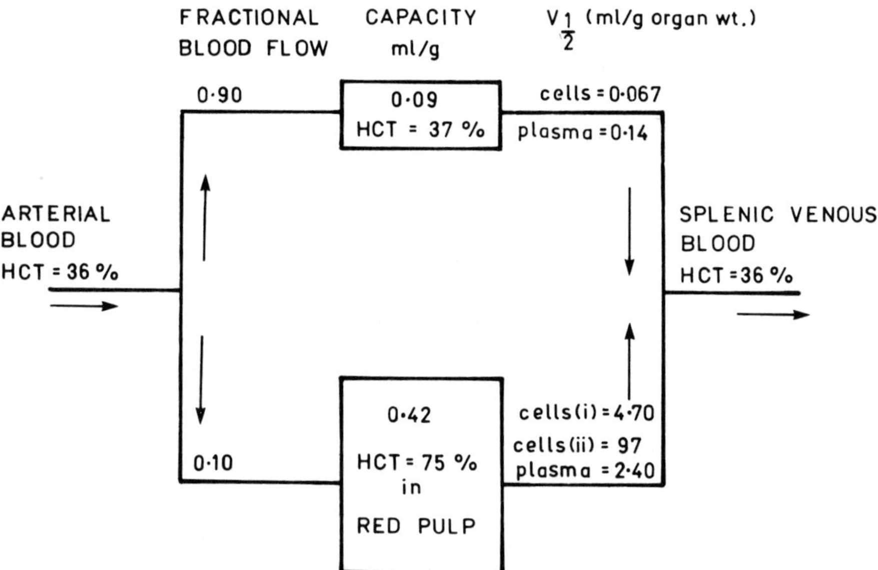

Fig. 8. Compartment model for distribution of whole blood in spleen (see text). Nine-tenths of splenic arterial blood passes through smaller compartment that contains blood of similar hematocrit (37%). One-tenth of total blood flow passes through red pulp (the major compartment) that contains blood of hematocrit 75%. From Levesque and Groom [47]: Washout kinetics of red cells and plasma from the spleen. Am J Physiol 231:1665–1671, 1976.

After isolation of the spleen and cannulation of the artery and vein, the inflow was occluded and the organ was permitted to empty its contents passively, via the venous cannula, into a series of 2 ml syringes. When the outflow ceased, 0.5 μg of norepinephrine in 0.1 ml saline was injected via a small polyethylene tube, inserted through the arterial cannula to the level of the bifurcation, to induce active contraction of the spleen. Further samples were then collected, until the outflow ceased, and the hematocrit values of all samples were then measured. These hematocrit values were plotted as a function of the cumulative volume of blood drained (Fig. 9:lower). The initial hematocrit was comparable to that of femoral venous blood but, as drainage continued, the hematocrit rose gradually, reaching a final value of 78.5%. This value is very similar to the hematocrit value of 75% determined from the washout experiments for the larger of the two vascular compartments, and the agreement lends support to the hypothesis that this larger compartment does indeed represent the blood in the red pulp.

We are now in a position to review the question posed earlier in this paper, and to answer fairly confidently that there are indeed morphological counterparts to the compartments of the simple equivalent model, derived from washout kinetics, for the storage and transit of red cells in the spleen. In the process of discovering the answer, we have learned much about the retention of immature and abnormal cells in the spleen and the means by which this is brought about. There never was much doubt that cellular deformability is very important for the safe transit of red cells through a capillary network, but these investigations appear to suggest that the surface properties of the cells may be of importance also. At least, the spleen appears to think so.

Physical Environment in the Splenic Pulp: Harmful to Red Cells, or Not?

It has been estimated that of the 120 days of a human red cell's life, two days are spent in the spleen [14]. This is because the red cells move so slowly through the red pulp. Our own investigations have confirmed, quantitatively, that the mean transit time of cells through the fast compartment in the spleen is comparable to that through the vascular bed of skeletal muscle, whereas that of normal, mature red cells through the pulp is 70 times greater. However, for immature and abnormal cells in the pulp the transit time would be 1,400 times greater. The cells in the pulp are thus exposed to the local splenic environment for long periods. Furthermore, since the hematocrit in the red pulp is so very high, the cells could be competing for substrate under conditions of inadequate supply.

It has been suggested that, quite apart from the activity of the phagocytic cells, prolonged exposure to the physical environment in the splenic pulp may itself be harmful to red cells, particularly to those that are abnormal [14, 26, 54, 74]. Red cells from the pulp and the splenic vein certainly show increased osmotic and mechanical fragility [19] and have a lower sodium—potassium ratio [59]. Many of the changes occuring in the sequestered cells are similar to

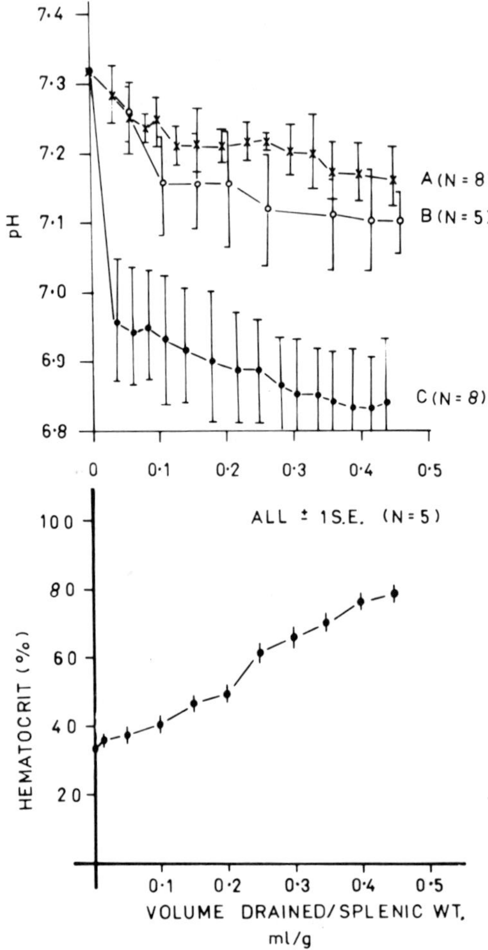

Fig. 9. pH and hematocrit of successive samples of venous blood drained from cat spleen after clamping arterial inflow. With no stasis pH fell to 7.16 (group A), with stasis during cannulation procedures (8–10 minutes) pH fell to 7.10 (group B), and after stasis for 60 minute pH fell to 6.83 (group C). High hematocrit values at end show that blood came from red pulp. From Levesque and Groom [48] : pH environment of red cells in the spleen. Am J Physiol 231:1672–1678, 1976.

those resulting from incubation of blood at 37°C in vitro [52, 72] , and such changes have been designated by the term "conditioning" — ie, modification of red cells for further challenge within the spleen or the peripheral circulation [26] .

What is known, definitively, about the environmental conditions within the pulp? The answer has to be that little is known, but much has been surmised. The generally accepted hypothesis is that, as a consequence of stasis, the PO_2, glucose concentration, and pH must all be low, creating a hostile environment for red cells. For instance, a pH as low as 6.8 [54] would lead to increased mean cellular volume [55] and membrane rigidity [43], increased blood viscosity at low shear rate, denoting increased aggregability [17, 54], decreased cell filterability, and increased osmotic fragility [54]. Such a series of physical changes occurring in an acidic environment might well contribute to metabolic changes and lead to the shortened life-span of cells incubated within the pulp [26, 72].

In practice, it is very difficult to test the above hypothesis, because it has not been possible to collect blood from the splenic pulp without contamination from blood in the rapidly flowing parts of the splenic circulation [38, 60]. This objection applies when blood is collected from the cut surface of the freshly excised spleen [59] and also when samples are withdrawn from the splenic pulp by tissue aspiration [70]. The splenic drainage procedure that we have described appears to provide, for the first time, a means of sampling blood from the red pulp selectively, since expulsion of blood from the fast-transit pool is completed ahead of that of blood from the slow-transit pool in the red pulp.

The pH values of successive blood samples, collected anerobically during splenic drainage, are shown in Figure 9 (upper) as a function of the cumulative volume of blood drained [48]. In eight experiments side vessels were cannulated so that blood flow through the spleen was not interrupted until the start of venous sampling (curve A). In five experiments the main vessels themselves were cannulated, causing an 8–10-minute period of stasis before sampling began, and in eight further experiemnts splenic blood flow was occluded completely for one hour prior to drainage. The mean hematocrit values in these same experiments are shown in Figure 9 (lower). It is clear that when normal blood flow through the spleen was not interrupted at any time prior to the drainage procedure, the pH of the final sample was 7.16. After eight to ten minutes of stasis the final pH was 7.10, whereas after stasis for one hour the corresponding value was 6.83. The drainage procedure took about five to six minutes from start to finish, so that the final blood sample in curve A was collected after the arterial inflow had been occluded for this period of time. On the basis of this, and from a comparison of curves A and B, we may estimate the pH value of blood in the red pulp of the spleen, perfused normally in situ, to be 7.20.

From other experiments we were able to show [48] that the pH within the red pulp results from the interplay of two separate factors: 1) pH-determining elements of the splenic tissue itself that buffer at 6.8, and 2) the buffering provided by red blood cells passing through the pulp. Under normal conditions the resultant pH is 7.2, but when complete stasis of blood occurs this pH falls progressively toward that of the fixed structures of the spleen — viz 6.8. Measurements of blood gases and glucose concentration (Fig. 10) in samples collected anerobically during splenic drainage suggest that the O_2 tension and glucose concentration in the red pulp are normally not less than 60% of the corresponding

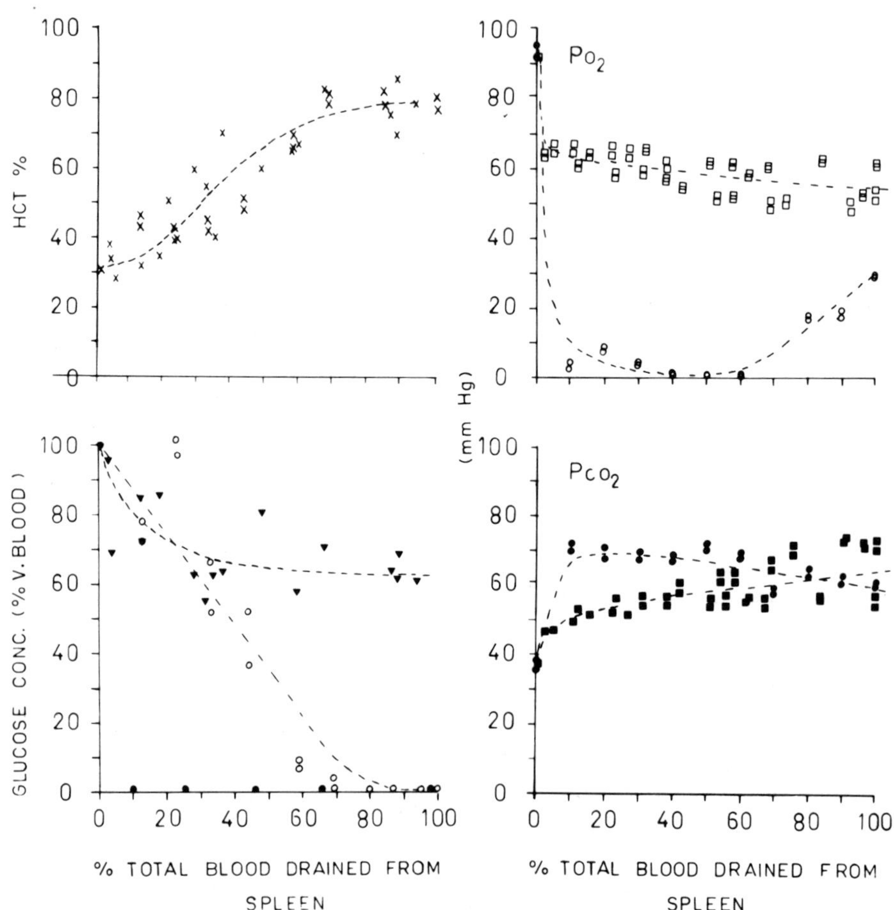

Fig. 10. Hematocrit, glucose concentration (% of that in venous blood), PO_2 and PCO_2 in successive samples of blood collected from the spleen of cat, during drainage with the arterial inflow occluded (see text). Abscissa: cumulative volume of splenic blood drained (% of total) up to the point when each sample was taken. Spleens with no prior interruption of flow: glucose ▾, PO_2 ▫, PCO_2 ▪. Occlusion for 20 minutes: glucose, ○. Occlusion for 60 minutes: glucose ●, PO_2 ○, PCO_2 ●. From Groom et al [27]: Flow stasis, blood gases and glucose levels in the red pulp of the spleen. In "Oxygen Transport to Tissue." III. Adv Exp Med Biol 94:567–572, 1977.

values in peripheral venous blood [27]. However, when blood flow is occluded the glucose concentration in splenic blood falls very rapidly indeed, the concentration in the pulp reaching zero in less than 20 minutes. At the same time the O_2 tension falls, though much more slowly; thus, after stasis for one hour, near-zero values are obtained except for a rise to 30 mm Hg in the last quarter of the blood drained from the spleen.

These results indicate that the commonly accepted notion of a hostile environment for red cells in the pulp of the normal spleen by virtue of the very low values of pH, O_2 tension, and glucose concentration which exist there is not, in fact, true. However, when normal blood flow through the spleen is arrested a hostile environment for red cells does develop in the pulp quite rapidly, the principal stress being substrate deprivation. Is it possible that in pathological conditions the trapping of abnormal red cells in the pulp could bring about a similar situation?

The Spleen: Reservoir or Internment Camp for Red Cells?

What is the purpose of the contractility of the spleen in so many species? The classical view, following the beautiful experiments and suggestions of Barcroft [1], is that the splenic pulp acts as a reservoir from which blood of high hematocrit can be discharged into the circulation, in order to increase the venous return and oxygen-carrying capacity of blood during an emergency. This explanation is not entirely satisfying, however, particularly as regards the high hematocrit of blood in the red pulp. What would be the disadvantage of having these red cells in the general circulation all the time? Why have them tucked out of the way in the spleen? The usual reply is that the viscosity of the circulating blood is thereby maintained at a lower value while the body is at rest, thus minimizing the work of the heart. It is difficult for a biophysicist to get too enthusiastic about this idea. The peripheral hematocrit rises only by about ten percentage units when the normal relaxed spleen contracts and, unless the hematocrit was unduly high to begin with, the increase in blood viscosity is comparatively small.

We wish to propose an alternative hypothesis, related to the role of the spleen as a filter for removing unsuitable red cells from the blood. During Wolrd War II, a continuous trickle of refugees from France and the low countries succeeded in reaching the east coast of Britain in small row-boats. Most of these people were perfectly genuine refugees, but there were some who were in fact espionage agents posing as refugees. Every refugee was therefore sent to an internment camp for a period of intensive questioning, after which he was either released or detained to await His Majesty's pleasure. It seems to us that the red pulp of the spleen constitutes an internment camp "par excellence" for blood cells, consisting as it does of a sponge-like matrix filled with slowly flowing blood. Any cells traversing the pulp have a high probability of coming into direct contact with the complicated mesh of reticulum fibers [68]; in fact, the organ seems almost to be specialized so as to have a large contact surface area for blood cells. Normal, mature red cells appear not to adhere to these structures, but immature cells, which have not yet lost completely the adhesiveness they possessed during earlier development in the bone marrow, are retained thereby. Our hypothesis, then, is that during exercise all red cells are needed to carry O_2 to the tissues and the spleen remains in the contracted state; during rest however, when the metabolic demands for O_2 are much lower, the spleen dilates, withdrawing large numbers of cells from the general circulation and screening them for undesirable surface properties. One would expect this function to be important, since all cells comprising the tissues and organs of the body adhere together strongly, except for those which are to be transported and must therefore acquire a different set of surface characteristics from the rest.

SEQUESTRATION OF HEAT-TREATED RED CELLS DURING A SINGLE TRANSIT THROUGH THE SPLEEN

Clinicians have for many years made use of altered red cells in order to study, to a limited degree, the splenic uptake function in man [13, 42, 71]. What is done is to label these cells with ^{51}Cr and to follow, as a function of time after their injection into the circulation, the radioactivity in samples of peripheral venous blood or the accumulation of the radioactivity in the spleen. The first of these methods allows the rate of decay of the number of altered cells in the blood to be measured, and the results are often expressed in terms of the disappearance half-time, $t\frac{1}{2}$ [51, 71]. The second method is semi-quantitative at best, because of the problems inherent in determining the radioactivity in the spleen by surface counting. The rate of removal of altered red cells from the blood by the spleen, in intact subjects, depends not only on the efficiency of trapping during a single passage but also on the splenic blood flow, expressed as the fraction of the subject's total blood volume passing through the organ per minute. We wished to study the transit and sequestration of altered red cells in the spleen during a single passage, as a means of determining what alterations in red cell properties the body considers most undesirable. For this purpose we have injected unlabeled red cells into the isolated, Ringer-perfused, cat spleen preparation described earlier in this chapter [49].

Each spleen was perfused at constant pressure with one liter of oxygenated (5% CO_2 in O_2), buffered Ringer's solution to wash out most of the red cells contained therein. At that stage, the color of the spleen was a good indication that most of the cells had in fact been removed from the organ. Samples of the outflow were then collected for measurement of the "background" or preinjection cell concentration. A small bolus of a red cell suspension, containing 2.0×10^6 cells, was then injected quickly into the arterial cannula close to the spleen itself; in some experiments the injectate also contained 0.1 μCi of radio-iodinated (^{125}I) human serum albumin. At the same time, rapid sequential sampling of the venous outflow was begun, this being continued until 50 ml of perfusate had passed through the organ from the instant of injection. This was done to ensure that the cellular concentration in the outflow had returned to its preinjection value before the end of the sampling period. In every experiment the first injection contained normal autologous red cells; ie, it constituted a control run. The second injection contained "altered" red cells (foreign cells or autologous cells collected or treated as described below), and in some experiments this was repeated once or twice more. All these samples were then assayed individually for their concentrations of red cells and, where appropriate, of albumin.

We used six different groups of altered red cells, as explained later. Our experimental approach was to examine very thoroughly the transit and retention of one particular group of cells in the spleen — viz, heat-treated (HT) cells [49] — and then to compare with this the results obtained using the other five groups. The HT cells were prepared by incubating, at $49.5°C$ for 20 minutes, "washed" (3X) autologous red cells suspended in Ringer's solution at a hematocrit of 35%. This treatment produces cells which, although of various sizes and shapes, are mainly spherocytic; they are less deformable, because of their spheroidal shape, and are osmotically and mechanically more fragile than normal red cells. It is not known whether changes in cell surface charge also occur as a result of heat treatment. This procedure is the one most used clinically for studying cellular retention in the spleen.

The outflow concentrations of HT cells and normal red cells, together with those of ^{125}I-albumin, are shown in Figure 11. It is clear that the peak cellular outflow for HT cells was delayed in comparison with that for normal cells and that the return to background concentration occurred sooner. In addition, the area under the curve was substantially less for HT cells than for normal cells, although the same number of cells was injected in both cases. Obviously HT cells are slowed in their passage through the spleen, and many do not reach the venous outflow at all; ie, they become trapped in the spleen. By contrast, there was no delay of the peak albumin concentration when HT cells, instead of normal cells, were injected.

The concentration curves of Figure 11 may be integrated, with respect to the outflowing volume of perfusate, to give the cumulative outflow of cells and plasma from the spleen. Further, since the total amounts injected into the arterial inflow were also measured (by making comparable injections into a beaker containing a known volume of Ringer's solution and measuring the resulting concentrations), the cumulative outflows could then be expressed as percentages of the injected material. It is clear, from Figure 12, that 90% of normal cells and 96% of the accompanying albumin had reached the outflow when 35 ml of perfusate had been collected (2 minutes, approx); in fact, no additional cells

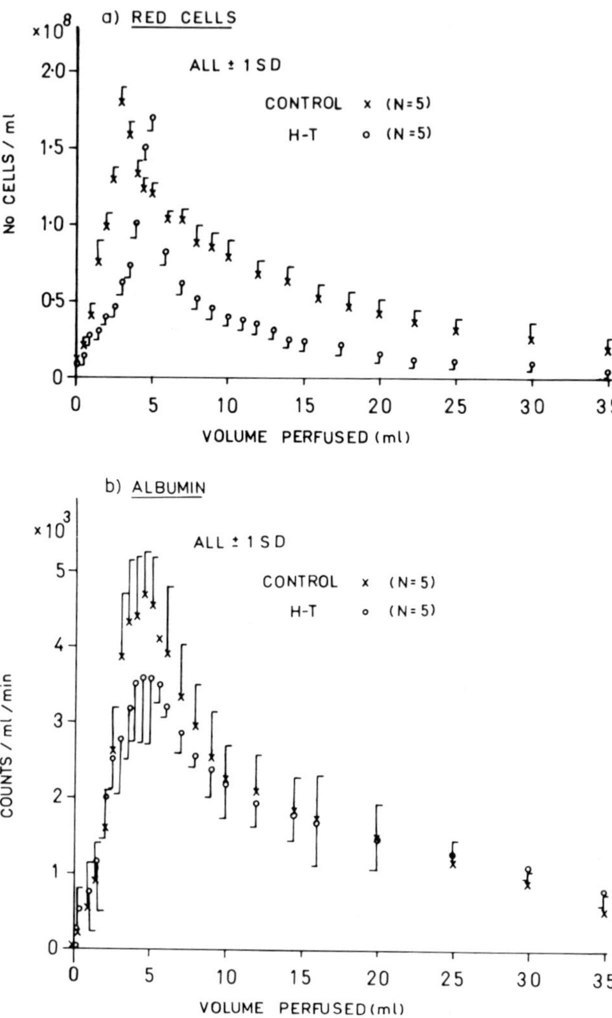

Fig. 11. Outflow concentrations of (a) red cells, and (b) ^{125}I-albumin, following a slug-injection into the splenic artery of ^{125}I-albumin plus either normal or HT cells; isolated spleens perfused with Ringer's solution at constant pressure. The peak concentration for HT cells occurs only after a larger volume of perfusion than that observed for normal cells (control), whereas the albumin curves show little difference. From Levesque and Groom [49] : Sequestration of heat-treated, autologous red cells in the spleen. J Lab Clin Med 90:666–679, 1977.

were recovered in the next 15 ml of perfusate, but the recovery of albumin rose to 98%. In the case of HT cells, however, only 57% recovery was obtained at both 35 and 50 ml perfusate; the corresponding values for the accompanying albumin were 87% and 90%, respectively.

We showed earlier that 90% of both cells and plasma entering the spleen travels via the fast pathway and the remainder travels very slowly via the red pulp; presumably this explains why 10% of normal cells were missing from the outflow in the present experiments. The HT cell values for percentage outflow were 0.65 times those for normal cells, suggesting that 35% of the HT cells were trapped during a single transit through the spleen. Clearly some of the plasma label was also held up, in its passage through the spleen, as a result of the trapping of HT cells. This was surprising, and suggested that areas of the pulp might be completely obstructed by the trapped HT cells. We therefore repeated these experiments, perfusing the spleen under constant flow conditions instead of constant pressure. A considerable rise in perfusion pressure occurred as a result of the injection of HT cells, and the percentage trapping of cells and plasma was reduced from 35% and 10%, respectively, to corresponding new values of 10% and zero. Obviously the rise in pressure was forcing more HT cells through the spleen.

Subsequent experiments, in which HT cells were injected into the splenic artery in vivo, followed one hour later by cell washout during Ringer perfusion, yielded the surprising result that only one-half of the normal red cells in the spleen were

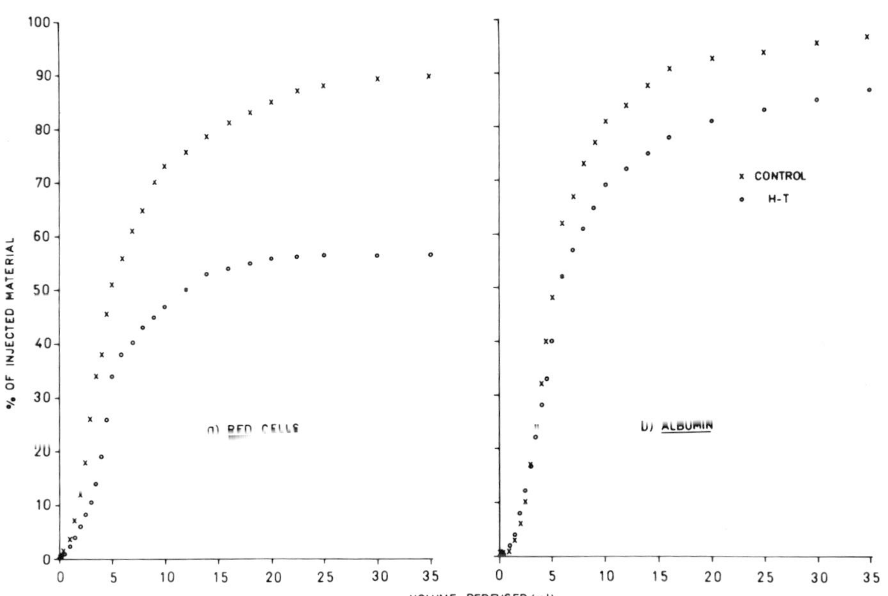

Fig. 12. Cumulative outflow vs volume of perfusate for the curves of Figure 11. Ninety percent of normal cells were recovered in the volume of perfusate collected, compared with only 56% of HT cells (a). A small amount of plasma trapping occurred with the HT cells only (b). From Levesque and Groom [47]: Sequestration of heat-treated, autologous red cells in the spleen. J Lab Clin Med 90:666–679, 1977.

free to circulate any longer [49]. The number of HT cells injected was equivalent to only 0.8% of the splenic red cell volume, yet this caused the immobilization of 50% of the cells in the pulp (Table III). This was confirmed qualitatively by the fact that HT cell-injected spleens remained a deep red color throughout the entire perfusion period, whereas the control spleens became extremely pale. Our results showed, however, that all the plasma remained free to circulate, albeit much of it slowly.

Earlier we asked the question: Is it possible that in pathological conditions the trapping of abnormal red cells in the splenic pulp could occlude blood flow and lead to low values of pH, O_2 tension, and glucose concentration, ie, create a "hostile" environment for red cells? We can now answer with certainty that red cell flow is indeed completely occluded under these conditions in much of the splenic pulp. However, the plasma continues to circulate, though at a much reduced rate. The major question, therefore, still remains unanswered, although it could now be rephrased, as follows: Is it possible, after obstruction of red cell flow in the splenic pulp, that transport of dissolved substances via the plasma might be maintained at a rate sufficient to prevent deleterious changes in red cells from taking place?

TRANSIT AND RETENTION OF "ALTERED" RED CELLS IN SPLEEN VERSUS SKELETAL MUSCLE

The foregoing studies on the trapping of HT cells in the spleen provided a starting point for comparing the retention of different types of "altered" cells. We have used the following cells to explore the effects of changes in bulk properties (deformability, shape, and size) and surface properties (adhesiveness) on trapping in both the spleen [46] and skeletal muscle [3]. For the studies

TABLE III. Circulating Red Cell and Plasma Volumes of the Rapidly and Slowly Exchanging Splenic Vascular Compartments (Fig. 8), Before and After the Injection of 1.57×10^9 HT Cells. Mean values ± SE

Circulating volume (ml/gm)	Controls	After HT cells
Rapidly exchanging		
Plasma	0.111 ± 0.003	0.105 ± 0.010
Red cells	0.046 ± 0.005	0.034 ± 0.003
Slowly exchanging		
Plasma	0.058 ± 0.002	0.074 ± 0.007
Red Cells	0.308 ± 0.008	0.156 ± 0.019

Modified from Levesque and Groom [49]: Sequestration of heat-treated autologous red cells in the spleen. J Lab Clin Med 90:666–679, 1977.

involving skeletal muscle an isolated Ringer-perfused preparation of cat gastrocnemius was used, from which the blood had already been washed out [28].

1) Neuraminidase-treated cells: Neuraminidase removes the sialic acid portions from the glycoproteins on the red cell surface, and in that process the electric charge on the cell is lost also [12]. Aggregation of red cells in the presence of dextrans increases when surface charge is lost [7, 36], although adhesion to protein-coated glass plates is not enhanced thereby [58]. Mean cellular volume and deformability are unchanged [18]. For our experiments one ml of packed cells was incubated for 60 minutes at 37°C with 30 units of Vibrio Cholerae neuraminidase [23].

2) N-ethyl-maleimide (NEM)-treated cells: NEM reacts with the sulfhydryl groups of several enzyme systems [35]. Cells incubated for 45 minutes at 37°C at a concentration $8-16$ μM/ml are of normal size, shape and fragility; they are selectively removed by the spleen and subsequently phagocytosed [34].

3) Glutaraldehyde-treated cells: Cells fixed in 1% glutaraldehyde solution become rigid without change of shape [40, 53]. They are highly resistant to deformation and to mechanical or osmotic lysis.

4) Cells from the splenic pulp: Red cells from the pulp contain a high proportion of immature cells [67]. Although these cells are found adhering to the fine structures of the pulp, they have a higher surface charge than mature cells [15, 76]. For the present investigation we used the final 3 ml of blood obtained from the spleen by means of the drainage method described earlier.

5) Foreign red cells: Cells freshly taken from a healthy human donor were used. Although these are larger than cat red cells (diameters 7.8 μm vs 5.7 μm; volumes 96 μm^3 vs 57 μm^3), they would, presumably, retain their deformability. We considered it possible that they might be "recognized" as foreign bodies and retained in the microcirculation by some adhesion process.

In all cases the organ was perfused with Ringer's solution under constant pressure conditions. For spleen, the driving pressure (ΔP) was 75 cm $H_2$0 (approx). However, in the case of skeletal muscle, a much lower value, 20 cm H_2O, was eventually selected in order to give as much opportunity as possible for "abnormal" cells to remain trapped in the vasculature. For each individual injection of "abnormal" cells, the number recovered at the outflow was expressed as a percentage of the number injected. Mean values of % recovery were then computed (Table IV) for each separate type of cell used, and the results in spleen vs skeletal muscle were compared.

It is clear from Table IV that marked retention in the spleen occurred for each of the six types of "abnormal" cell used [46]. However, only in the case of glutaraldehyde-fixed cells was the recovery always, and unequivocally, zero. Heat-treated cells were retained least of all. Cells collected previously from the splenic pulp and cells treated with NEM or neuraminidase all showed similar

TABLE IV. Recovery of "Abnormal" Red Cells (% Cells Injected ± SE) After a Single Transit Through Spleen or Skeletal Muscle of the Cat

Type of cells	Spleen	Skeletal muscle
Heat treated	57 ± 6*	93.8 ± 4.4**
Neuraminidase treated	31 ± 11*	96.8 ± 4.0**
NEM treated	28 ± 8*	not studied
Glutaraldehyde treated	zero*	21.4 ± 2.8*
From splenic pulp	34 ± 12*	97.0 ± 1.3**
Foreign cells (human)	14 ± 5*	102.8 ± 3.8**

Difference from 100%: *significant (P < 0.001)
**not significant (0.05 < P < 0.60)

Modified from Levesque MJ [46]: phD. Thesis, 1977; Bowden TJ [3], PhD. Thesis, 1978. University of Western Ontario.

degrees of trapping (70% approx). Human cells were retained to a considerable extent (85%). It appears, therefore, that changes in either the surface properties of red cells (NEM- or neuraminidase-treated autologous cells, and foreign cells) or cellular deformability (HT cells, glutaral dehyde-treated cells) will lead inevitably to sequestration in the spleen. We have studied only the initial process of trapping; what happens to the trapped red cell thereafter may depend on the precise way in which its properties differ from those of normal, mature cells. Thus immature cells collected previously from the splenic pulp are again retained, in this case for maturation before being released into the circulation. Presumably most of the other types of "abnormal" cells are likely victims for phagocytosis.

Our initial premise in these investigations was that by observing the spleen at work, as the body's filter for abnormal red cells, we should be able to determine quite clearly what cellular properties are important for red cells to function properly in the circulation. The unequivocal answer seems to be that deformability and a suitable set of surface properties are both important. It is therefore surprising to find (Table IV) that only deformability seems to matter where the transit of red cells through skeletal muscle is concerned [3]. Thus recovery of glutaraldehyde-treated cells from the venous outflow of muscle was only 21%, whereas that of each other type of cell studied was not significantly different from 100%. This seems to indicate that the surface properties of red cells are not important for safe transit through the microcirculation.

How is it, then, that the spleen appears to differ from skeletal muscle on this vital issue? One answer could be that they differ only in degree. The slug-injection method is not able to assess very well the cells that have longer transit times through an organ, whereas the washout method is extremely sensitive in this respect. Thus cells from the splenic pulp, of which a high proportion are reticulocytes, appear not to be sequestered at all in skeletal muscle, as judged from the present experiments. However, by the washout method, it was shown previously that reticulocytes are cleared from skeletal muscle much more slowly than mature red cells [28], and we attributed this to adherence of reticulocytes to vessel walls. Of course, it is also possible that this difference derives from the presence or absence of high molecular weight plasma proteins which influence adhesion. The reticulocytes were already present (and adhering?) in muscle during the blood-perfused steady state prior to washout, whereas in the slug-injection experiments the "abnormal" cells were suspended in Ringer's solution, after being washed three times, and injected into a Ringer-perfused muscle. However, this same procedure did not appear to prevent the spleen from retaining abnormal cells.

It seemed possible that altered red cells might be slowed down in their passage through skeletal muscle, even though all reach the venous outflow successfully. However, only in the case of HT cells was there any evidence of this, and a change in deformability of these cells is probably the key factor involved, not a change in surface properties. It certainly appears that surface properties are not of great consequence in regard to the transit of red cells through the microcirculation in skeletal muscle. Why then should the spleen sequester cells with altered surface properties? If the primary function of the spleen, as part of the reticuloendothelial system, is to filter out garbage from the blood, and if it does this by recognizing the abnormal surface properties of cell fragments, could it then be that we are merely fooling the spleen by changing the surface properties of cells so that it treats perfectly good red cells as garbage?

ACKNOWLEDGMENTS

The reader will now appreciate how it has come about that one whose research interest is "oxygen transport to tissue in skeletal and cardiac muscle" has developed, for several years past, a parallel interest in "the physiological role of the spleen with respect to red blood cells." These are the respective titles of the grants awarded to me by the Ontario Heart Foundation and the Medical Research Council of Canada, and I gratefully acknowledge their support.

It is a great pleasure to pay tribute to my colleagues, Drs. Seh-Hoon Song, Murina Levesque, and Timothy Bowden, who worked, as graduate students, on successive phases of this investigation. The friendship and labors of these people are deeply appreciated.

REFERENCES

1. Barcroft J, Stephens JG: Observations on the size of the spleen. J Physiol (Lond) 64:1–22, 1927.

2. Berendes M: The proportion of reticulocytes in the erythrocytes of the spleen as compared with those of circulating blood, with special reference to hemolytic states. Blood 14:558–563, 1959.

3. Bowden TJ: Storage and transit of red blood cells in skeletal muscle of the cat. Ph.D. Thesis, University of Western Ontario, 1978.

4. Braasch D, Jenett W: Erythrocyte flexibility, hemoconcentration and blood flow resistance in glass capillaries with diameters between 6 and 50 microns. Pfluegers Arch 302:245–254, 1968.

5. Branemark P-I: Rheological aspects of low flow states. In Shepro D, Fulton GP (eds): "Microcirculation as Related to Shock." New York: Academic Press, 1968, pp 161–180.

6. Branemark P-I: "Intravascular Anatomy of Blood Cells in Man." Basel: S. Karger, 1971.

7. Brooks DE: Red cell interactions in low flow states. In Grayson J, Zingg W (eds): "Microcirculation." New Yrok: Plenum Press, 1976, pp 33–52.

8. Canham PB, Burton AC: Distribution of size and shape in populations of normal human red cells. Circ Res 22:405–422, 1968.

9. Chien S, Usami S, Dellenback RJ, Gregersen MI: Blood viscosity influence on erythrocyte deformation. Science 157:827–829, 1967.

10. Chien S, Usami S, Dellenback RJ, Gregersen MI, Nanninga B, Guest MM: Blood viscosity influence of erythrocyte aggregation. Science 157:829–831, 1967.

11. Chien S, Usami S, Taylor HN, Lundberg JL, Gregersen MI: Effect of hematocrit and plasma proteins on human blood rheology at low shear rates. J Appl Physiol 21:81–87, 1966.

12. Cook GMW, Heard DH, Seaman GVF: Sialic acids and the electrokinetic charge of the human erythrocyte. Nature 191:44–47, 1961.

13. Crome P, Mollison PL: Splenic destruction of Rh-sensitized and heated red cells. Br J Haematol 10:137–157, 1964.

14. Crosby WH: Normal functions of the spleen relative to red blood cells: A review. Blood 14:399–408, 1959.

15. Danon D, Marikovsky Y: Difference de charge électrique de surface entre erythrocytes jeunes et ages. Compt Rend Acad Sci 253:1271–1272, 1961.

16. Danon D, Marikovsky Y: Determination of density distribution of red cells. J Lab Clin Med 64:668–674, 1964.

17. Dintenfass L, Burnard E: Effect of hydrogen ion concentration on in vitro viscosity of packed red cells and blood at high hematocrits. Med J Aust 1:1072–1074, 1966.

18. Durocher JR, Payne RC, Conrad M: Role of sialic acid in erythrocyte survival. Blood 45:11–20, 1975.

19. Emerson CP, Shen SC, Ham TH, Fleming EM, Castle WB: Studies on the destruction of red blood cells. IX. Quantitative methods for determining the osmotic and mechanical fragility of red cells in the peripheral blood and splenic pulp; the mechanism of increased hemolysis in hereditary spherocytosis (congenital hemolytic jaundice) as related to the function of the spleen. Arch Intern Med 97:1–38, 1956.

20. Eriksson E, Lisander B: Low flow states in the microvessels of skeletal muscle in cat. Acta Physiol Scand 86:202–210, 1972.

21. Fung Y-C: Red blood cells and their deformability. Ch 12. In Kaley G, Altura BM (eds): "Microcirculation," vol 1. Baltimore, Maryland: University Park Press, 1977, pp 255–278.

22. Ganzoni A, Hillman RS, Finch CA: Maturation of the macroreticulocyte. Br J Haematol 16:119–135, 1969.

23. Gattegno L, Bladier D, Cornillot P: Aging in vivo and neuraminidase treatment of rabbit erythrocytes: Influence on half-life as assessed by ^{51}Cr-labelling. Hoppe-Seylers Z Physiol Chem 356:391–397, 1975.

24. Geiger HB, Song SH, Groom AC: Release of red cells from the slowly exchanging splenic pool after noradrenaline administration. Can J Physiol Pharmacol 54:477–484, 1976.
25. Goldsmith HC: Deformation of human red cells in tube flow. Biorheology 7:235–242, 1971.
26. Griggs RC, Weisman R Jr, Harris JW: Alterations in osmotic and mechanical fragility related to in vivo erythrocyte aging and splenic sequestration in hereditary spherocytosis. J Clin Invest 39:89–101, 1960.
27. Groom AC, Levesque MJ, Brucksweiger D: Flow stasis, blood gases and glucose levels in the red pulp of the spleen. In Silver IA, Erecinska M, Bicher HI (eds): "Oxygen Transport to Tissue-III." New York: Plenum Press, 1977, pp 567–572.
28. Groom AC, Song SH, Campling B: Clearance of red blood cells from the vascular bed of skeletal muscle with particular reference to reticulocytes. Microvasc Res 6:51–62, 1973.
29. Groom AC, Song SH, Lim P, Campling B: Physical characteristics of red cells collected from the spleen. Can J Physiol Pharmacol 49:1092–1099, 1971.
30. Ham TH, Dunn RF, Sayre RW, Murphy JR: Physical properties of the red cells related to effects in vivo. I. Increased rigidity as measured by viscosity of cells altered by fixation, sickling and hypertonicity. Blood 32:847–861, 1968.
31. Harris JH, Kellermeyer RW: "The Red Cell." Cambridge, Massachusetts: Harvard University Press, 1970.
32. Hillman RS: Characteristics of marrow production and reticulocyte maturation in normal man in response to anemia. J Clin Invest 48:443–453, 1969.
33. Holzbach RT, Shipley RA, Clark RE, Chudzik EB: Influence of spleen size and portal pressure on erythrocyte sequestration. J Clin Invest 43:1125–1135, 1964.
34. Jacob HS, Jandl JH: Effects of sulfhydryl inhibition on red blood cells. I. Studies in vivo. J Clin Invest 41:1514–1523, 1962.
35. Jacob HS, Jandl JH: Effects of sulfhydryl inhibition on red blood cells. II. Mechanisms of hemolysis. J Clin Invest 42:779–792, 1962.
36. Jan K-M, Chien S: Role of surface charge in red blood cell interactions. J Gen Physiol 61:638–654, 1973.
37. Jandl JH: The agglutination and sequestration of immature red cells. J Lab Clin Med 55:663–681, 1960.
38. Jandl JH: Hereditary spherocytosis. In Beutler E (ed): "Hereditary Disorders of Erythrocyte Metabolism." New York: Grune & Stratton, 1967, p 209.
39. Jay AWL: Viscoelastic properties of the human red cell membrane. I. Deformation, volume loss and rupture of red cells in micropipettes. Biophys J 13:1166–1182, 1973.
40. Jay AWL, Canham PB: The sedimentation of individual glutaraldehyde-fixed biconcave human erythrocytes. J Cell Physiol 80:367–372, 1972.
41. Jay AWL, Rowlands S, Skibo L: The resistance to blood flow in the capillaries. Can J Physiol Pharmacol 50:1007–1013, 1972.
42. Kimber RJ, Lander H, Robson HN: The sequestration of NEM-treated red cells in normal and abnormal subjects: A test of splenic uptake function. J Lab Clin Med 63:951–966, 1965.
43. LaCelle PL: Alteration of the deformability of the erythrocyte membrane in stored blood. Transfusion 9:238–245, 1969.
44. Landaw SA, Russell ES, Bernstein SE: Splenic destruction of newly-formed red blood cells and shortened erythrocyte survival in mice with congenital microcytosis. Scand J Haematol 7:516–524, 1970.
45. Leblond PF, LaCelle PL, Weed RI: Cellular deformability: A possible determinant of the normal release of maturing erythrocytes from the bone marrow. Blood 37:40–46, 1971.

46. Levesque MJ: Sequestration of normal and abnormal red cells by the spleen. Ph.D. Thesis, University of Western Ontario, 1977.

47. Levesque MJ, Groom AC: Washout kinetics of red cells and plasma from the spleen. Am J Physiol 231:1665–1671, 1976.

48. Levesque MJ, Groom AC: pH environment of red cells in the spleen. Am J Physiol 231:1672–1678, 1976.

49. Levesque MJ, Groom AC: Sequestration of heat-treated, autologous red cells in the spleen. J Lab Clin Med 90:666–679, 1977.

50. Levesque MJ, Groom AC: Effects of pH and flow rate on the release of "bound" red cells from the splenic pulp. Can J Physiol Pharmacol 56:260–268, 1978.

51. Marsh GW, Lewis SM, Szur L: The use of ^{51}Cr-labelled heat-damaged red cells to study splenic function. I. Evaluation of method. Br J Haematol 12:161–166, 1966.

52. Mayman D, Zipursky A: Hereditary spherocytosis: The metabolism of erythrocytes in the peripheral blood and in the splenic pulp. Br J Haematol 27:201–217, 1974.

53. Morel FM, Baker RE, Wayland H: Quantification of human red blood cell fixation by glutaraldehyde. J Cell Biol 48:91–100, 1971.

54. Murphy JR: The influence of pH and temperature on some physical properties of normal erythrocytes and erythrocytes from patients with hereditary spherocytosis. J Lab Clin Med 69:758–775, 1967.

55. Murphy JR: Erythrocyte osmotic fragility and cell-water: Influence of pH and temperature. J Lab Clin Med 74:319–324. 1969.

56. Opdyke DF, Apostolico R: Splenic contraction and optical density of blood. Am J Physiol 211:329–334, 1966.

57. Piomelli S, Lurinsky G, Wasserman LR: The mechanism of red cell aging. I. Relationship between cell age and specific gravity evaluated by ultracentrifugation in a discontinuous density gradient. J Lab Clin Med 69:659–674, 1967.

58. Ponder E: The stickiness of red cells and ghosts. In Sawyer W (ed): "Biophysical Mechanisms in Vascular Homeostasis." New York: Appleton-Century-Crofts, 1965, pp 53–59.

59. Prankerd TAJ: Studies on the pathogenesis of haemolysis in hereditary spherocytosis. Q J Med 29:199–208, 1960.

60. Prankerd TAJ: The spleen and anemia. Br Med J 2:517–524, 1963.

61. Rand RP, Burton AC: Mechanical properties of the red cell membrane. I. Membrane stiffness and intracellular pressure. Biophys J 4:115–135, 1964.

62. Robertson JS: Theory and use of tracers in determining transfer rates in biological systems. Physiol Rev 37:133–154, 1957.

63. Snook T: A comparative study of the vascular arrangements in mammalian spleens. Am J Anat 87:31–78, 1950.

64. Song SH, Groom AC: Storage of blood cells in spleen of the cat. Am J Physiol 220:779–784, 1971.

65. Song SH, Groom AC: The distribution of red cells in the spleen. Can J Physiol Pharmacol 49:734–743, 1971.

66. Song SH, Groom AC: Immature and abnormal erythrocytes present in the normal, healthy spleen. Scand J Haematol 8:487–493, 1971.

67. Song SH, Groom AC: Sequestration and possible maturation of reticulocytes in the normal spleen. Can J Physiol Pharmacol 50:400–406, 1972.

68. Song SH, Groom AC: Scanning electron microscope study of the splenic pulp in relation to the sequestration of immature and abnormal red cells. J Morphol 144:439–452, 1974.

69. Sorbie J, Valberg LS: Splenic sequestration of stress erythrocytes in the rabbit. Am J Physiol 218:647–653, 1970.
70. Vaupel P, Wendling P, Thome H, Fischer J: Atemgaswechsel un Glucoseaufnahme der menschlichen Milz in situ. Klin Wochenschr 55:239–242, 1977.
71. Wagner HW Jr, Razzak MA, Gaertner RA, Caine WP, Feagin OT: Removal of erythrocytes from the circulation. Arch Intern Med 110:90–97, 1962.
72. Weed RI, LaCelle PL, Merrill EW: Metabolic dependence of red cell deformability. J Clin Invest 48:795–809, 1969.
73. Weiss L: A scanning electron microscope study of the spleen. Blood 43:665–691, 1974.
74. Weiss L, Tavassoli M: Anatomical hazards to the passage of erythrocytes through the spleen. Semin Hematol 7:372–380, 1970.
75. Wennberg E, Weiss L: The structure of the spleen and hemolysis. Ann Rev Med 20: 29–40, 1969.
76. Yaari A: Mobility of human red blood cells of different age groups in an electric field. Blood 33:159–163, 1969.

13

Summary Remarks: In Vitro Erythrocyte Characteristics

Geoffrey V. F. Seaman

The main thrust of the discussions on in vitro erythrocyte characteristics have centered on three principal areas:

1) The rheological properties of blood and the relations between red blood cell membrane mechanical properties and red cell deformability;

2) mechanisms of red blood cell aggregation and the role of this and deformability in microcirculatory flow; and

3) the organization of the erythrocyte membrane at the molecular level and how some of the molecular interactions are expressed at the cellular and blood system levels.

The physicochemical properties of blood or red cell suspensions can be examined at three levels of organization, namely the molecular, cellular, and the blood system levels. Some of the properties that fall under each of these categories are listed in Table I. Phenomenological biophysics, or macrobiophysics, treats the blood system as a continuum characterized by experimental parameters, without consideration of structure at the cellular or molecular level. An example of this approach is the measurement of apparent blood viscosity. Micro- or cellular biophysics, examines blood at the cellular level and considers the averaged properties of whole cells or plasma. Examples of cellular biophysics would be the measurement of the electrokinetic charge of cells [45], and the intrinsic mechanical properties of individual red cells [31]. In molecular biophysics and biochemistry attempts are made to analyze the molecular basis of a variety of properties observed at the cellular or phenomenological level. Examples include the characterization of N-acetylneuraminic acid as the component mainly responsible for red cell surface charge [15, 20], the identification of the erythrocyte D-glucose transport system [26], and the characterization of antigen sites involved in specific cellular aggregation (agglutination).

Erythrocyte Mechanics and Blood Flow, pages 261–271

TABLE I. Physicochemical Properties of Blood and Red Cell Suspensions

| Level of Organization | | |
Molecular	Cellular	Blood system
Enzymatic properties	Membrane mechanical properties	Viscous properties
Microviscosity	Deformability	Heat and gas transport
Functional groups	Geometry	Flow behavior in capillary beds
Sites of specific recognition	Surace charge Membrane particle distribution	Cell-cell interactions

Before covering the principal issues discussed on in vitro erythrocyte characteristics it is proposed to consider the nature of the red blood cell and how many of its properties may change as a result of isolation procedures, the length and conditions of storage, in vivo aging, experimental manipulations, or modifications. The vast literature on red blood cells speaks to the deceptive ease with which they can be isolated from blood in large quantities. Experimental artifacts arise easily in the field of red cell research, and the problems identified in the recent literature [5, 26, 46] should serve as a reminder to the unwary.

Under physiological conditions the principal function of the erythrocyte is as a carrier of oxygen. The oxygen affinity of the hemoglobin is influenced by the level of 2,3-diphosphoglycerate (2,3-DPG). Increases in the intra-erythrocytic 2,3-DPG reduce the oxygen affinity of hemoglobin, and decreases in 2,3-DPG have the opposite effect [28]. Hemoglobin is not an essential structural component of the cell membrane [57], although it has been reported to interact electrostatically with the membrane [47]. One may anticipate therefore that the properties of hemoglobin will contribute additively to the red cell system rather than interactively or synergistically.

The normal human erythrocyte has a life-span of 110 to 120 days, with a relatively low level of random destruction [3]. The effete red cell is thought to be sequestered and subsequently broken down by the phagocytic cells of the reticuloendothelial system [42], although the precise mechanism involved has yet to be established.

A very large data base has been accumulated on the biochemical and physico-chemical properties of red cells. This is due in considerable measure to their ready availability and the presence of two desirable features — namely, a limited potential for repair or regeneration of cellular structures, and also the absence of membranous intracellular compartments. However many of the properties have been described in terms of mean values for a whole population of red cells. If selection of the population occurs during experimental manipulations or isolation procedures, for example, centrifugal methods will tend to separate young from old red cells [18, 34], then the mean values obtained will not be representative for the whole cell population.

In some studies the time required for isolation of cells and other manipulations may be important since a variety of changes can occur in red cells during their storage in vitro [52, 56]. In the case of whole blood, breakdown of the leukocytes with the release of proteases and lysosomal hydrolases [25] affords the possibility of enzymatic modification of the surface properties of the red blood cells. Autolytic changes are also possible since protease activity has been found to be associated with the membrane of the human erythrocyte [2]. There is also evidence for transglutaminase-catalyzed crosslinking of red cell protein components, which may be a biochemical cuase of irreversible membrane stiffening [51].

Isolation procedures of red blood cells from whole blood can also lead to the loss of more labile components from the cells during the washing procedure or, if a low ionic strength wash medium is used, to the adsorption of components from the plasma. Mollison and Polley [33] found that human red cells incubated with their own "deionized" serum at an ionic strength < 0.025 gmole liter^{-1} adsorb various immunoglobulins when complement components of $C'4$ and $C'3a$ are both present on the red cells. Changes in the surface properties of the cells can be recognized biologically, either as a shortened life-span on reinfusion or in vitro by an appropriate erythrophagocytosis test system employing homologous macrophages in culture [27].

If red cell ghosts are to be prepared from intact red cells, then the issue of loss of components or structural changes in the peripheral zone of the membrane becomes of even greater experimental relevance. There are possibly as many different types of ghosts as there are ways of hemolyzing red cells [39, 44]. Two types of ghosts prepared in hypotonic media, "white ghosts" and "resealed ghosts," are popularly used in red cell studies. White ghosts are essentially free of hemoglobin and largely devoid of intracellular contaminants. Such ghosts are used widely for biochemical work, including analysis of various membrane constituents. In work dealing with resealed ghosts the primary emphasis is on function rather than extreme purity of the membrane. One aim is usually to restore original membrane structure, especially transport function, after sealing. Aside from transport studies resealed ghosts find use in experiments on the sidedness of effects of chemical and/or enzymatic modification on membrane structure and function.

Reorganization of at least parts of the membrane has been demonstrated for the phospholipids of the ghost membrane by means of the lipid-soluble reagent 2:4-dinitrofluorobenzene (DNFB). DNFB labels only part of the phosphatidylethanolamine. Little if any of the phosphatidylserine is dinitrophenylated in intact cells, whereas in white ghosts there is a more complete modification of both types of phospholipids. Thus, although no appreciable loss of lipid occurs during the preparation of the ghosts, a significant structural change has taken place. Along these lines, the distribution of certain membrane components may be changed as a result of interaction with membrane reagents — for example, aldehyde fixatives or polycations such as poly-L-lysine [40].

In the field of red cell enzymology the removal of leukocytes and platelets from whole blood is of critical importance since the activities of a variety of enzymes are many-fold higher in the non-red cellular elements of blood [6]. A further incompletely resolved complication is the extreme difficulty encountered in removing reticulocytes, which if present tend to cloud quantitative studies on the enzymology of mature red cells [1]. The activity of acetylcholinesterase, for example, has been used as a measure of reticulocytosis [43], and the activity of this enzyme in the reticulocyte has been estimated to be three times that of the average mature erythrocyte [1].

Procedures in which red cells are subjected to relatively low shear stresses can result in irreversible changes in some of their physicochemical properties [35, 37], as is also the case for some mechanically fragile proteins [53]. Consequently care must be exercised to ensure that stresses of sufficient magnitude to alter surface properties of the cell membrane or other physicochemical characteristics are not applied. Care must be exercised also in modification procedures that alter cell surface charge [45], since such a change will lead to appreciable alterations in local pH — such as surface pH [23]. A process such as this potentially could result in the activation and/or deactivation of membrane-associated enzymes and be accompanied by degradative processes in the cellular peripheral zone.

BLOOD RHEOLOGY AND CELL DEFORMABILITY

Steady Flow Theology of Blood

Dr. Herbert Meiselman points out several principal conclusions that have been derived from macrorheological studies on blood; namely, these:

1) The macroscopic flow properties of normal human blood are those which are characteristic of a non-Newtonian pseudoplastic fluid at lower rates of shear but Newtonian at high rates of shear ($> 1,000$ sec^{-1}) (measurements made in cone-plate or couette instruments whose geometry is large compared with the size of red cells). The issue of whether blood has a yield stress has not yet been resolved [14].

2) The increase in apparent blood viscosity at lower shear rates is primarily related to fibrinogen-induced red cell aggregation (rouleaux formation).

3) In contrast, non-aggregating suspensions, such as red cells in serum or isotonic buffer, have much lower apparent viscosities at the lower shear rates, and the systems are almost Newtonian in behavior.

4) Packed cell volume (PCV) is the suspension property that most influences both the apparent viscosity and the extent of non-Newtonian behavior. At high rates of shear the viscosity is an exponential function of PCV.

5) Red cell deformability also plays an important role in the rheological properties of cell suspensions. Aldehyde-fixed red cell suspensions display greatly increased viscosities in comparison to normal erythrocyte suspensions and also behave as Newtonian fluids over a wide range of shear rates [8].

Microrheological Behavior of Cells

Cellular deformability is assessed in several different ways, including 1) measurement of volumetric flow rates of cell suspensions through various types of filters, having pores in the size range 3–8 μm diameter [12, 54]; 2) determination of the negative pressure required to draw a cell into a micropipette or release it again [19, 41, 55]; 3) assessment of fluid mechanical deformation of red cells [17, 24]; or 4) viscometric measurement of the relative viscosity of the cell suspension at high rates of shear [55].

The microsieving technique (1) and viscometric measurement technique (4) are used to measure the average deformability for a cell population, whereas the micropipette method (2) and fluid mechanical deformation procedure (3) are used to measure the properties of individual cells.

Recently a centrifugal method for determining red cell deformability has been developed by Corry and Meiselman [16] in which the cells are centrifuged through buffer into a glutaraldehyde solution.

Dr. Meiselman also discussed the roles of metabolic depletion and cell morphology in determining the rheological properties of cell suspensions. Besides the normal biconcave discoid shape, red cells can assume a wide range of shapes. Bessis [4] has described these erythrocyte forms, of which two are of especial interest. The echinocyte resembling a sea urchin is a spiculated (crenated) sphere, and the stomatocyte is an indented cup-shaped cell. Many agents convert normal red cells into either echinocytes or stomatocytes, suggesting that the shape of the cell depends on the balance of two opposing forces These observations have been explained by the bilayer couple hypothesis of Sheetz and Singer [50].

Dr. Meiselman pointed out that both the flow behavior [30] and membrane mechanical properties of ATP-depleted human erythrocytes [31] and fresh human red cells were very similar, contrary to earlier reports [55]. Using human red cells suspended in buffer, Meiselman [29] has studied the rheological effects of the discocyte–echinocyte transformation induced under isovolumic conditions by nitrophenols or sodium salicylate. He found an increased apparent viscosity at low rates of shear for suspensions of echinocytes. Restoration of the biconcave discoid shape in the presence of the echinocytic agent resulted in the return of the rheological behavior of the suspension toward normal. It appears that the changes in flow properties are related to an increased cell-cell interaction brought about by the echinocyte form.

MECHANISMS OF RED BLOOD CELL AGGREGATION

Dr. Donald Brooks addressed red cell aggregation mainly from the point of view of aggregation produced by macromolecules. Erythrocytes suspended in simple ionic media probably remain disaggregated because of the ionic double layer that surrounds each cell and which cannot be easily penetrated by other double layers. Aggregation can occur if very strong short-range forces are present, which are able to overcome the electrostatic repulsion, or if attractive forces are present that can act at distances greater than the effective extension of the ionic double layers [7]. Because of their size, macromolecules are often able to adsorb simultaneously onto two adjacent cell surfaces at separation distances where double layer repulsion is small. This type of macromolecular bridging is thought to account for the red cell rouleaux formation observed in the presence of fibrinogen or plasma and for the aggregation produced by higher molecular weight neutral polymers such as the dextrans [7]. Extensive use has been made of the dextran-erythrocyte interaction in model studies designed to evaluate the relative roles of macromolecular bridging forces, electrostatic repulsive forces, and mechanical shearing forces in red cell aggregation and disaggregation [11, 13]. The cell membrane is an important element in this aggregation reaction since it forms the substrate to which the bridging macromolecules adsorb. The dextran-membrane interaction is weak and probably nonspecific, unlike lectin-induced aggregation [36]. The deformability of the normal red cell membrane permits large areas of close membrane opposition thus allowing extensive intercellular bridging by macromolecules.

Dextran imparts an unexpected increase in the surface electrostatic potential to cells with which it is in contact. The increase in surface electrostatic potential (zeta potential) is a function of both concentration and molecular weight of the dextran. Electrophoretic data for red cells in dextran are usually expressed in terms of relative zeta potential, Z, where Z is the zeta potential in the presence of polymer divided by that in its absence. The zeta potential increase results from an expansion of the ionic double layer produced by exclusion of ions by adsorbed dextran segments. Since the cell surface charge does not change as a result of polymer adsorption, the ion exclusion produces an increase in surface potential. These findings imply that the erythrocyte disaggregation observed at high dextran concentrations arises from enhanced electrostatic repulsion between the cells.

In order to elucidate mechanisms of red cell aggregation, it is desirable to have a quantitative measure of the degree of aggregation present in a given suspension. Viscometric measurements may be used to provide a relative estimate of the degree of aggregation present in a suspension. Under certain conditions the ratio of the apparent viscosities of aggregated and disaggregated suspensions of equal cell volume concentration should provide an index of the extent of aggregation present. If the viscosities of the continuous phase differ in the suspensions to be compared, the ratio of relative viscosities in the absence of cellular deformability should provide the required information. Since normal erythrocytes are deformable, the contribution of deformation must be minimized by examination of the systems at very low rates of shear. Thus, the relative viscosity ratio, R, will provide a quantitative index of the degree of cell aggregation when evaluated at a sufficiently low rate of shear [9].

Dr. Brooks pointed out that cell aggregation under a shear regime will often be different from the conditions of static aggregation, and furthermore, that under appropriate circumstances, irreversible time-dependent changes may occur. He illustrated this point by describing recent experiments in which he has been able to demonstrate a shear-induced aggregation of red cells by fairly high levels of concanavalin A at relatively high rates of shear (50 sec^{-1}). The shear appears to be producing changes in the red cells that allow interaction with the concanavalin A. This is yet another example of the experimental conditions producing irreversible changes in the system under examination.

MOLECULAR ORGANIZATION OF THE ERYTHROCYTE MEMBRANE

Dr. Michael Sheetz presented the bilayer and shell model for the erythrocyte membrane and compared it with the current fluid mosaic model of membrane structure. The experimental information on which the shell model is based was obtained by Dr. Sheetz and coworkers [48, 49] using dissection of the intact erythrocyte membrane with the non-ionic detergent Triton. The shell model consists of three basic units — the membrane bilayer, the shell of inner peripheral proteins, and the links between the shell and bilayer. The membrane bilayer comprises membrane lipids, glycophorin, and band 3. The shell proteins consist of spectrin, actin, and components 4.1 and 4.9, with component 2.1 and possibly other proteins such as band 4.1 or 4.2 acting as links holding the proteins between the bilayer and shell together.

Spectrin is a major component of the erythrocyte shell and plays a major role in maintaining the structure since its removal results in loss of the reticular structure of the shell [49]. The shape and deformability of the red cell depends on the properties of the cytoskeleton adherent to the inner surface of the bilayer and spectrin and actin have important functions in controlling these properties. It has been suggested also that spectrin controls the asymmetric distribution of red cell lipids such as phosphatidylethanolamine and phosphatidylserine [22]. Extraction of lipids and lipid-associated proteins does not alter the integrity or morphology of the erythrocyte cytoskeleton, suggesting that they are not essential structural components and confirming that a peripheral protein complex or shell exists as a structure independent of the bilayer.

ROLE OF SPECTRIN IN RED CELL DEFORMABILITY

Dr. Alfred Greenquist described a variety of experiments that support the view that spectrin is involved in the control of red cell shape. He reported that both heating and metabolic depletion result in decreased erythrocyte deformability and spectrin extractability from the membrane [32], which experimental studies suggest are interrelated phenomena. Dr. Greenquist also reported that membranes from erythrocytes heated to 50°C showed a marked decrease in the endogenous phosphorylation of spectrin. In addition, during metabolic depletion when ATP has dropped to about 15% of normal, spectrin undergoes rapid dephosphorylation. In parallel with these changes in spectrin the cells are transformed from biconcave discoids to progressive varieties of spiculated discoids (echinocytes I, II, and III) and finally to smooth spheroidal forms termed spheroechinocytes.

In mice with the genetic trait Sph/Sph, in which spectrin is virtually absent, the cells are spheroidal and membranes isolated from such murine red cells are unstable and readily fragmented. In the human disease, hereditary spherocytosis, the erythrocytes are less deformable and isolated membranes also exhibit an alteration in the spectrin phosphorylation reaction [21].

PROTEIN-PROTEIN INTERACTIONS IN THE RED BLOOD CELL MEMBRANE

Dr. J. Palek detailed studies devised to gain understanding of spectrin organization by its alteration at the molecular level. The technique employed was that of intermolecular disulfide coupling of the nearest membrane protein neighbors followed by two-dimensional PAGE, which provided evidence for crosslinking of actin, spectrin, spectrin-actin, and crosslinking between band 2 and 4.9. He reported also that red cell ATP depletion under aerobic conditions results in the spontaneous formation of several protein complexes that appear to have arisen from intermolecular disulfide couplings because they were reversed by dithiothreitol reduction and their formation did not occur under anaerobic conditions. Subsequent exposure of anaerobically ATP-depleted red cell ghosts to catalytic oxidation with $CuSO_4$ and O-phenanthroline produces similar protein complexes. One of the large complexes was shown to contain spectrin polypeptides 1 and 2 as the prominent components. Formation of this large spectrin-rich complex in ATP-depleted red cells was completely prevented by maintaining red cell ATP at normal levels. In ATP-depleted red cells but not fresh cells a complex of a similar molecular size was produced by presumably crosslinking the neighbor protein amino groups with glutaraldehyde, suggesting that the ATP depletion produces a rearrangement of the spectrin with enhanced contacts between the individual polypeptides, which permits more ready crosslinking into a large aggregate. The salient features of Dr. Palek's current work have been reviewed recently [38].

SUMMARY

While much remains to be done before a complete understanding of the molecular basis of the membrane mechanical properties of red cells is achieved or the phenomenological properties of blood systems predicted from their cellular characteristics, progress has been substantial. An encouraging situation is the increasing awareness that we are dealing with a dynamic system, with a large complement of enzymes, the properties of which are time dependent and, in addition, are a marked function of the method of preparation and the manipulations that have been carried out. Investigators dealing with red cells should perhaps bear in mind the words of Humpty Dumpty to Alice when he said in a rather scornful tone, "it means just what I choose it to mean — neither more nor less" [10]. Many of the apparent discrepancies and disagreements have arisen because individuals have carried out different experiments which they think to be identical to those conducted by others.

ACKNOWLEDGMENTS

This work was supported by grants HL 18284 and HL 19255 from the U.S. Public Health Service.

REFERENCES

1. Allison AC, Burn GP: Enzyme activity as a function of age in the human erythrocyte. Br J Haematol 1:291–303, 1955.
2. Ballas SK, Burka ER: Protease activity in the human erythrocyte: Localization to the cell membrane. Blood 53:875–882, 1979.
3. Berlin NI, Berk PD: The biological life of the red cell. In Surgenor D MacN (ed): "The Red Blood Cell," Vol 2, New York: Academic Press, 1975, pp 957–1019.
4. Bessis M: Red cell shapes: An illustrated classification and its rationale. In Bessis M, Weed RI, Leblond PF (eds): "Red Cell Shape." New York: Springer-Verlag, 1973, pp 1–25.
5. Beutler E, Kuhl W: Human erythrocyte membrane enzymes. Blood 51:367–368, 1978.
6. Beutler E, West C, Blume K-G: The removal of leukocytes and platelets from whole blood. J Lab Clin Med 88:328–333, 1976.
7. Brooks DE: Red cell interactions in low flow states. In Grayson J, Zingg W (eds): "Microcirculation," Vol 1, New York: Plenum 1976, pp 33–52.
8. Brooks DE, Goodwin JW, Seaman GVF: Interactions among erythrocytes under shear. J Appl Physiol 28:172–177, 1970.
9. Brooks DE, Goodwin JW, Seaman GVF: Rheology of erythrocyte suspensions: Electrostatic factors in the dextram-mediated aggregation of erythrocytes. Biorheology 11:69–77, 1974.
10. Carroll L (alias Dodgson CL): "Through the Looking Glass," Ch 6, 1872.
11. Chien S: Electrochemical interactions between erythrocyte surfaces. Thromb Res (Suppl II)8:189–202, 1976.
12. Chien S, Luse SA, Bryant CA: Hemolysis during filtration through micropores: A scanning electron microscopic and hemorheologic correlation. Microvasc Res 3:183–203, 1971.

13. Chien S, Sung LA, Kim S, Burke AM, Usami S: Determination of aggregation force in rouleaux by fluid mechanical technique. Microvasc Res 13:327–333, 1977.

14. Cokelet GR, Smith JH: The effect of concentric-cylinder viscometer gap size on the experimental rheological properties of blood. Biorheology 10:51–56, 1973.

15. Cook GMW, Heard DH, Seaman GVF: Sialic acids and the electrokinetic charge of the human erythrocyte. Nature 191:44–47, 1961.

16. Corry WD, Meiselman HJ: Centrifugal method of determining red cell deformability. Blood 51:693–701, 1978.

17. Corry WD, Meiselman HJ: Deformation of human erythrocytes in a centrifugal field. Biophys J 21:19–34, 1978.

18. Danon D, Marikovsky Y: Determination of density distribution of red cell population. J Lab Clin Med 64:668–674, 1964.

19. Evans EA, Hochmuth RM: Membrane viscoelasticity. Biophys J 16:1–11, 1976.

20. Eylar EH, Madoff MA, Brody OV, Oncley JL: The contribution of sialic acid to the surface charge of the erythrocyte. J Biol Chem 237:1992–2000, 1962.

21. Greenquist AC, Shotat SB: Phosphorylation in erythrocyte membranes from abnormally shaped cells. Blood 48:877–886, 1976.

22. Haest CWM, Plasa G, Kamp D, Deuticke B: Spectrin as a stabilizer of the phospholipid asymmetry in the human erythrocyte membrane. Biochim Biophys Acta 509:21–32, 1978.

23. Hartley GS, Roe JW: Ionic concentrations at interfaces. Trans Faraday Soc 36:101–109, 1940.

24. Hochmuth RM, Mohandas N: Uniaxial loading of the red-cell membrane. J Biomech 5:501–509, 1972.

25. Janoff A: Neutrophil proteases in inflammation. Annu Rev Med 23:177–190, 1972.

26. Kahlenberg A, Zala CA: Reconstitution of D-glucose transport in vesicles composed of lipids and intrinsic protein (Zone 4.5) of the human erythrocyte membrane. J Supramol Struct 7:287–300, 1977.

27. Kay MMB: Mechanism of removal of senescent cells by human macrophages in situ. Proc Natl Acad Sci USA 72:3521–3525, 1975.

28. MacDonald R: Red cell 2,3-diphosphoglycerate and oxygen affinity. Anaesthesia 32:544–553, 1977.

29. Meiselman HJ: Rheology of shape-transformed human red cells. Biorheology 15:225–237, 1978.

30. Meiselman HJ, Baker RF: Flow behavior and ATP-depleted human erythrocytes. Biorheology 14:111–126, 1977.

31. Meiselman HJ, Evans EA, Hochmuth RM: Membrane mechanical properties of ATP-depleted human erythrocytes. Blood 52:499–504, 1978.

32. Mohandas N, Greenquist AC, Shohet SB: Effects of heat and metabolic depletion on erythrocyte deformability, spectrin extractability and phosphorylation. In Brewer G (ed): "The Red Cell." New York: Alan R. Liss, Inc., 1978, pp 453–472.

33. Mollison PL, Polley MJ: Uptake of γ-globulins and complement by red cells exposed to serum at low ionic strength. Nature 203:535–536, 1964.

34. Murphy JR: Influence of temperature and method of centrifugation on the separation of erythrocytes. J Lab Clin Med 82:334–341, 1973.

35. Nanjappa BN, Chang H-K, Glomski CA: Trauma of the erythrocyte membrane associated with low shear stress. Biophys J 13:1212–1222, 1978.

36. Nicolson GL: The interactions of lectins with animal cell surfaces. In Bourne GH, Danielli JF (eds): "International Review of Cytology," Vol 39. New York: Academic Press, 1974, pp 89–190.

37. Ohshima N: Hemorheological effects of mechanical factors on blood cell damage in extracorporeal circulation: Significance of subhemolytic trauma. Biorheology 15: 295–302, 1978.

38. Palek J, Liu S-C: Dependence of spectrin organization in red blood cell membranes on cell metabolism: Implications for control of red cell shape, deformability, and surface area. Semin Hematol 16:75–93, 1979.

39. Ponder E: "Hemolysis and Related Phenomena." New York: Grune & Stratton, 1948.

40. Pricam C, Fisher KA, Friend DS: Intramembranous particle distribution in human erythrocytes: Effects of lysis, glutaraldehyde, and poly-L-lysine. Anat Rec 189: 595–607, 1977.

41. Rand RP, Burton AC: Mechanical properties of the red cell membrane. Biophys J 4:115–135, 1964.

42. Riflind RA: Destruction of injured red cells in vivo. Am J Med 41:711–723, 1966.

43. Sabine JC: Erythrocyte cholinesterase titers in hematologic disease states. Am J Med 27:81–96, 1959.

44. Schwoch G, Passow H: Preparation and properties of human erythrocyte ghosts. Mol Cell Biochem 2:197–218, 1973.

45. Seaman GVF: Electrokinetic behavior of red cells. In Surgenor D MacN (ed): "The Red Blood Cell," Vol 2. New York: Academic Press, 1975, pp 1135–1229.

46. Seaman GVF, Knox RJ, Nordt FJ, Regan DII: Red cell aging. I. Surface charge density and sialic acid content of density-fractionated human erythrocyte. Blood 50:1001–1011, 1977.

47. Shaklai N, Yguerabide J, Ranney HM: Interaction of hemoglobin with red blood cell membranes as shown by a fluorescent chromophore. Biochemistry 16:5585–5592, 1977.

48. Sheetz MP, Sawyer D: Triton shell of intact erythrocytes. J Supramol Struct 8:399–412, 1978.

49. Sheetz MP, Sawyer D, Jackowski S: The ATP-dependent red cell membrane shape change. A molecular explanation. In Brewer G (ed): "The Red Cell." New York: Alan R. Liss, Inc, 1978, pp 431–450.

50. Sheetz MP, Singer SJ: Biological membranes as bilayer couples. A molecular mechanism of drug-erythrocyte interactions. Proc Natl Acad Sci USA 71:4457–4461, 1974.

51. Siefring GE, Apostol AB, Velasco PT, Lorand L: Enzymatic basis for the Ca^{2+}-induced crosslinking of membrane proteins in intact human erythrocyte. Biochemistry 17:2598–2604, 1978.

52. Sirs JA: Effects of storage on the respiratory function and flexibility of red blood cells. Blood Cells 3:409–423, 1977.

53. Tirrell MV: Stress-induced structure and activity changes in biologically active proteins. J Bioeng 2:183–193, 1978.

54. Usami S, Chien S, Bertles JF: Deformability of sickle cells as studied by microsieving. J Lab Clin Med 86:274–279, 1975.

55. Weed RI, LaCelle PL, Merrill EW: Metabolic dependence of red cell deformability. J Clin Invest 48:795–809, 1969.

56. Weed RI, LaCelle PL, Udkow M: Structure and function of the red cell membrane: Changes during storage. In Greenwalt TJ, Jamieson GA (eds): "The Human Red Cell in Vitro." New York: Grune & Stratton, 1974, pp 65–82.

57. Weed RI, Reed CF, Berg G: Is hemoglobin an essential structural component of human erythrocyte membranes? J Clin Invest 42:581–588, 1963.

14

Summary Remarks: In Vivo Erythrocyte Behavior

B. W. Zweifach

The principal thrust of investigations in blood rheology is to provide quantitative expressions for the physical variables that control the perfusion of the microcirculatory apparatus. Investigations of this kind center about the interactive behavior of the red blood cells (RBC) since they make up some 45% of the blood volume and their dynamic behavior to a considerable degree determines the effectiveness of the blood flow through the microvascular system. It, therefore, becomes of singular importance that studies of blood and blood cells outside of the body should pay particular attention to the need to duplicate their natural environment within the body and to focus on forces of the same magnitude as those encountered in the intact microcirculation.

In terms of optimal experimental design, it would be desirable to obtain the necessary measurements on the living system under well-controlled conditions, but except in rare instances this is not possible. The effects of red cell aggregation on blood flow have been simulated through the use of a variety of viscometers that at best approach macrovessel flow conditions. Measurements that define red cell deformability in terms of the modulus of elasticity of the cell surface are even more difficult to extrapolate to the living system in anything but general terms.

A useful pragmatic approach would be to obtain a numerical estimate of cell deformability in vitro on normal red cells that have been subjected to known physicochemical alterations, or on abnormal cells from diseased individuals, and then to perfuse suspensions of such cells through the capillary network of well-established tissue preparations — eg, mesentery or skeletal muscle. Among the parameters that can be measured in such an experiment are pressure distribution, spacing of RBC in the capillaries proper, segmental blood flow, and blood cell distribution within the network. Such observations could then be

Erythrocyte Mechanics and Blood Flow, pages 273–281
© 1980 Alan R. Liss, Inc., 150 Fifth Avenue, New York, NY 10011

used to weight the relative contribution of specific aberrations in RBC material properties to the dynamics of microvascular flow. The extent to which changes in the rheologic characteristics of red cells will affect the ability of microvascular perfusion to meet tissue needs can be measured by the extent to which tissue pO_2 is maintained as monitored by use of microelectrodes.

The purpose of the present overview is to discuss the relevance or limitations of in vitro models of microvascular behavior to the living microcirculation. With this objective in mind, it can be seen that intravital microscopy serves a unique purpose by identifying those physical and functional variables that are basic to the problem and their relative importance. The relationship of in vivo modeling can best be dealt with under several headings.

MICROVASCULAR SYSTEM AS A WHOLE

For many years, direct intravital microscopy has served as the backbone of the approach to the microcirculatory system. With the availability of newer video techniques and improved electronic instrumentation, it has become possible to record on-line the basic physical parameters involved in microvascular flow. With the increased use of rheological concepts and methods, it has become possible to carry the analysis of microcirculatory behavior a step further by identifying the separate contributions of the fluid and cellular components of the blood.

Direct microscopy brings into view several basic aspects of microvascular behavior: 1) the actual flow rate in successive segments; 2) the distribution of blood cells within the network; and 3) the flow in the narrow capillaries where red cell deformation is necessary. Early rheologic studies were concerned primarily with the red cell aggregation phenomenon. Because of the difficulty of fitting in vitro derived quantitative expressions into comparable facets of microcirculatory behavior, there has been a tendency for physiologists to dismiss blood rheologic factors as having only a secondary impact on flow under normal conditions. In vivo measurements of pressure-flow relationships in the microvascular network for the most part neglect the viscosity term in the Poiseuille-Hagen relationship. The assumption is made that under normal conditions the viscosity term remains fairly constant in vessels below $25-30$ μm in diameter.

The two rheologic aspects that have received major attention are the establishment of the material properties of the red blood cell as a means of characterizing their flow deformability during passage of the blood through the capillary-sized vessels, and the delineation of the factors that increase the tendency for RBC to form aggregates in the bloodstream. Both facets of the problem are key determinants of the effectiveness of microvascular flow, particularly under pathological conditions. As noted above, intravital microscopy brings out the

range of deformability actually encountered, as well as the varying degrees of
red cell aggregation that can occur under specific circumstances. The direct
visual approach, however, has its limitations and must be buttressed by the
formulation of methods that describe these phenomena in a quantitative framework that in turn makes it possible to compare in vitro with in vivo measurements. Such a comparison has not always been possible. As in vitro rheological
experiments have become more sophosticated, their precise relationship to the
mechanics of blood flow in the living microcirculation becomes more difficult
to identify.

More recently, in vitro work has dealt with the deformability of the red blood
cell as reflected by its material properties. Included has been work on cells with
abnormal shapes (spherocytes, acanthocytes, sickle cells). In terms of the in vivo
circulation, little has been established aside from the obvious implication that
bizzare-shaped cells are less flexible and frequently become entrapped and plug
the entrance into side branchings.

There has been a trend for blood rheologists to rely almost exclusively on in
vitro models to obtain new data. It is obvious that in vitro test systems must become much more sophisticated before they can match currently available in vivo
data. Not only is there a need to put together an in vitro test system that
matches closely what is encountered in a particular vascular bed, but even more
critical is the need to have sufficient understanding of the actual system so as to
be in a position to expand the model without neglecting essential elements of the
system. Once equations of flow have been set up, parallel experiments should be
conducted, using both the living microcirculation and the model system as test
objects; eg, introduce cells with different viscoelastic moduli into both set-ups
and compare the effects on flow, resistance, etc.

In the same context, it can be appreciated that studies of the in vivo microvascular system make it possible to set up guidelines for structural boundary
values, as well as the order of magnitude of the forces involved in perfusion of
the terminal vascular bed. Unless the two approaches continue to interlock, both
will become progessively concerned with minutiae or gross approximations of
structure-function relationships.

All too often, models are set up without questioning the purpose of the new
model. Basically, such constructs should provide new insight or embellish the
information that is already on hand. In the case of the microcirculation, the
most common reason that has been advanced for the adoption of an analytical
approach is the relative inaccessibility of the network so that key information
cannot be obtained by other than in vitro models. In many ways, recent measurements of the viscoelastic properties of the red cell obtained by the micropipette
suction procedure are more compatible with recent advances in the biochemical
makeup of the cell membrane than with the flow characteristics of the blood
through the capillary network.

VESSEL GEOMETRY

Most models of the microvessels assume the vessel can be approximated by a tube of uniform diameter. Although in a general sense, such an approach can serve as a useful first step, modifications must be introduced to bring it closer to reality as the questions posed become more complex. Inspection of photographs of living vessels indicate that the influence of a number of ancillary factors must be taken into account.

As a rule, individual capillaries are some 200–250 μm in length and their diameter is not uniform but increases by 20–25% from the proximal to the distal end. Furthermore, being a mosaic of endothelial cells, the vessel has an irregular contour with the nuclear portion of the cell bulging into the lumen and the overlapping edges of continuous cells frequently disposed as flat projections into the lumen. Irregularities of this kind in narrow vessels must have a significant effect on many calculations including the blood cell velocity profile and the applicability of the lubrication layer concept.

Many of the precapillary feeding vessels are narrower than the true capillaries, particularly at their point of origin from the parent arteriole. This narrowed neck-like portion of the precapillary has a lumen diameter of some 3 μm and may be about 20–40 μm long, so that a considerable resistance to flow is imposed at the point of entry. In estimating the effect of vessel diameter on flow, most calculations assume the cross-section is circular. Here again, the narrowed neck region has a folded endothelium and a very irregular contour. In addition, capillary branching points may have an elliptical cross-section.

The collective effect of the above set of physical variants probably brings about significant departures from Poiseuille flow conditions. In the literature when in vivo and in vitro data are compared, close agreement is accepted when the two sets of data fall within ± 100% to 200% of each other. Yet in the living circulation, variations of this magnitude represent the actual range of physiological adjustments.

BRANCHING CONFIGURATIONS

As a result of the high shear rates that prevail on the arterial side of the microcirculation, the red cells lose their biconcave shape and take on an array of distorted profiles. In the larger arterioles, where the diameter is sufficient to allow the cross-sectional deployment of several red cells, these bizzare shapes are the result of hydrodynamic forces imposed by the high flow rate. After some six to eight side branches are given off, the terminal portion of the arterioles is narrowed to the point that the red blood cells traveling at a high rate become further distorted by the bending and buckling needed to accomodate the larger red cell to the narrow lumen of the precapillary arterioles. The terminal arterioles thus originate as an 18–20 μm vessel and become tapered as successive precapillary offshoots are given off until at its distal end the arteriole is only 6–8 μm wide.

Along the course of the terminal arterioles some 7–10 branches of capillary size are distributed at right angles to the parent stem. RBC are swept into the offshoots in accord with the prevailing pressure drop and become grossly deformed as they pass through these junctional orifices. Both the tapering and entry problems of this kind are difficult to duplicate in vitro.

As has been repeatedly noted, the number of red blood cells in the capillary sized vessels per unit volume of blood is much smaller than in the large feeding vessels. Values as low as 8% have been calculated for capillary hematocrit. The phenomenon of red "cell separation" in situ and distribution within the microvascular network appears to be determined more by the prevailing pressure gradients and by the relative dimensions of the two branches than by the material property of the red cell. As indicated in cases where the branch is less than 8–10 μm in diameter, entry conditions may require considerable deformation of the red cell for passage into the capillary branch. Under such circumstances, the material properties of the red cell become a more important determinant of their distribution. Under normal conditions, however, it is the net driving pressure that is the major factor sweeping the red cell into a given branch. As is the case in other related problems, there is a need to relate cell separation in the microcirculation to both intravascular and to vascular determinants.

ENDOCAPILLARY LAYER

Much has been made of the Lighthill-Fitzgerald model based on hydrodynamic theory of a thin lubrication layer between the moving red cells and the endothelial surface in capillaries to facilitate the passage of the deformed red cells in these narrow channels. Little or no attention has been given to the fact that the luminal surface of the endothellum is lined with a glycoprotein similar to the calyx present on the surface of other cells and that this material serves as a mechanical lubricant reducing frictional interaction with the bloodstream. Failure of this boundary is more likely to occur through an erosion or deficiency of this lining material, rather than through a physical narrowing of the space between RBC and the endothelial surface. Experiments where flow is measured through glass tubes do not duplicate these surface conditions. Blood cells adhere to glass, whereas they do not interact with the luminal surface of the endothelium. Experiments in which blood or blood cells are perfused through glass microtubes must take this important aspect into account.

RBC DEFORMABILITY

Early rheologic analyses were in the main concerned with the flow properties of the blood as a suspension of red cells and dealt with factors such as the number of cells (hematocrit), the effect of changes in shear rate, and the size of the tube or vessel. Subsequently, attention was focused on the red cell shape and the extreme deformability of these cells. The latter two features have been analyzed recently in abnormal or pathological conditions where the red blood cell assumes an unusual shape, is more fragile, and is more easily hemolyzed.

The data presented at this meeting dealt largely with the material properties of the RBC which in terms of microvascular perfusion affect primarily cell deformability. This physical phenomenon will have its greatest impact on the flow through vessels of capillary dimensions — where the RBC are larger than the vessels and must be deformed to permit blood to flow from the arterial to the venous side of the network. The material properties of red cells probably also change with age, despite no apparent change in shape. In some diseases the cells may become spherical (spherocytosis), or assume bizzare shapes, as in sickle cell disease or thalassemia.

A key feature with respect to red blood cells is the constant volume to surface area relationship, because of which surface area cannot be increased under conditions of deformation. Apparently, this constraint holds for both leucocytes and red blood cells. Because of the RBC viscoelastic properties, changes in cell behavior must be considered from both a static and a dynamic interactive mode.

When red cells assume bizzare shapes or are transformed into spherocytes, they are believed to be much less deformable and, hence, unable to be distributed uniformly into the all-important exchange portion of the capillary network. As discussed in the present meeting, in vitro experiments with micropipettes show no substantial difference in the viscoelastic properties of pathological cells in patients with sickle cell disease, diabetes, spherocytosis, or acanthocytosis. Yet such cells in the living microcirculation represent a physical hazard, since they become lodged in branchings and block blood flow to localized areas of tissue. The in vitro data obtained on the material properties of such cells are of limited practical value in quantifying their impact on microvascular perfusion. As is the case for other rheological measurements based on in vitro test systems, what is needed here, for example, would be some estimate of the extent to which tissue perfusion has been impaired as reflected by a change in tissue oxygenation. Such tests could be made in animal preparations in which the pathological RBC have been substituted for normal cells by intravenous injection.

Pathological red cells have been recognized by changes in their size or in departures from the biconcave disc shape. Red blood cells differ in size under normal conditions with the difference presumably a function of the age of the cell. It is not known whether differences in size are associated with differences in the deformability of the cell. This aging characteristic has been proposed as a mechanism for the detection and removal of "old" or "dying" red blood cells.

Data based on the ability of RBC to penetrate artificial membranes with 1–3 μm pores have been used as an index of RBC deformability. Such experiments tend to minimize the importance of entry conditions and emphasize the cell deformation within the capillary proper. This concept is in conflict with in vivo data obtained as blood crosses the narrow junctions of the offshoot with the terminal arterioles, showing that both pressure and flow are lowered abruptly in the neck-like entrance into the capillary-sized vessels. Conditions governing the relative impact of these two factors require clarification.

In the present state of our knowledge, analysis of the material properties of RBC would seem to have a greater relevance to the basic biochemical defect responsible for their unusual shape — eg, a change in the spectrin content or makeup, as in spherocytosis.

RED CELL AGGREGATION

Diabetes mellitus in man is associated with severe angiopathies, particularly in the microcirculatory system. Direct examination of the capillary flow in the bulbar conjunctiva and retina of the eye has indicated that red cell aggregation is consistently present in diabetics. Although some in vitro work with viscometers has supported this concept, the in vivo and in vitro observations are difficult to compare with one another. There is as yet no acceptable method for expressing numerically the severity of the aggregation phenomenon in vivo. Presumably such a trend would be exaggerated either by a reduced shear rate or by a change in the vessels proper. It is known that specific changes in red cell hemoglobin occur during the development of the disease, and it would be useful to relate the changes in red cell flow properties with the biochemical evidence of hemoglobin alteration.

The ineffective perfusion of the capillary network in diabetes has also been attributed to changes in RBC deformability on the basis of filtration of red cell suspensions across 1–2 μm pores in special filters and the movement of such blood cells in capillary-sized tubes. More convincing evidence in this regard could be obtained by direct microscopy of capillary blood vessels under conditions where both pressure and RBC velocity were being measured. Animal models of diabetes would be well suited for this purpose.

In view of the need for glucose by most cells of the body, an insulin deficiency should affect both vascular endothelium and leucocytes. Electron microscope studies of wound healing in diabetic animals show an impaired deposition of collagen and a defect in endothelial cell replacement. Intra vital preparations of the microcirculation could be used to study the vascular and blood cell response to localized injury as a basis for evaluating both the severity of the condition and the response to therapy.

CELL ENVIRONMENT FOR IN VITRO STUDIES

In vitro measurement of red blood cell properties are carried out with the cells immersed in some arbitrary solution, usually a balanced electrolyte mixture that may or may not contain a colloid such as albumin. In view of the fact that deformability of the cell is a reflection primarily of its outer membrane complex, enviornmental conditions can be an important determinant of the red cell response to such tests. As discussed at this meeting the surface membrane complex contains an inner layer of spectrin combined with contractile proteins and other membrane constituents. These delicate relationships are easily disturbed by interfering with ion fluxes or active transport mechanisms.

Presumably the optimal "natural" environment for red blood cells is the blood plasma, so that even solutions that contain albumin as a colloid may still be deficient. In fact there is evidence that extravasation of the red cell into the interstitium by itself is sufficient to change the physical properties of the cell and to lead to its ingestion by macrophages. The fact that alterations either in the electrolyte composition or in the plasma protein content of the medium used to suspend RBC have no measurable effect on viscoelastic properties, as measured by micropipettes or by viscometers, does not by itself prove anything with respect to the behavior that such treated cells would exhibit in an in vivo system.

RECOGNITION OF ABNORMAL RBC

Dr. Groom focused attention on the mechanism by which "dying" or abnormal RBC are recognized and removed from the active circulation. Previous assumptions were based on the inability of the red cell to be deformed beyond the point where its surface area was increased, so that an "aged" red blood cell would be detected because it was less deformable and would hemolyze when forced through fine cell separations in structures, such as the spleen. The data presented by Dr. Groom did not support such a mechanism, but suggested instead that some surface property was involved, as in the case of sequestration of lymphocytes or other leucocytes by the endothelium in particular organs, such as lymph nodes.

Relatively few in vivo experiments have dealt with blood rheology because of the difficulty in obtaining sufficiently accurate measurements of the driving force (Δ P), red cell volume, vessel diameter, and red cell velocity at the same time. The few that have been carried out, on the whole provide data that approximate in vitro values for stress-strain relationships. Material properties of the red cell cannot as yet be approximately from in vivo observational data.

Changes in red cell shape in vitro and the formation of rouleaux in vivo and in vitro depend upon the adsorption of macromolecules to the red cell surface. In experimental procedures used to study isolated systems or red blood cell sus-

pensions, the cells are thoroughly washed to remove platelets and plasma proteins. During this process, variable amounts of materials absorbed to the red cell
surface are undoubtedly washed off or altered, so that even well-controlled in
vitro experiments deal with a built-in abnormality.

RHEOLOGICAL EFFECTS OF LEUCOCYTES

Leucocytes that become lodged inside a blood capillary change blood flow
drastically; pressure rises in the proximal end to arteriolar levels and leads to net
filtration. Such leucocytes take some four to five seconds to traverse the length
of the average capillary. Red cells do not readily enter such an obstructed
capillary until the leucocyte is moved down into the venular tributaries. From
the point of view of postcapillary resistance, the adhesion and rolling of leucocytes along the venule wall represent a substantial increase in the resistance to
blood flow. Even a slight increase in this trend can affect both capillary flow
and pressure. The term "postcapillary resistance" is somewhat ambiguous, since
the first postcapillaries and their confluent collecting venules (20–25 μm wide)
are not contractile. The "R" term thus encompasses the number of vessels at this
cross-section, the effect of a decreased shear on blood viscosity, the effect of
leucocyte palisading in the postcapillaries, and possibly even a retrograde effect
on the number of capillaries being perfused with blood. All of these factors remain to be evaluated by a suitable combination of in vivo and in vitro techniques.

SUMMARY COMMENT

Each of three principal approaches used for microvascular research — intravital microscopy, physical modeling, and theoretical hydrodynamics — tends to
remain self-contained, although all rely on one another for definitive data. Unless
the in vivo and ex vivo approaches remain mutually self-supporting, it will be
difficult to maintain the initial momentum developed during the past ten years.
Of particular importance is the need for both groups to recognize the significance
of the questions that remain to be answered. A complete synthesis of the living
system is obviously impossible. An attractive compromise would be to continue
to incorporate into in vivo experiments new data as they are derived from in vitro
analogues so as to extract hitherto unattainable numerical relationships.

ACKNOWLEDGMENTS

This work was supported by a grant from the USPHS (HL-10881).

Index